Uni

T0229996

Springer
Berlin
Heidelberg
New York
Barcelona
Hong Kong
London
Milan
Paris
Singapore
Tokyo

ÉCOLE POLYTECHNIQUE

The Ecole Polytechnique, one of France's top academic institutions, has a longstanding tradition of producing exceptional scientific textbooks for its students. The original lecture notes, the *Cours de l'Ecole Polytechnique*, which were written by Cauchy and Jordan in the nineteenth century, are considered to be landmarks in the development of mathematics.

The present series of textbooks is remarkable in that the texts incorporate the most recent scientific advances in courses designed to provide undergraduate students with the foundations of a scientific discipline. An outstanding level of quality is achieved in each of the seven scientific fields taught at the *Ecole*: pure and applied mathematics, mechanics, physics, chemistry, biology, and economics. The uniform level of excellence is the result of the unique selection of academic staff there which includes, in addition to the best researchers in its own renowned laboratories, a large number of world-famous scientists, appointed as part-time professor or associate professor, who work in the most advanced research centers France has in each field.

Another distinctive characteristics of these courses is their overall consistency; each course makes appropriate use of relevant concepts introduced in the other textbooks. This is because each student at the Ecole Polytechnique has to acquire basic knowledge in the seven scientific fields taught there, so a substantial link between departments is necessary.

The distribution of these courses used to be restricted to the 900 students at the Ecole. Some years ago we were very successful in making these courses available to a larger French-reading audience. We now build on this success by making these textbooks also available in English.

Michel Demazure

Bifurcations and Catastrophes

Geometry of Solutions to Nonlinear Problems

Translated from the French
by David Chillingworth

With 56 Figures

Springer

Michel Demazure

Cité des Sciences et de l'Industrie
30 av. Corentin-Cariou
75019 Paris, France
e-mail: m.demazure@cite-sciences.fr

David Chillingworth

Department of Mathematics
University of Southampton
Southampton SO17 1BJ, UK
e-mail: drjc@maths.soton.ac.uk

The original French edition was published in 1989 by Ellipses under the title
Géométrie – Catastrophes et Bifurcations

Mathematics Subject Classification (1991): 58F14, 58A05, 58F21, 58C27, 58C28, 34C35

Library of Congress Cataloging-in Publication Data

Demazure, Michel.
 [Géométrie-catastrophes et bifurcations. English]
 Bifurcations and catastrophes : geometry of solutions to nonlinear problems / Michel
Demazure translated from the French by D.J.R. Chillingworth.
 p. cm. -- (Universitext)
 Originally published: Géométrie-catastrophes et bifurcations, 1989.
 Includes bibliographical references and index.
 ISBN 3540521186 (softcover : alk. paper)
 1. Bifurcation theory. 2. Catastrophes (Mathematics) 3. Differentiable manifolds. I.
Title. II. Series.

QA380 .D46 2000   99-053117
515'.35--dc21

ISBN 3-540-52118-6 Springer-Verlag Berlin Heidelberg New York

© Springer-Verlag Berlin Heidelberg 2000
Printed in Germany

Typesetting: By the author using a Springer T$_E$X macro package
Cover design: *design & production* GmbH, Heidelberg

Printed on acid-free paper   SPIN: 10017546   41/3143at – 5 4 3 2 1 0

Table of Contents

Introduction

In this introduction we try to give some idea of the motivation and content of the course of lectures on which this book was based. Most of the points mentioned will be discussed in the text, but some of them are referred to merely in order to indicate possible extensions.

1. At the origin of the word geometry is the word *measure*. However, even when we are aiming for an explicit numerical description, the study of *form* necessarily has to precede the detailed construction and hence the measurement.

As an example, take the simple problem of drawing the *apparent outline* of a surface S (given by an equation $f(x, y, z) = 0$) with respect to projection into the xy plane. This is the curve C which is the projection into the xy plane of the curve D consisting of points (x, y, z) of S where the tangent plane is vertical, and therefore given by the equations

$$f(x, y, z) = 0, \quad f'_z(x, y, z) = 0.$$

The naïve approach to this question is to trace out the curve point by point after dividing up the xy plane into small squares. In doing this we may obtain a very inaccurate result for the following reason: tracing the curve C point by point assumes that it is regular, whereas – and this is a fundamental observation – this curve almost certainly has cusp points (we shall see why later). Therefore we need to start differently, by studying the form of this curve and specifically determining its cusp points and their tangents, before we complete the picture by drawing (point by point) the regular branches that connect them.

If we now want to find out what happens when we deform the surface S (by simply changing the chosen projection, for example) we must again start by investigating how the cusp points move. In fact in the general case they move regularly as long as two of them do not meet; after such a collision there is a 'change of form', a 'change of state' as a physicist would say, or, as Thom[1] says, a *catastrophe*.

---

[1] René THOM, French mathematician, born 1923, Fields Medal 1958, gave the name *catastrophe theory* to a collection of definitions and theorems that are part of the theory of singularities of differentiable maps. This term has had an astonishing treatment in the media (see Sect. 5.12).

2. We meet phenomena like this in the related context of *Morse*[2] *theory.* Here is a simple example.

Again in three-dimensional space, consider a (compact) surface S, given as above by an equation $f(x, y, z) = 0$, and imagine trying to discover the form of S by studying its sections by planes $z = a$. In this way we would obtain S as a union of 'curves' $C(a)$, where each $C(a)$ is defined in the plane $z = a$ by the equation $f(x, y, a) = 0$. The curves $C(a)$ are regular at all points (x, y, a) at which $f'_x(x, y, a)$ and $f'_y(x, y, a)$ do not vanish simultaneously: in fact this is a consequence of the *Implicit Function Theorem*.

In general (here, as previously, the justification for this 'in general' appeals to what is called the Transversality Theorem) the 'bad' points are finite in number and are the *critical points*; there must be some of these, if only the maxima and minima of the function z on S. The corresponding values of a are the *critical values*. Among the essential points of Morse theory is that, roughly speaking, on the one hand "nothing happens between the critical values" as significant changes in form of the curves $C(a)$ take place only for critical values (and, for those values, at the critical points), while on the other hand the global form of the surface S can be reconstituted starting from knowledge of the critical points alone (see Figs. 0.1a,b). Here also, if we deform the surface S then the critical points and critical values move continuously as long as two critical points do not collide.

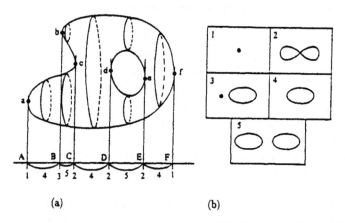

(a)   (b)

Fig. 0.1. (a) Critical points and critical values, (b) the five forms of level curves

3. In these two examples we see a general philosophy emerging. To explain this we use the first example, which is somewhat richer. The object is to study the projection map $(x, y, z) \mapsto (x, y)$ from the surface S to the plane P with equation $z = 0$, going from the general to the particular.

---

[2] Marston MORSE (1892-1977), American mathematician, originator (together with his compatriot Hassler WHITNEY who will feature significantly in this text) of a large number of the ideas that we shall be encountering.

a) Above a general point p of P there is a finite number (possibly zero) of points of S, all with non-vertical tangent plane; as the point p varies each of its inverse images moves continuously (this is once again the Implicit Function Theorem); this 'general' part of the surface can be described as a union of *leaves* that can each be parametrized in the form $z = g(x, y)$.

b) When the point p reaches the apparent outline C at a general point of C the simplest of the catastrophes occurs, namely the *fold* (Fig. 0.2a): two inverse images of p coalesce at a point of S where the tangent plane is vertical, that is at a point of the curve D.

c) When the point p is even more special, that is when it is the projection of a point (x, y, z) of D with vertical tangent and which therefore, as can easily be checked, satisfies the equations

$$f(x, y, z) = 0, \quad f'_z(x, y, z) = 0, \quad f''_{zz}(x, y, z) = 0,$$

then three inverse images of p come together and p is a cusp point of C. This is the second catastrophe in order of complexity, the *cusp* (Fig. 0.2b).

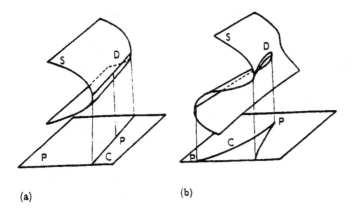

(a)   (b)

Fig. 0.2. (a) Fold, (b) cusp

d) If we have one more dimension, for example if S depends on time t and so is in fact given by $f(x, y, z, t) = 0$, we obtain the third catastrophe – called the *swallowtail* – when, for a particular value of t, two cusp points coalesce, with four inverse images.

These examples exhibit the fundamental characteristics of *singularities*; they are in general unavoidable, they are stable and above all they are 'structural': it is they that 'carry' the form of the geometric objects. This explains why differential geometry has evolved historically from the study of regular situations to the study of singularities.

4. In a general way which needs to be made precise in each instance, a geometric object decomposes into *strata*, the situation in the interior of each stratum being regular (that is to say technically validated by the Implicit Function Theorem), and the passage from one stratum to another taking place via one of the elementary catastrophes listed by Thom.

Moreover, when the given object is sufficiently general in its class this decomposition into strata and the catastrophes conforms to the usual dimensional intuitions: each new condition introduced translates into a drop in dimension by 1 of the corresponding stratum (surface, curve, point). This is a more difficult result, based on the Transversality Theorem. We can give an idea of what is involved using the previous example. While it is fairly clear that for general functions f, g and h the three equations $f = 0$, $g = 0$, $h = 0$ define isolated points, it is not immediately obvious why this should also be the case for the equations $f = 0$, $f'_z = 0$, $f''_{zz} = 0$. This is just the type of situation that the said theorem deals with.

A third step is the proof of the stability of the whole analysis under small perturbations. This is even more delicate, and we shall return to it later.

It is one of the central tenets of Thom's philosophy that these phenomena are a part of our everyday observation. It is the case, for example, with the luminous caustics that we can see each morning in our cup of tea: we immediately notice that first of all there have to be cusp points and that secondly their general form is essentially independent of the experimental conditions.

5. Furthermore, the very notion of *stability* is fundamental and it conditions every instance of mathematical modelling. Since the adjective 'stable' is used classically in many contexts, not always compatible with each other, from now on we shall say *structurally stable* to refer to properties not of a particular configuration but of the system as a whole. Thus we shall speak of stable equilibrium or stable orbit, but of a structurally stable differential equation. This notion will be made more precise at the appropriate time.

Now we give an example that uses a theory so simple as to be indisputable, namely that of a spherical mirror. From a certain point of view this may nevertheless seem a little suspect because spherical mirrors do not exist; anything that we can or could ever manufacture would be only approximately spherical. This implies that all that could ever be observed would be the (structurally) stable properties of spherical mirrors, that is to say those properties that remain approximately valid for approximately spherical mirrors. It is curious to note that an argument of this kind – which today seems imposed from the outset on anyone wishing to formulate any physical law whatsoever – leads directly to the present-day mathematical definition of continuity, whereas the historical development of this notion was long and difficult[3]. We remark in

---

[3] This definition (due in fact to BOLZANO) is named after Cauchy. Augustin CAUCHY (1789-1857) established several fundamental notions of classical analysis in his

passing, without elaborating the point, that the feature of being 'evident *a posteriori*' is common to many scientific concepts.

In parentheses we give, following Arnol'd[4], an amusing illustration of this 'philosophy'. Consider an experiment that consists of releasing a disturbance at the centre of a circular bowl, for example by letting a drop fall into the centre of a cup of tea. Theory says that the circular waves will propagate out from the point of impact, reflect on the boundary, and then reconverge. And this is indeed what is observed, although it is clearly impossible not only to obtain a circular cup but also to hit the centre exactly. In fact, if we consider a circle as an ellipse with coincident foci and then slightly separate the foci we observe that the deformation of the circumference is of second order relative to the separation of the foci, so that conversely a circle can be regarded as an ellipse whose foci are a pair of points, very close to each other and positioned symmetrically about the centre of the circle, but chosen arbitrarily according to the demands of the problem. Therefore what happens in reality is that the drop falls a little to one side of the (alleged) centre of the (alleged) circle and, as must happen in a genuinely elliptic mirror, the waves converge at the other focus, namely the symmetric point with respect to the centre.

6. We continue with these observations on structural stability, but now we consider the opposite case in which the situation that we aim to study really is not structurally stable, the neighbouring situations sharing common properties which are different from those of the initial situation. These are therefore the properties that will in fact be observed.

This remark enables us to understand the phenomenon called *symmetry-breaking* which occurs when the initial situation has symmetry properties which are not preserved under perturbation. Here is a simple example, to which we shall return in Sect. 5.9. It concerns finding the equilibrium states of a bar constrained to move in a vertical plane, freely jointed at its foot and kept in a vertical position by two symmetrical springs (see Fig. 0.3a). Calculations reveal what is called a *bifurcation*. When the spring compression is sufficiently strong the vertical position is the only equilibrium state and it is stable. On the other hand, when the compression is weak the vertical position is unstable and two stable equilibrium states appear symmetrically placed with respect to the vertical. This passage from one regime to another (the 'bifurcation') happens for a given parameter value. Naturally, this is not what is observed (see Fig. 0.3b). In fact the condition of exact symmetry of the two springs cannot be satisfied in practice because *it is not structurally stable*.

---

course (1821) at the Ecole Polytechnique. For further details see [HM], Vol.1, pp.336-345, or [MA], Chapt.2.

[4] Vladimir ARNOL'D, Russian mathematician born in 1937, published many works, all fascinating, on the theory of singularities and its applications, notably in mechanics. Their influence on this text is undeniable; they are recommended reading, although it is necessary to be aware of some ideosyncracies in terminology or orthography (particularly in proper names) and the unconstructively polemical (and sometimes seemingly abusive) nature of certain claims of priority.

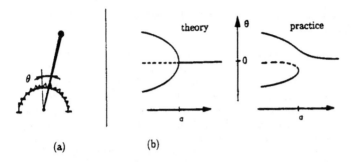

(a)   (b)

Fig. 0.3. (a) Bar supported by symmetric springs, (b) broken symmetry in the bifurcation

7. However, this in turn opens up another argument: how on earth can we effectively distinguish between harmless simplifying assumptions and those which make the model not structurally stable and therefore quite unrealistic for reasons that we have just seen?

To explain how this question can be answered we have to start by specifying the context a little more accurately. Consider a system (mechanical, physical, chemical, ...) described by a certain number of *state variables* satisfying certain characteristic relations (algebraic or differential, for example) in which some *control variables*[5] are involved in a known way (such as the spring control in the previous example), while an unspecified number, quite probably infinite, of *hidden parameters* are involved in unknown ways which in particular reflect the 'imperfections' of the concrete physical object that is modelled by our abstract system. In other words, all that we know is that our system is near, and as close as we wish (or rather as we are prepared to pay for), to the theoretical system described by our state and control variables and by the relations that connect them. Now, if this theoretical situation is structurally stable, the real system will have behaviour that is close (and even as close as we wish etc. ...) to the theoretical behaviour. On the other hand, if the theoretical system is too special to be structurally stable (supposing that our two springs are exactly symmetric) then the hidden parameters come into play in unpredictable ways; in that case it usually turns out to be enough to incorporate new control variables (so that instead of considering the difference between our two springs as a hidden variable, we take account of it) in order to recover a structurally stable system.

Consequently the structurally unstable symmetric system with one control variable

$$\ddot{\theta} = \sin\theta - a\theta,$$

which (in suitable units) models the bar held by springs that are assumed to be perfectly symmetric, has to be replaced by the asymmetric system with

---

[5] One contemporary French philosopher uses the excellent terminology *dynamic variables* and *strategic variables*.

two control variables

$$\ddot{\theta} = \sin\theta - a\theta + b,$$

which can be shown to be structurally stable and which therefore correctly takes account of the real situation, whatever (small) hidden perturbation it may be subjected to.

8. In the same spirit, we can present the problem of *linearization* of a differential system as follows. Consider the system

$$(S) \qquad\qquad \frac{dx}{dt} = f(x,y), \quad \frac{dy}{dt} = g(x,y),$$

where the vector field

$$X : (x,y) \mapsto (f(x,y), g(x,y))$$

vanishes at the origin, and also the linearized system

$$(L) \qquad\qquad \frac{dx}{dt} = ax + by, \quad \frac{dy}{dt} = cx + dy,$$

with

$$f(x,y) = ax + by + \cdots, \quad g(x,y) = cx + dy + \cdots.$$

We wish to compare the trajectories of the system (S) in a neighbourhood of 0 with those of the linear system (L). We see the resemblence to the earlier discussion: with (S) being close to (L), and the nearer to the origin the closer it is to (L) (by Taylor's formula), can we deduce that the trajectories of (S) are close to those of (L)? We are dealing with a 'structural stability' property of (L), and the answer is given by the Hartman-Grobman Theorem (in the special case of dimension 2 as here, a slightly stronger theorem applies[6]): if the matrix $\left(\begin{smallmatrix} a & b \\ c & d \end{smallmatrix}\right)$ has no purely imaginary eigenvalue then in a neighbourhood of the origin the system (S), which is a perturbation of (L), has its trajectories close to those of the latter.

On the other hand, in the case when these eigenvalues are purely imaginary the trajectories of (L) are circles centred at 0, while those of (S) may have very different global form however close (S) is to (L). In fact in this case (L) is already structurally unstable within linear systems (see Fig. 0.4).

9. These reflections naturally lead us to hope that at the end of the day structural stability will be a common occurrence. To be more precise, in order for the philosophy sketched in Para. 7 above to work effectively we would want it to be the case that structurally stable sytems could be found in the neighbourhood of any system under consideration. In mathematical terms, the structurally stable systems ought to be *dense* in the set of systems, or, to put it another way, a system chosen at random (a so-called 'generic' system)

---

[6] For these two theorems see Sects. 8.7 and 8.11.

Fig. 0.4. Example of a structurally unstable linear system

ought to be structurally stable. This is what is called somewhat pompously the 'structural stability hypothesis'. We shall see later what we should think of it.

Meanwhile, there is something that obviously has to be made precise in all that we have been saying so far: what do we understand technically by 'nearby properties', 'situation that does not change', 'dense', 'generic', etc.? Take the example of Hartman's Theorem mentioned above. It can be stated as follows: if the linear system (L) has no purely imaginary eigenvalue (in short, if it is 'hyperbolic') then for every system (S) with linearization (L) there is a (fairly regular) transformation of the xy plane in a neighbourhood of 0 which takes the trajectories of (L) to those of (S) preserving the parametrization by time.

Clearly the only content of a theorem of this kind is the assertion of regularity of the transformation: the more regularity we assert, the more significant the theorem becomes and the more worthwhile the operation of linearization will be in practice, but of course the lower the likelihood of the theorem being true. In the precise case of linearization the answer is a little subtle: without additional hypotheses we cannot obtain very much regularity and we can not even ensure differentiability of the transformation, but only a somewhat weaker property (except as already indicated in dimension 2; but even in this case the existence of second derivatives is not guaranteed). On the other hand, with an additional assumption on the eigenvalues (absence of 'resonances'[7], which is a 'generic' assumption) we can assert the infinite differentiability of the transformation: this is Sternberg's Theorem.

10. In these few rather informal paragraphs we have met the essential keywords of the subject: *transversality, genericity, structural stability, linearization, bifurcations, catastrophes, ...* There are at least two missing. The first is *dissipative*; in fact we shall be saying nothing serious about non-dissipative or *conservative* systems whose study is more complicated and involves tools of infinite subtlety. The absence of the other word is more curious. Is this really a text on *geometry*? The question raises another: what is geometry? It is not easy to answer this, and the answers change over time. Nevertheless geometric objects exist, and the things we have been briefly talking about above certainly belong to the universe of geometry. Moreover, it is clear that

---

[7] See Sects. 8.10 and 8.11.

differentiable objects and their singularities are omnipresent in nature and also in the theoretical models that have been developed to take account of them. This is much less clear for the students of today than for their elders. Teaching geometry at university level nowadays requires picking up the threads of a broken tradition. In particular, the absence of prior knowledge in certain areas (particularly in algebra) and the general state of teaching in schools rule out whole sections of geometry (algebraic geometry, finite geometries, ...). In my view the themes developed here have the advantage of being relevant to both the content and outlook of other texts as well as to current movements in contemporary ideas (in bifurcation, turbulence and so on). The later chapters will lead us to living questions on which research is active and fruitful and which would have seemed totally far-fetched just a few years ago.

In fact a title more in line with the content of this book could have been: 'geometric methods in the study of singularities and bifurcations', or (why not ?) 'from Poincaré to Smale and beyond' ....

11. However, even with the best intentions in the world, it is not possible to do mathematics without technique. The essential characteristic of true statements is that they possess a proof. We would like to be able to delegate to specialists alone the task of checking the existence and validity of proofs. To a large extent this is possible, but there are of course several drawbacks. First of all (and this is the case for the easiest ones) the most economical way to understand the meaning and the limitations of an assertion is often to read a proof of it. But there are more important considerations: in order to be able to apply the conclusions of a theorem it is necessary for the hypotheses to be satisfied (although it would seem that this obvious fact has not penetrated the whole of science, especially the 'inexact'[8] sciences). It is quite common, when confronted with a concrete problem, to find a theorem which 'almost applies'. This is why in order to master a proof it is usually necessary to have a clear idea of the status of the hypotheses that it contains: are they there for circumstantial reasons (convenience of exposition, simplification of the proof, sympathy for the reader ...) or because they are essential? Often a small modification of the premises entails only a minor change in the consequences (could we then talk about 'stable' or 'generic' results? Unfortunately, a result picked at random has little chance of being interesting! ...); sometimes, on the other hand (could we call these 'critical' results?) a slight weakening of the hypotheses turns into complete collapse of the conclusions. The role of *counterexamples* is precisely to set boundaries on the possible, and it is not due to perversity (or at least not totally) that mathematical texts tend to exhibit monsters[9].

---

[8] So called in opposition to the 'inhumane' sciences ... . On the abuses of pseudo-mathematical modelling in the social sciences, see [BE].

[9] Hermite said, no doubt with a touch of humour, that he "turned away in fear and horror from this lamentable plague of continuous functions that have no

As far as the theorems in this text are concerned, they come in three forms. Some of them, usually the most elementary ones, are accompanied by a complete proof. For others, just the main ideas of the proof are indicated. Finally, for the last type which are too difficult or which appeal to notions too far afield, no indication of the proof is given. Note incidentally that the sign □ marks the end of a proof (or its absence when following directly after the statement).

12. Prerequisites. This book assumes a basic knowledge of linear algebra, general topology and functions of several variables. To help the reader, some of the most important results needed are recalled in the text. Apart from the notational items listed below, these results are placed where they arise most naturally rather than being lumped together in a special section that would inevitably be indigestible. This is the reason why it sometimes happens that we use an auxiliary notion that is not taken up again until later; in that case use of the Index should enable the reader to find the corresponding reference easily.

In places we use small type to give commentaries or include additional observations that may help some readers to make connections between the text and knowledge they may already have from elsewhere. For example, a few variants of terminology are noted in this way. Those who find that these inclusions add to the difficulty can ignore them, at least on the 'first reading'. In the footnotes there is some brief biographical information about mathematicians whose work is quoted. The last section of each chapter aims to give a slightly more global historical overview; this may also contain some biographical pointers. Symmetrically, in the first introductory section of each chapter there is an attempt to motivate or justify the subjects treated in the chapter and the approach chosen. Other comments of this nature can also be found scattered in the text.

13. References. Within each chapter the number of that chapter is to be understood. Thus to refer to Proposition 2.4 in Chapter 3 we shall say "by Proposition 2.4" or "by Proposition 3.2.4" according to whether we are inside Chapter 3 or not. Numbered formulae are referred to in the same way, with the difference that the number is placed in parentheses as in "formula (2.4)" or "formula (3.2.4)", or perhaps just (2.4) or (3.2.4).

Throughout the text the capital letters in square brackets (for example [HM]) refer to the bibliography. The works included vary greatly in length and level. Some of them such as [AS], [A3], [A6] are elementary introductions, while others such as [AA], [BL], [HS], [IR], [PS] are at a level comparable to the present text. The monographs [AM], [A5], [AG], [HA], [GH], [KH], [MV],

---

derivatives", but it was necessary to exhibit some of them if only to put an end to the string of false proofs, often underwritten by famous names, of the fact that every continuous function was differentiable.

[PM], [RO] give an idea of more advanced developments and in some cases the present state of research.

14. I have benefitted from the help of several colleagues in the preparation of this text. I especially thank Marc Chaperon as well as Jean-Pierre Bourguignon, Marc Giusti and Jean Lannes who have shared with me many valuable observations. Several errors in the earlier (French) edition were detected by students in the class of 1984, among whom I must give a special mention to Max Bezard.

1 Local Inversion

1.1 Introduction

This chapter is concerned with the following problem : how can we conveniently recognize when a map from (an open subset of) one vector space into another is invertible, and what regularity can we hope for in the inverse map? In fact it is very rare to be able to prove that the map is globally invertible, and we have to restrict ourselves to a 'local' statement.

The basic idea behind this type of statement is *linearization*: if the map is sufficiently close to an invertible linear map then it ought itself to be invertible, at least locally. This leads automatically to an assumption about the tangent linear map and hence to a statement in terms of derivatives. We shall take this opportunity to revise quickly the rudiments of differential calculus of several variables, mainly in order to fix the vocabulary and notation.

Another theme that merges with the above throughout this chapter is that of *curvilinear coordinates*. We start with a finite-dimensional vector space E, an open set U in E and a basis $(e_1, \ldots, e_n)$ for E. Each element ξ of U can be written uniquely as $\xi_1 e_1 + \cdots + \xi_n e_n$; we denote the maps $\xi \mapsto \xi_1, \ldots, \xi \mapsto \xi_n$ defined on U by $x_1, \ldots, x_n$ (the basis dual to the given basis). The functions $x_1, \ldots, x_n$ are then the components of a map from U into $\mathbf{R}^n$ which is the restriction of a vector space isomorphism from E to $\mathbf{R}^n$; we say that they form a system of linear coordinates on U. It is helpful to be able to use more general coordinate systems that may be better adapted to the situation we want to study, and in particular to any symmetries that may be present (such as polar or spherical coordinates, for example). A *system of generalized coordinates* on U is a sequence of n real functions $x_1, \ldots, x_n$ on U that define a bijection f, say, from U to a subset U' of $\mathbf{R}^n$. Naturally, if we wish to work with such a coordinate system we have to impose some regularity conditions.

The weakest condition is the following : U' is open in $\mathbf{R}^n$, and f and f^{-1} are continuous (that is, f is a *homeomorphism*). This gives a very flexible notion, but it is too weak for many applications since *a priori* the map f could transform two curves that are tangent into two curves that are not tangent, which would rule out the use of such coordinates for studying contact between two curves, for example. Therefore we have to insist that f and f^{-1} be sufficiently differentiable (we say that f has to be a *diffeomorphism*) ; for most applications we can impose the strongest version of this, namely that

f and f^{-1} be infinitely differentiable. Note incidentally that the expression 'curvilinear coordinates' comes from the fact that straight lines in $\mathbf{R}^n$ then correspond to curves in U.

In fact the situation is a little more subtle. Take the example of polar coordinates : it is not possible to find a map $(x, y) \mapsto (r, \theta)$ or a map $(r, \theta) \mapsto (x, y)$ which works for the whole domain under consideration (the (x, y) plane with the origin removed, or the (r, θ) plane). This leads us to the notion of a *local* system of (curvilinear) coordinates, and it is here that we meet again the theme of local invertibility.

To end these brief comments we note that the fact that we have considered coordinates as functions on the domain U, a point of view in a sense dual to the usual one in elementary calculus, will turn out to be very convenient in practice as (we hope) the reader will later agree.

We now give a brief description of the plan of this chapter. A preliminary statement, which is very useful in its own right, deals with the case of perturbing the identity by adding a term which is 'small' in a suitable sense; this is the subject of Sect. 2. In Sect. 3 we introduce the notions of partial derivative and strictly differentiable function, and in Sect. 4 we prove the *Local Inversion Theorem* in very general form. After discussing the notion of functions of class C^r in Sect. 5 we deduce in Sect. 6 the Local Inversion Theorem for C^r functions. In Sect. 7 we introduce the notion of 'curvilinear coordinates'. The last section gives some historical comments.

The Local Inversion Theorem has many useful variants (the Rank Theorem, the Implicit Function Theorem, ...). In general these are easier to understand from the geometric point of view, which is why we shall wait until we are able to use the language of *submanifolds* in the next chapter before dealing with them.

1.2 A Preliminary Statement

1.2.1. As an introduction to what follows, recall the following elementary result : if u is a continuous[10] endomorphism of the complete normed vector space E with $\|u\| < 1$, then the endomorphism $Id_E + u$ is invertible with inverse $Id_E - u + u^2 - u^3 + \cdots$, where obviously we write u^2 for $u \circ u$, etc.. The series converges absolutely since $\|u\| < 1$. Let us now present things in a slightly different way and look for the inverse of $Id_E + u$ in the form $Id_E + v$. Writing I for the identity map Id_E on E for ease of notation, we thus have to solve $(I + u) \circ (I + v) = I$, or in other words $x + v(x) + u(x + v(x)) = x$ or

$$v(x) = -u(x + v(x)), \quad x \in E. \tag{2.1}$$

---

[10]This condition is automatically satisfied if the space under consideration is finite-dimensional. This is essentially the case which interests us ; nevertheless the general case does not present any additional difficulty and is extremely useful.

However, since u is by assumption *linear* (this is the first time that property is used) we can write the above as $v = -u - u \circ v$. By the method of successive approximation we let

$$v_0 = 0, \ v_1 = -u, \ v_2 = -u - u \circ v_1 = -u + u^2, \ \ldots, \ v_n = -u - u \circ v_{n-1},$$

from which we find $v_n = -u + u^2 - u^3 + \cdots + (-1)^n u^n$ and $v = \lim v_n = \sum_{n=1}^{\infty} (-1)^n u^n$, where the v_n form a Cauchy sequence because $\|v_n - v_{n-1}\| \leq \|u^n\| \leq \|u\|^n$.

Written in this way, the proof is easily adapted to the nonlinear case. Fix a complete normed linear space E and for $r > 0$ and $\lambda > 0$ let $L_{r,\lambda}$ denote the set of maps u from the open ball $B_r = \{\, x \in E \mid \|x\| < r \,\}$ into E such that

$$u(0) = 0 \qquad (2.2)$$

and which are *Lipschitz* with Lipschitz constant λ, that is they satisfy

$$\|u(x) - u(y)\| \leq \lambda \|x - y\|, \ x \in B_r, \ y \in B_r. \qquad (2.3)$$

Lemma 1.2.2. *Suppose $\lambda < 1$ and let $r' = (1 - \lambda)r$, $\lambda' = \lambda/(1 - \lambda)$. Let $u \in L_{r,\lambda}$. Then there exists a unique element v of $L_{r',\lambda'}$ such that*

$$v(x) = -u(x + v(x)), \ x \in B_{r'}. \qquad (2.4)$$

Note first of all that the right hand side of the above does make sense because for $x \in B_{r'}$ we have $\|x + v(x)\| \leq (1 + \lambda')\|x\| \leq (1 + \lambda')r' = r$. Next observe that for $v \in L_{r',\lambda'}$ and $x \in B_{r'}$ we have

$$\|v(x)\| = \|v(x) - v(0)\| \leq \lambda' \|x - 0\| < \lambda' r' = \lambda r.$$

First we fix x in $B_{r'}$ and prove the existence of a unique y in B_r with

$$y = -u(x + y). \qquad (2.5)$$

The uniqueness results from the fact that if y and y' both satisfy (2.5) then $\|y - y'\| = \|u(x + y') - u(x + y)\| \leq \lambda \|y - y'\|$ and so $\|y - y'\| = 0$. As before, we define a sequence of elements of $B_{\lambda r}$ by $y_0 = 0$, $y_1 = -u(x), \ldots$ and

$$y_n = -u(x + y_{n-1}). \qquad (2.6)$$

This makes sense because if we suppose that $\|y_{n-1}\| < \lambda r$ then we deduce that $\|x + y_{n-1}\| < r' + \lambda r = r$ and $\|u(x + y_{n-1})\| < \lambda r$. For $n \geq 2$ we have

$$\|y_n - y_{n-1}\| = \|u(x + y_{n-1}) - u(x + y_{n-2})\| \leq \lambda \|y_{n-1} - y_{n-2}\|;$$

from this and the initial estimate $\|y_1 - y_0\| = \|u(x)\| \leq \lambda \|x\|$ we deduce

$$\|y_n - y_{n-1}\| \leq \lambda^n \|x\|.$$

Hence the y_n form a Cauchy sequence in the complete metric space E. Let

$$y = \lim_{n \to \infty} y_n \in E.$$

Since $\|u(x) - u(x')\| \leq (1+\lambda)\|x - x'\|$ the map u is continuous and (2.5) follows from (2.6) on passing to the limit. It remains to prove that the map $x \mapsto y$ does belong to $L_{r',\lambda'}$. First, it is clear that for $x = 0$ we have $y_n = 0$ for all n and so $y = 0$. Consider a second element x' of $B_{r'}$, and construct the sequence y'_n and its limit y' just as above. We shall show that $\|y - y'\| \leq \lambda'\|x - x'\|$, which will complete the proof. For all $n \geq 1$ we have

$$\|y_n - y_{n'}\| = \|u(x + y_{n-1}) - u(x' + y'_{n-1})\| \leq \lambda\|x - x'\| + \lambda\|y_{n-1} - y'_{n-1}\|;$$

since $\|y_0 - y'_0\| = 0$ it follows by induction that for all n we have

$$\|y_n - y'_n\| \leq (\lambda + \cdots + \lambda^n)\|x - x'\| \leq \lambda'\|x - x'\|$$

and on taking the limit as $n \to \infty$ we obtain the desired inequality.   □

Proposition 1.2.3. *Let r and λ be real numbers with $r > 0$ and $0 < \lambda < 1$. Let f be a map from B_r into E such that*

$$\|f(x) - f(x') - (x - x')\| \leq \lambda\|x - x'\|$$

for x and x' in B_r. Let $r' = (1 - \lambda)r$, $\lambda' = \lambda/(1 - \lambda)$, $U' = f^{-1}(B_{r'}) \cap B_r$ and $V' = B_{r'}$. Then U' is open, f induces a continuous injective map from U' into $V' = B_{r'}$, the inverse bijection $g : V' \to U'$ is continuous, and we have

$$\|g(y) - g(y') - (y - y')\| \leq \lambda'\|y - y'\|$$

for y and y' in V'.

Define a map $u : B_r \to F$ by $f(x) = x + u(x)$. With the notation above we thus have $u \in L_{r,\lambda}$ and the lemma implies that there exists $v \in L_{r',\lambda'}$ with $v(x) + u(x + v(x)) = 0$ for all $x \in B_{r'}$. Since

$$(1 - \lambda)\|x - x'\| \leq \|f(x) - f(x')\| \leq (1 + \lambda)\|x - x'\|$$

the map f is continuous and injective. Since f is continuous, U' is open. Define $g : V' \to E$ by $g(y) = y + v(y)$; for y in V' we have $g(y) \in B_r$ and $f(g(y)) = g(y) + u(g(y)) = y + v(y) + u(y + v(y)) = y$. Hence g maps V' into U' and f and g are mutually inverse bijections between U' and V'. By construction, g is continuous and satisfies the stated inequality, which completes the proof.   □

Note that if f is linear we may take r to be arbitrarily large, say $r = +\infty$, and $\lambda = \|f - 1\|$. We then recover precisely the statement at the beginning of this section; we obtain moreover that $\|f^{-1} - 1\| \leq \|f - 1\|/(1 - \|f - 1\|)$, an estimate which can also be derived from the construction of f^{-1} as the sum of a geometric series.

1.3 Partial Derivatives. Strictly Differentiable Functions

1.3.1. Let E be a vector space with U a subset of E, let F be a normed vector space and let $f : U \to F$ be a map. Let a be a point in U and let v be a vector in E. Consider the *partial function* $t \mapsto f(a + tv)$ which is defined on the set I of those $t \in \mathbf{R}$ such that $a + tv$ belongs to U. If I contains an interval $(-\varepsilon, \varepsilon)$ and if

$$\frac{1}{t}(f(a + tv) - f(a))$$

has a limit in F as t tends to 0, this limit is called the *partial derivative* (or the partial derivative vector) of f at a with respect to v. If this derivative exists for every $a \in U$ we say that the partial derivative of f with respect to v exists on U and we write

$$(L_v f)(a) = \lim_{t \to 0} \frac{1}{t}(f(a + tv) - f(a)) \in F. \tag{3.1}$$

For any real number λ we see at once that $L_{\lambda v} f = \lambda L_v f$.

Naturally, if E is finite-dimensional and a basis has been chosen for F, then all these definitions immediately reduce to the analogous definitions for the components of f, which are scalar functions on U.

More generally, if u is a continuous linear form on F then $L_v(u \circ f) = u \circ L_v f$; the above is the special case when we take as u the coordinate forms on F.

1.3.2. Traditionally, when E is finite-dimensional and a basis $(e_1, \ldots, e_n)$ has been chosen for E together with its dual basis $(x_1, \ldots, x_n)$ such that

$$\xi = \langle \xi, x_1 \rangle e_1 + \cdots + \langle \xi, x_n \rangle e_n, \quad \xi \in E,$$

we write $\frac{\partial f}{\partial x_i}$ instead of $L_{e_i} f$. This classical notation has a serious disadvantage of which it is important always to be aware : if $L_e f$ is well defined in terms of f and e then $\frac{\partial f}{\partial x}$ does not depend only on the choice of f and of $x \in E^*$; but in fact $\frac{\partial f}{\partial x_i}$ depends more on the choice of the remaining x_j with $j \neq i$! For example, if E is a plane with basis $\{i, j\}$ consider the new basis $\{i' = i + j, \ j' = j\}$, so that $xi + yj = x'i' + y'j'$ with $x' = x$, $y' = y - x$. We have then $x' = x$ but $\frac{\partial}{\partial x'} = L_{i'} \neq L_i = \frac{\partial}{\partial x}$, and conversely $y' \neq y$ but $\frac{\partial}{\partial y'} = L_{j'} = L_j = \frac{\partial}{\partial y}$. Thus $\frac{\partial y'}{\partial x'}$ is either 0 or -1 according to the context. This abuse of notation is precisely the same one that occurs in writing $x_1 = e_1^*$.

The notion of partial derivative can be generalized to that of *Lie derivative*[11]. Let X be a map from U into E (a 'vector field' on U). The Lie

---

[11]The Norwegian mathematician Sophus LIE (1842-1899), who was professor at the University of Christiania (now Oslo) from 1872 to 1899, produced major work on infinite groups that was unrecognized during his lifetime. The Lie derivative is known by many names, notably *particular derivative* in mechanics and *convective derivative* in a slightly more general context.

derivative $L_X f$ of f with respect to the field X is the map of U into F defined, when it exists, by

$$(L_X f)(a) = \lim_{t \to 0} \frac{1}{t}\big(f(a + tX(a)) - f(a)\big) \in F. \tag{3.2}$$

For any scalar function g on U we then have

$$L_{gX} f = g L_X f. \tag{3.3}$$

When X is the constant vector field with value v we recover the partial derivative introduced above.

1.3.3. Let E and F be two normed vector spaces with norms denoted by $x \mapsto \|x\|_E$ and $y \mapsto \|y\|_F$. Let U be an open set in E with a belonging to U and let $f : U \to F$ be a map. We say that f is *strictly differentiable* at the point a if there exists a continuous [12] linear map $\phi : E \to F$ having the following property : for h and k in E, small enough so that $a + h$ and $a + k$ belong to U, define the element $\alpha(h, k)$ of F by

$$f(a + h) - f(a + k) = \phi \cdot (h - k) + \alpha(h, k); \tag{3.4}$$

then as $\sup(\|h\|_E, \|k\|_E)$ tends to zero we have

$$\lim \|\alpha(h, k)\|_F / \|h - k\|_E = 0. \tag{3.5}$$

Let $v \in E$; taking $h = tv$ with t real and sufficiently small and $k = 0$ we obtain

$$\phi \cdot v = \lim_{t \to 0} \frac{1}{t}\big(f(a + tv) - f(a)\big) = (L_v f)(a)$$

and $\phi \cdot v$ is the partial derivative of f at a with respect to v. Thus the linear map ϕ is determined uniquely; we call it the *tangent map* to f at a, or the *derivative* of f at a, and denote it by

$$f'(a) \in L(E; F). \tag{3.6}$$

Thus for every $v \in E$ we have

$$f'(a) \cdot v = (L_v f)(a). \tag{3.7}$$

In particular, if f is strictly differentiable at a then the map which associates to each element v of E the partial derivative of f at a with respect to v is *linear and continuous*.

The natural setting for the above definitions is that of *affine* spaces. Indeed, if E and F are two affine spaces with E^{tr} and F^{tr} their respective vector spaces of translations, and if f is a map from an open set U of E into F, the formulae

---

[12]Recall that if E and F are finite-dimensional then every linear map from E to F is continuous.

above remain meaningful when h,k and v are taken as elements of E^{tr} and $f'(a)$ is a linear map from E^{tr} into F^{tr}.

If f is strictly differentiable at every point of U we have the *derived* map or *derivative* map

$$f' : U \to L(E; F)$$

which associates to each point a of U the linear map $f'(a)$.

In infinite dimensions this notion of derivative is generally called the *derivative in the sense of Fréchet*. There is a slightly weaker notion, which differs from this one for more general spaces, called the *derivative in the sense of Gâteaux*.

1.4 The Local Inversion Theorem: General Statement

Theorem 1.4.1. *Let E and F be two complete normed vector spaces with U an open subset of E. Let a be a point of U, and let f be a map from U into F which is strictly differentiable at a. Suppose that the linear tangent map $f'(a)$ is an isomorphism[13] from E onto F. Then there exists an open set U' in U containing a and an open set V' of F containing $f(a)$ with the following properties:*

a) *f induces a continuous bijection from U' onto V',*
b) *the inverse map : $V' \to U'$ is continuous,*
c) *g is strictly differentiable at $f(a) \in F$; its derivative is $f'(a)^{-1} \in L(F; E)$.*

We begin the proof with a small digression. Let G be a third complete normed vector space and $u : F \to G$ an isomorphism. Write $g = u \circ f : E \to G$. Then (by the theorem on differentiation of compositions : see 5.4) g is strictly differentiable at a with derivative $u \circ f'(a)$ and it is elementary to verify that proving the theorem for $f : U \to F$ is tantamount to proving it for $g : U \to G$. Now, we may take u to be the isomorphism $F \to E$ inverse to $f'(a)$, and this reduces us to proving the theorem in the particular case where $F = E$ and where $f'(a) = Id_E$. Since translations in E and F are obviously harmless, we may also replace f by $x \mapsto f(x + a) - f(a)$. Thus we have now reduced the situation to the case where $F = E$, $f'(a) = Id_E$, $a = 0$ and $f(a) = 0$, which is what we shall assume from now on.

Provisionally fix λ with $0 < \lambda < 1$. By definition of strict differentiability there exists $r > 0$ such that the open ball B_r is contained in U and for x and x' in B_r we have

$$\|f(x) - f(x') - (x - x')\| \leq \lambda \|x - x'\|.$$

---

[13]The expression "isomorphism of E onto F" means "bijection from E onto F with continuous inverse"; the latter property is a consequence of the former when E and F are finite-dimensional.

We can therefore apply Proposition 2.3 and deduce the existence of open sets U' and V' having the required properties a) and b). Moreover, if we put $r' = (1 - \lambda)r$ and $\lambda' = \lambda/(1 - \lambda)$ we then have

$$\|g(y) - g(y') - (y - y')\| \leq \lambda'\|y - y'\|$$

for y and y' in $B_{r'}$. Since $\lambda' = \lambda/(1 - \lambda)$ and λ could be chosen to be arbitrarily small, this implies that g is strictly differentiable at the point 0, with derivative Id_E. The proof is therefore complete.   □

1.5 Functions of Class C^r

Let E be a finite-dimensional vector space, with U an open subset of E and f a map from U into a normed vector space F. Let $(e_1, \ldots, e_n)$ be a basis for E, with dual basis $(x_1, \ldots, x_n)$. Equip E with any norm (since it is well known that they are all equivalent), for example

$$\|\xi_1 e_1 + \cdots + \xi_n e_n\| = \sup_i |\xi_i|.$$

Likewise, give $L(E; F)$ the norm

$$\|\phi\| = \sup_i \|\phi(e_i)\|_F.$$

Proposition 1.5.1. *The following conditions are equivalent:*

(i) *the partial derivatives $L_{e_i} f = \frac{\partial f}{\partial x_i}$ exist and are continuous on U;*
(ii) *the map f is strictly differentiable at every point of U and the derivative map $a \mapsto f'(a)$ from U into the normed space $L(E; F)$ is continuous.*

Since $(\partial f/\partial x_i)(a) = f'(a) \cdot e_i$, property (i) follows immediately from (ii). Suppose conversely that (i) is satisfied; we shall prove a result that implies (ii), namely that for every $a \in U$ the map f is strictly differentiable at a with derivative $f'(a)$ equal to

$$(\xi_1, \ldots, \xi_n) \mapsto \xi_1 \frac{\partial f}{\partial x_1}(a) + \cdots + \xi_n \frac{\partial f}{\partial x_n}(a).$$

Thus, fix a point a of U, let $\varepsilon > 0$ be fixed, and choose $r > 0$ with $a + B_r \subset U$ and

$$\|\frac{\partial f}{\partial x_i}(a + x) - \frac{\partial f}{\partial x_i}(a)\|_F \leq \varepsilon, \quad x \in B_r, \quad i = 1, \ldots, n. \tag{5.1}$$

We prove a preliminary result : let $u \in B_r$ and $i \in [1, \ldots, n]$ and let α be a real number such that $u + \alpha e_i \in B_r$; we then have

$$\|f(a + u + \alpha e_i) - f(a + u) - \alpha \frac{\partial f}{\partial x_i}(a)\|_F \leq |\alpha|\varepsilon. \tag{5.2}$$

To see this, let g be the map

$$t \mapsto f(a + u + t\alpha e_i) - \alpha t \frac{\partial f}{\partial x_i}(a).$$

On the interval $[0, 1]$ this map admits a continuous derivative, namely $g'(t) = \alpha \frac{\partial f}{\partial x_i}(a + u + t\alpha e_i) - \alpha \frac{\partial f}{\partial x_i}(a)$; by (5.1) we have $\|g'(t)\| \leq |\alpha|\varepsilon$ for all t in $[0, 1]$, and so $\|g(1) - g(0)\| \leq |\alpha|\varepsilon$ – which is precisely (5.2).

Having established this, now let $h, k \in B_r$ and let

$$u_0 = h, \ u_1 = h + (k_1 - h_1)e_1, \ u_2 = u_1 + (k_2 - h_2)e_2, \ \ldots, \ u_n = k.$$

Applying (5.2) for $u = u_{i-1}$ and $\alpha = k_i - h_i$ we obtain

$$\|f(a + u_i) - f(a + u_{i-1}) - (k_i - h_i)\frac{\partial f}{\partial x_i}(a)\|_F \leq |k_i - h_i|\varepsilon,$$

and hence by summation

$$\|f(a + k) - f(a + h) - \sum_{i=1}^{n}(k_i - h_i)\frac{\partial f}{\partial x_i}(a)\|_F \leq n\varepsilon\|h - k\|_E.$$

But this means precisely that f is strictly differentiable at a and that $f'(a)$ is as stated. $\qquad\square$

1.5.2. We thus see that the condition (i) is independent of the choice of basis $(e_1, \ldots, e_n)$. Functions which satisfy this condition are said to be of *class C^1*. If f is of class C^1 the derived map $f' : U \to L(E; F)$ is continuous (we also say of class C^0) ; by induction we say that f is of class C^r (where r is an integer ≥ 1) if f' is of class C^{r-1}. Since the coefficients of f' are the $\partial f/\partial x_i$ this means also that the $\partial f/\partial x_i$ are of class C^{r-1}. Applying the proposition again we deduce that saying that f is of class C^r is the same as saying that for every integer $k \in \{1, \ldots, r\}$ and every sequence $\{i_1, \ldots, i_k\}$ of elements of $\{1, \ldots, n\}$ the iterated partial derivative $L_{e_{i_1}} \cdots L_{e_{i_k}} f$, which we also write as

$$L_{e_{i_1}} \cdots L_{e_{i_k}} f = \frac{\partial^k f}{\partial x_{i_1} \cdots \partial x_{i_k}} = f_{x_{i_1} \ldots x_{i_k}},$$

exists and is continuous on U. This tells us too (*Schwarz'* [14] *Lemma*) that the above iterated partial derivatives are symmetric functions of the indices $(i_1, \ldots, i_k)$.

We say that f is of class C^∞ if it is of class C^r for every r, that is if all its iterated partial derivatives exist and are continuous on U.

When E is infinite-dimensional we take the property (ii) of 5.1 as the definition of functions of class C^1. We then define the functions of class C^r, $r \in [1, \infty]$ as above.

---

[14]Hermann Amandus SCHWARZ (1843-1921) succeeded Weierstrass in 1892 at the University of Berlin.

As a matter of curiosity, note that in (ii) the continuity assumption is superfluous: if f is strictly differentiable at every point of U then f' is continuous and f is of class C^1. Of course, if we delete the adjective *strictly* the statement becomes false: the function $x \mapsto x^2 \sin(1/x)$, extended to have the value 0 at 0, is differentiable everywhere but its derivative is not continuous at 0. This function is not strictly differentiable at 0.

1.5.3. We now look at the two particular cases of the above that arise when one of the spaces is $\mathbf{R}$. First of all consider an open interval I of $\mathbf{R}$ and a map γ of I into a finite-dimensional vector space E (so γ is a *parametrized arc in E*). Then at a point t of I the derivative $\gamma'(t)$ (in the sense above) is the linear map $\lambda \mapsto \lambda \dot{\gamma}(t)$ of $\mathbf{R}$ into E, where $\dot{\gamma}(t)$ is the usual derivative vector of γ at the point t. To avoid confusion we shall systematically use the notation $\dot{\gamma}(t)$ for the usual derivative vector, and we shall call it the *velocity vector* of γ at time t (and at the point $\gamma(t)$).

In dual fashion, consider an open set U of E, a point a of U and a map $\phi : U \to \mathbf{R}$. Then the derivative $\phi'(a)$ is a linear map from E to $\mathbf{R}$, and therefore an element of the dual space E^* of E. The map $U \to E^*$ defined in this way is called the *differential* of ϕ and is denoted by $d\phi$. If a basis $(e_1, \ldots, e_n)$ is chosen for E, with dual basis $(x_1, \ldots, x_n)$, the differential of the function x_i is given by the formula

$$dx_i(a) \cdot \xi = \xi_i = \langle \xi, x_i \rangle,$$

and is thus the constant map with value x_i. Hence we obtain the classical expression from differential calculus

$$d\phi = \frac{\partial \phi}{\partial x_1} dx_1 + \cdots + \frac{\partial \phi}{\partial x_n} dx_n$$

relating the maps $d\phi$ and $dx_1, \ldots, dx_n$ from U into E^* with the functions $\frac{\partial \phi}{\partial x_i}$ on U.

Furthermore, when the space E is equipped with a euclidean scalar product $(\xi \mid \eta)$ we can associate to the linear form $d\phi(a)$ the vector $\mathrm{grad}(\phi)(a)$ such that

$$(\mathrm{grad}(\phi)(a) \mid \eta) = d\phi(a) \cdot \eta, \qquad \eta \in E.$$

In particular, the vector $\mathrm{grad}(f)(a)$ is orthogonal to the kernel of the linear form $df(a)$. Letting a vary, we obtain a vector field $\mathrm{grad}(f)$ on U, called the *gradient* of f (relative to the given euclidean structure on E). In an orthonormal basis the components of $\mathrm{grad}(\phi)$ are the partial derivatives $\partial \phi / \partial x_i$.

1.5.4. Let E' be another finite-dimensional vector space, with U' an open set in E' and g a map from U' to U. Then if f and g are of class C^1 the composition $f \circ g$ is of class C^1 and in $L(E; F)$ we have the *formula for differentiating composed functions (maps)*

$$(f \circ g)'(a) = f'(g(a)) \circ g'(a), \qquad a \in U'. \tag{5.3}$$

Indeed, it can be verified immediately from the definition (see (3.4) and (3.5)) that $f \circ g$ is strictly differentiable at every point a of U' and has derivative $f'(g(a)) \circ g'(a)$. However, this expression depends continuously on a and so $f \circ g$ is actually of class C^1. It follows by induction that if f and g are of class C^r, $r \in [1, \infty]$ then the map $f \circ g$ is also of class C^r.

We now look at some particular cases of the formula for differentiating composed functions. Let E, F, U and f be as above. First of all consider a parametrized arc $\gamma : I \to U$ in U. The two sides of the equation $(f \circ \gamma)'(t) = f'(\gamma(t)) \circ \gamma'(t)$ are linear maps from $\mathbf{R}$ into E, and when applied to the element 1 of $\mathbf{R}$ they give

$$(f \circ \gamma)'(t) = f'(\gamma(t)) \cdot \dot{\gamma}(t). \tag{5.4}$$

Dually, let ϕ be a function on F (or on an open subset of F containing $f(U)$). We have

$$d(\phi \circ f)(x) = d\phi(f(x)) \circ f'(x), \tag{5.5}$$

which means that the element $d(\phi \circ f)(x)$ of E^* is the image of the element $d\phi(f(x))$ of F^* by the linear map from F^* to E^* that is the dual of $f'(x) : E \to F$.

If E and F each have a euclidean structure then $\mathrm{grad}(\phi \circ f)(x) \in E$ is the image of $\mathrm{grad}(\phi)(f(x)) \in F$ under the linear map $F \to E$ that is adjoint to $f'(x)$.

On composing these two situations we obtain

$$(\phi \circ f \circ \gamma)'(t) = \langle f'(\gamma(t)) \cdot \dot{\gamma}(t), d\phi(f(\gamma(t))) \rangle. \tag{5.6}$$

In the particular case when we take ϕ to be a coordinate linear form y_i and γ to be a 'straight line' $t \mapsto a + te_j$ we recover the fact that the matrix for the linear map $f'(a)$ is formed from the partial derivatives at a of the components of f: if we write $f(x) = \sum f_j(x)e_j$ then

$$f'(a) \cdot e_i = \sum_j \frac{\partial f_j}{\partial x_i} e_j.$$

Finally, taking $E = F$ and $f = Id_E$ gives the pretty relationship

$$\frac{d(\phi \circ \gamma)}{dt}(t) = \langle \dot{\gamma}(t), d\phi(\gamma(t)) \rangle. \tag{5.7}$$

1.6 The Local Inversion Theorem for C^r Maps

Theorem 1.6.1. *Let E and F be two finite-dimensional vector spaces with U an open subset of E, let a be a point of U and let $f : U \to F$ be a map of class C^r, $r \in [1, \infty]$. Suppose the linear map $f'(a)$ is bijective. Then there exists an open set U' of U containing a and an open set V' of F containing $f(a)$ such that f induces a bijection from U' onto V' and the inverse map is also of class C^r.*

Since the map $x \mapsto f'(x) \in L(E; F)$ is continuous, $f'(x)$ is bijective for x sufficiently close to a (observe, for example, that the determinant of $f'(x)$ with respect to fixed bases of E and F depends continuously on x). Restricting U if necessary, we may therefore suppose that $f'(x)$ is bijective for all x in U. Moreover, since f is strictly differentiable at a and $f'(a)$ is invertible we can apply Theorem 4.1 and obtain an open subset U' of U containing a and an open set V' of F containing $f(a)$ such that f induces a bijection from U' to V' with continuous inverse g. But for every $x \in U'$ the same theorem, applied this time at the point x, shows that g is strictly differentiable at the point $f(x)$, with derivative $f'(x)^{-1}$. Hence for all $y \in V'$ we have

$$g'(y) = f'(g(y))^{-1}. \tag{6.1}$$

Since g and f' are continuous it follows from this that g' is continuous, which means that g is of class C^1. Likewise, if g is assumed to be of class C^s with $s < r$ it follows from (6.1) that g' is also of class C^s and so g is of class C^{s+1}. Thus by induction we see that g is of class C^r.   □

If we replace "finite-dimensional spaces E and F" by "complete normed spaces E and F" and "$f'(a)$ is bijective" by "$f'(a)$ is an isomorphism" the theorem remains true, with the same proof.

1.6.2. In order to make efficient use of the above theorem we introduce some terminology. Let E and F be two finite-dimensional vector spaces with U, V open subsets of E, F respectively, and $f : U \to V$ a map. We say classically that f is a *homeomorphism* if f is bijective, and if f and its inverse map $f^{-1} : V \to U$ are continuous. For $r \in [1, \infty]$ we say analogously that f is a *diffeomorphism of class C^r* if f is bijective and if f and f^{-1} are of class C^r. We say that f is a *local diffeomorphism* of class C^r at the point a of U (or, more correctly, in a neighbourhood of a) if there exist open sets U' of E and V' of F with

$$a \in U' \subset U, \quad f(a) \in V' \subset V,$$

such that f induces a diffeomorphism of U' onto V'.[15]

---

[15]It follows trivially from these definitions that a diffeomorphism is a local diffeomorphism at every point. The converse is true in dimension 1 but false in dimension 2 as the example of the complex exponential shows.

Suppose that f is a local diffeomorphism at a and with the above notation let $g : V' \to U'$ be the inverse map to $f : U' \to V'$. Differentiating the relation $g(f(x)) = x$ we obtain $g'(f(x)) \circ f'(x) = Id_E$; differentiating the relation $f(g(y)) = y$ we likewise obtain $f'(g(y)) \circ g'(y) = Id_F$. Thus if $x \in U'$ and $y \in V'$ correspond (so $y = f(x)$ and $x = g(y)$) the linear maps $f'(x) : E \to F$ and $g'(y) : F \to E$ are mutually inverse. In particular, $f'(a)$ is bijective. The theorem above may therefore be restated in the following form:

Theorem 1.6.3. *Let $r \in [1, \infty]$ and let f be a map of class C^r from U into F. In order for f to be a local diffeomorphism of class C^r at a it is necessary and sufficient that the tangent linear map $f'(a)$ be bijective.* $\qquad\square$

We often say (local) diffeomorphism instead of (local) diffeomorphism of class C^1. The expressions "diffeomorphism of class C^r" and "local diffeomorphism of class C^r" may appear ambiguous at first sight, but one of the consequences of the above is precisely that there is no possible error : if f is of class C^r and is a (local) diffeomorphism, then it is a (local) diffeomorphism of class C^r. We can see this by applying the theorem above in the two senses, one of the equivalent conditions being independent of r, or more directly by using the formula (6.1) as we did above.

1.6.4. One consequence of the easy implication in the above theorem is that non-empty open sets in $\mathbf{R}^n$ and $\mathbf{R}^m$ are not diffeomorphic if n is different from m . In fact it is already true that non-empty open sets in $\mathbf{R}^n$ and $\mathbf{R}^m$ are not homeomorphic if n is different from m ; this result, which is more difficult to prove than the previous one, is the *theorem of invariance of dimension* due, like the related theorem of 'invariance of domain', to Brouwer (1911; for the history of these theorems see [HM], vol. 2, pp. 228 and 236).

This question has played an important historical role. In 1874, Cantor[16], who the year before had proved the impossibility of constructing a bijection between $\mathbf{N}$ and $\mathbf{R}$, posed the problem of dimension. After having tried unsuccessfully for three years to prove the nonexistence of a bijection between $\mathbf{R}$ and $\mathbf{R}^n$ for $n > 1$, he succeeded, to his own amazement, to establish such a bijection : he wrote to Dedekind[17] "I see it but I don't believe it". In reply, Dedekind suggested that it ought to be feasible to prove the impossibility of a *bicontinuous* bijection between $\mathbf{R}^n$ and $\mathbf{R}^m$ for $n \neq m$. Finally on this subject we note that although the case $n = 1$ is easy (since it suffices to observe that the complement of a point in $\mathbf{R}$ is not connected, in contrast to the situation for $\mathbf{R}^m$ with $m > 1$), the general case requires more elaborate tools arising from *algebraic topology*.

---

[16]Georg CANTOR (1845-1918), German mathematician, professor at Halle from 1879 to 1905, founder of the theory of infinite sets.

[17]Richard DEDEKIND (1831-1916), German mathematician, professor at Brunswick Technical University from 1863 to 1894, was the inventor of (among other things) ideals and recursive functions.

1.6.5. Let us now return to our main theorem and apply it explicitly in the case where we have chosen bases in E and F (which are necessarily of the same dimension), which is the same as taking $E = F = \mathbf{R}^n$. Thus we consider an open set U in $\mathbf{R}^n$, a point $a = (a_1, \ldots, a_n)$ in U, and n functions $f_i(x_1, \ldots, x_n)$, $i = 1, \ldots, n$ of class C^r on U, with $r \in [1, \infty]$. The derivative of f at a is the linear map from $\mathbf{R}^n$ to $\mathbf{R}^n$ with matrix

$$(\frac{\partial f_j}{\partial x_i})(a_1, \ldots, a_n),$$

the *Jacobian*[18] *matrix* of $f_1, \ldots, f_n$ at a. If the *Jacobian determinant*

$$\frac{D(f_1, \ldots, f_n)}{D(x_1, \ldots, x_n)} = \det\left(\frac{\partial f_j}{\partial x_i}\right)_{i,j=1,\ldots,n} \tag{6.2}$$

is nonzero at a then in a suitable open subset U' of U containing a the equations

$$f_1(x_1, \ldots, x_n) = y_1, \quad \ldots \quad, f_n(x_1, \ldots, x_n) = y_n$$

have a unique solution for $(y_1, \ldots, y_n)$ close to

$$b = (b_1, \ldots, b_n) = (f_1(a_1, \ldots, a_n), \ldots, f_n(a_1, \ldots, a_n)).$$

Moreover, this solution may be written as

$$x_1 = g_1(y_1, \ldots, y_n), \ldots, x_n = g_n(y_1, \ldots, y_n),$$

where the functions $g_1, \ldots, g_n$ are of class C^r in a neighbourhood of b. Finally, the Jacobian matrices $(\frac{\partial f_i}{\partial x_i})$ and $(\frac{\partial g_i}{\partial y_j})$ are, at the corresponding points, inverse to each other. We customarily write $(\partial y_j/\partial x_i)$ and $(\partial x_i/\partial y_j)$ instead of $(\partial f_j/\partial x_i)$ and $(\partial g_i/\partial y_j)$, as being more aesthetic ... and ambiguous.

1.7 Curvilinear Coordinates

1.7.1. As above, let E be a finite-dimensional vector space with U an open subset of E. Consider a finite sequence $(\phi_1, \ldots, \phi_n)$ of functions of class C^∞ on U and let $f : U \to \mathbf{R}^n$ denote the map whose components are $\phi_1, \ldots, \phi_n$. We say that $(\phi_1, \ldots, \phi_n)$ is a *system of curvilinear coordinates* on U if $f(U)$ is an open set in $\mathbf{R}^n$ and f is a diffeomorphism of class C^∞ from U to $f(U)$. Likewise we say that $(\phi_1, \ldots, \phi_n)$ is a *system of local coordinates* at a point a of U (or, more correctly, in a neighbourhood of a) if f is a local diffeomorphism of class C^∞ at a, that is (recall the definitions) if there is an open subset U' of U containing a such that the (restrictions of the) ϕ_i form a

---

[18]Named after Jacob JACOBI (1804-1851), German mathematician, professor at Kœnigsberg from 1831 to 1848.

system of curvilinear coordinates on U'. A *system of local coordinates centred at a* is a system of local coordinates in a neighbourhood of a which vanish at a (which can always be arranged by replacing each ϕ_i by $\phi_i - \phi_i(a)$).

1.7.2. The main advantage of curvilinear or local coordinates is that they allow us to express properties of differential type just as well as with linear coordinates, while offering much greater flexibility. We give a few examples. With the notation above, suppose $(\phi_1, \ldots, \phi_n)$ is a system of local coordinates at a. Let u be a function defined on a neighbourhood of a, with values possibly in some other vector space F. Since in particular f is bijective on the neighbourhood U' of a we may express U in the form $v \circ f$ where v is a function of n variables with values in F defined on a neighbourhood of $f(a)$. Therefore for all $\xi \in E$ sufficiently close to a we have

$$u(\xi) = v(\phi_1(\xi), \ldots, \phi_n(\xi)),$$

which it is natural to write (forgetting ξ) as

$$u = v(\phi_1, \ldots, \phi_n)$$

(the "expression for u in the system of local coordinates $(\phi_1, \ldots, \phi_n)$"). Since f and f^{-1} are of class C^∞ and we have $u = v \circ f$ and $v = u \circ f^{-1}$, to say that u is continuous at a, of class C^r at a, continuous in a neighbourhood of a, etc. is equivalent to saying that the function v of n variables has the same property at the corresponding point $b = f(a)$. Likewise, since the derivative of u at a and the derivative of v at b are related by the formulæ $u'(a) = v'(b) \circ f'(a)$ and $v'(b) = u'(a) \circ f'(a)^{-1}$, to say that $u'(a)$ is injective, or surjective, or bijective, or of rank r, etc. is the same as stating the analogous property for the derivative $v'(b)$, a property which itself can be translated into properties of the partial derivatives of v or, given a basis for F, into properties of the Jacobian matrix of v. Thus, for example, the criterion for local invertibility given in 6.5 may be verified using the Jacobian determinant calculated in local coordinates just as well as in linear coordinates. This illustrates the following principle :

Fundamental principle. *All formulæ and statements in first order differential calculus are expressed in curvilinear coordinates exactly as in linear coordinates.*

1.7.3. In general we can reason directly in curvilinear (or local) coordinates as follows. To each system of curvilinear coordinates $(\phi_1, \ldots, \phi_n)$ on an open set in a vector space E we can associate in dual fashion a system of parametrized arcs playing the role of straight lines parallel to the coordinate axes. Fix a point a of U; the n 'coordinate' arcs $\gamma_1, \ldots, \gamma_n$ passing through a are defined as follows : each $\phi_i \circ \gamma_j$ is constant and equal to $\phi_i(a)$ for $i \neq j$, and $\phi_j \circ \gamma_j(t) = \phi_j(a) + t$. These arcs are transformed by the diffeomorphism

$f : U \to f(U) \subset \mathbf{R}^n$ that is associated to the curvilinear coordinate system into the n straight lines parallel to the coordinate axes and passing through $f(a)$. In particular, we have

$$\langle \dot{\gamma}_j(0), d\phi_i(a) \rangle = \delta_{i,j} \; ;$$

thus the $\dot{\gamma}_j(0)$ form a basis for E and the $d\phi_i(a)$ are the dual basis of E^*. In the special case when the ϕ_i are linear we are back in the familiar linear situation. By working with the ϕ_i and γ_j we can mimic all the computations that are done in linear coordinates. For example, if γ is a parametrized arc with $\gamma(0) = a$ we may write

$$\dot{\gamma}(0) = \sum_{j=0}^{n} a_j \dot{\gamma}_j(0) \qquad \text{with} \qquad a_j = \frac{d}{dt}(\phi_j \circ \gamma)(0).$$

Dually, if ϕ is a function defined on a neighbourhood of a then

$$d\phi(a) = \sum_{i=0}^{n} b_i d\phi_i(a) \qquad \text{with} \qquad b_i = \frac{d}{dt}(\phi \circ \gamma_i)(0).$$

In the same way we can calculate the tangent map to an arbitrary map whenever we have a system of local coordinates on the source and on the target. In fact, studying the differential properties of any map $f : U \to V$, where U, V are open subsets of E, F respectively, reduces to studying those of a real-valued function $\phi \circ f \circ \gamma$ of one variable, where ϕ is a function on (a subset of) V and γ is an arc in U ; the formula

$$\frac{d}{dt}(\phi \circ f \circ \gamma)(t) = \langle f'(\gamma(t)) \cdot \dot{\gamma}(t), d\phi(\gamma(t)) \rangle,$$

which we have already seen in 5.4, allows us to calculate the 'elements of the matrix' $\langle f'(a).\xi, \lambda \rangle$ where ξ belongs to E and λ to F^*.

Now return to the original situation. Let a be a point of the finite-dimensional vector space E, let $\phi_1, \ldots, \phi_n$ be functions (of class C^∞) defined on a neighbourhood of a and let $f : U \to \mathbf{R}^n$ be the associated map. The derivative of f at a is the linear map $E \to \mathbf{R}^n$ whose components are the linear forms $d\phi_i(a)$. To say that $f'(a)$ is bijective means therefore that the $d\phi_i(a)$ form a basis for E^*, so they are linearly independent and we have $\dim(E) = n$. Hence we obtain a new form of the Local Inversion Theorem :

Corollary 1.7.4. *In order for the n functions ϕ_i to form a system of local coordinates at a it is necessary and sufficient for the n linear forms $d\phi_i(a)$ to form a basis for E^*.*   □

Suppose more generally that the $d\phi_i$ are *linearly independent* (so we have $n \leq \dim(E)$). By the theorem on completing a basis, we can choose linear forms $u_1, \ldots, u_k$ with $k = \dim(E) - n$ which together with $(\phi_1, \ldots, \phi_n)$ form a basis for E^*. Since the u_j are their own differentials, the preceding corollary implies that $(\phi_1, \ldots, \phi_n, u_1, \ldots, u_k)$ is a system of local coordinates at a. Hence:

Corollary 1.7.5. *In order for the n functions ϕ_i to form part of a system of local coordinates at a it is necessary and sufficient for the n linear forms $d\phi_i(a)$ to be linearly independent in E^*.*   □

It is perhaps worth recalling that to say that the linear forms $d\phi_i(a)$ are linearly independent is equivalent to saying that the intersection of their kernels is a vector subspace of E of codimension n (*a priori* this codimension is $\leq n$). In the next chapter we shall return to consider the geometrical significance of this condition.

1.7.6. Retaining the above notation, let $h : U \to \mathbf{R}^n \times \mathbf{R}^k$ be the map whose components are $(\phi_1, \ldots, \phi_n, u_1, \ldots, u_k)$; this is a local diffeomorphism at a, and so there is an open set U' of U containing a and an open set V' of $\mathbf{R}^n \times \mathbf{R}^k$ such that h induces a homeomorphism (which we still call h) from U' onto V'. Therefore *the image of $f : U \to \mathbf{R}^n$ contains a neighbourhood of $f(a)$*. A noteworthy consequence of this is that if θ is a function of n variables such that $\theta(\phi_1(x), \ldots, \phi_n(x))$ vanishes for all x sufficiently close to a then θ is identically zero on a neighbourhood of $f(a)$: in the traditional terminology we say that the functions $\phi_1, \ldots, \phi_n$ are *functionally independent* at a (or in a neighbourhood of a).

We have in fact proved in passing that a map $U \to F$ of class C^∞ (clearly, C^1 would be enough) whose derivative at a is surjective is, in topological terms, *open* at a : the image of any sufficiently small open set in U that contains a is an open set in F. It is easy to deduce from this that if $f'(a)$ is surjective for every $a \in U$ then f is open (the image of every open set in U is open in F) and in particular $f(U)$ is open in F.

1.8 Generalizations of the Local Inversion Theorem

The Implicit Function Theorem seems to have been used without justification until Cauchy gave a proof of it in 1839 based on the existence theorem for solutions of differential equations. The method of proof by 'successive approximations', conceived by Liouville in 1837 in connection with the same existence theorem, was generalized and popularized by Picard (circa 1890) and then extended to the general context of complete metric spaces by Banach in his thesis in 1922.

To describe directions in which the theorem has since been generalized, we look again at the situation studied in this chapter. Consider two vector spaces E and F (finite-dimensional — or, more generally, complete normed spaces), an open set U in E, a map $f : U \to F$ and a point a in U. The general problem is to deduce from a hypothesis about the derivative $f'(a)$ (which is presumed to exist) some conclusion about the behaviour of f in a neighbourhood of a. Thus we have seen that if $f'(a)$ is bijective (in infinite dimensions we have to say "is an isomorphism") then the equation $f(x) = y$, where y is close to $f(a)$, has a unique solution x close to a. We also saw that if $f'(a)$ is surjective (in infinite dimensions a stronger property is required) then f is surjective in a neighbourhood of a and the equation therefore has solutions for y close to $f(a)$; consequently if $f'(a)$ is surjective for all a then the set of those y for which the equation has solutions is an open set. If we have additional information which enables us to assert that this set is closed (often obtainable from 'a priori inequalities') then we have a criterion for solubility of nonlinear equations ; suitably sharpened and generalized, this is the so-called 'continuity method' which is often used to solve nonlinear partial differential equations.

One can try to generalize the theorem in two directions. First, the assumptions on the derivative may be weakened and one then asks what can be deduced about the function. We shall see later the simplest cases of this situation (parametrizations of submanifolds, bifurcations of critical points, elementary catastrophes ...). Also, and this is essential for many applicatons, one can attempt to escape the limitation of normed spaces and take as E and F spaces of C^∞ functions, spaces of distributions, and so on. Unfortunately, this is much more difficult, and there is no real generalization of the Local Inversion Theorem which retains its power and simplicity. See the discussion of these questions by J. D. Bost [BO], who gives a good approximation to such a theorem as well as certain applications, notably the famous Kolmogorov–Arnol'd–Moser (KAM) theorem on invariant tori which plays an essential role in the theory of conservative dynamical systems. It involves results quite beyond the scope of this book, however.

The practical simplicity of the Local Inversion Theorem and the absence of a really convenient version in more general spaces partly explains the systematic use that is made of complete normed spaces in functional analysis.

2 Submanifolds

2.1 Introduction

Submanifolds are common and unavoidable objects. While in ordinary geometry they are low-dimensional, in mechanics they can be significantly more
complicated. For example, the space of configurations of a solid body with
one fixed point can be identified with the group $\mathbf{SO}(3)$ of rotations of $\mathbf{R}^3$,
which is a 3-dimensional submanifold of the space $\mathbf{R}^9$ of 3×3 matrices. The
configuration space of a (free) solid has three more dimensions due to translations; its phase space, taking account of velocities, has dimension 12. Likewise
the phase space for the three-body problem has dimension $3 \times 3 \times 2 = 18$.

Although it is natural to define a submanifold by equations in the ambient
space, we usually work on the manifold itself via (local) parametrizations. For
instance, it seems unreasonable to define the sphere $\mathbf{S}^2$ as the image of the
parametrization by latitude and longitude, or to define the rotation group as
the image of the parametrization by the Euler angles (mainly because of the
trouble caused by singularities of the parametrization at certain points). In
contrast, however, it is no joke calculating in the rotation group by dragging
around the nine matrix coefficients.

One of the reasons for being interested in parametrizations is that they
allow us to discard useless coordinates in the ambient space in order to concentrate on the submanifold itself. The natural end-result of this step is the
notion of an 'abstract' manifold defined without recourse to any particular
embedding in a space $\mathbf{R}^n$. Thus, to study the 2-dimensional torus which is
the natural space on which to define doubly periodic functions, it is simpler
to regard it as the quotient of $\mathbf{R}^2$ by the equivalence relation "$(x, x') \equiv (y, y')$
if $x - y$ and $x' - y'$ are integers" (which clearly brings out its group structure)
than to embed it in $\mathbf{R}^2 \times \mathbf{R}^2 = \mathbf{R}^4$ as the product $\mathbf{S}^1 \times \mathbf{S}^1$ of two circles or,
more economically but more perversely, in $\mathbf{R}^3$ as a torus of revolution.

Unfortunately, the fact that an approach is conceptually simpler does
not mean that it is easier to understand, and experience shows that there
is no best way to teach manifolds. On the one hand, we can choose (as we
do here) a presentation geared to the notion of submanifold of an ambient
vector space, which quickly gets down to the essentials, and then we find
ourselves in a straightjacket in which closely related statements cannot be

conveniently unified, and where the 'intrinsic' character of the notions introduced is disguised by the embeddings incorporated into the definitions (see
Proposition 8.1 for an example of this type of difficulty). On the other hand,
we could take the leisure to develop the intrinsic approach, which makes submanifolds of $\mathbf{R}^n$ into a rather artificial and somewhat shameful species of
manifold, and then spend time explaining that, despite appearances, we are
really talking about simple and natural objects. Given the present constraints
of time and space, we adopt the 'low profile' approach; this also offers the
symbolic advantage of breaking with the pernicious practice of subordinating
every concrete application, however elementary, to a deep understanding of a
theoretical apparatus which at the end of the day is quite out of proportion.

The contents of this chapter are as follows. The definition of submanifolds
and some examples take up Sects. 2 and 3, and tangent spaces are discussed
in Sect. 4. In Sects. 5 and 6 we introduce the convenient and simple notion
of transversality. Sect. 7 is devoted to the geometric version of the Implicit
Function Theorem. In Sects. 8 and 9 we introduce embeddings and the slightly
weaker notion of immersions. Sect. 10 contains a little topology : there we
discuss proper maps. In Sect. 11 we suggest how the intrinsic approach can
be worthwhile, and we conclude in Sect. 12 with some historical remarks.

The only tool that we really use is the Local Inversion Theorem. In fact,
this whole chapter can be considered just as a variation on that theme, with
the introduction of a geometric vocabulary that we have tried to keep down
to essentials.

2.2 Definitions of Submanifolds

Let E be a vector space of finite dimension $\dim(E) = n$ with V a subset of
E and a a point of V.

Definition 2.2.1. A set of C^∞ functions $\phi_1, \ldots, \phi_m$, defined on a neighbourhood of a, is said to form a *nondegenerate system of local equations for V at
a* if the following two conditions are satisfied :

a) there is an open set U of E containing a in which the ϕ_i are defined and
such that $V \cap U$ is the set of points $x \in U$ with $\phi_1(x) = 0, \ldots, \phi_m(x) = 0$;
b) the linear forms $d\phi_i(a)$ are linearly independent.

We express the *nondegeneracy* condition b) by saying that the functions
are *functionally independent* at a (see 1.7.6). Note immediately that if condition b) is true at the point a then it remains true in a neighbourhood of a.
Therefore we may assume that we have chosen U so that we have, as well as
condition a), the stronger condition b') :

b') for each $x \in U$ the linear forms $d\phi_i(x)$ are linearly independent.

We then say that the functions ϕ_i form a *nondegenerate system of equations for V in U*.

The nondegeneracy condition b) (or b')) is essential. Indeed, in the absence of such a condition the set of common zeros of a family of C^∞ functions can be extremely pathological (see later). We shall see below that the integer m is uniquely determined (there, too, condition b) is essential : consider the sum of the ϕ_i^2). It is called the *codimension* of V at a, and the integer $d = n - m$ is the *dimension* of V at a. We denote these by $\mathrm{codim}_a V$ and $\dim_a V$ respectively.

Definition 2.2.2. We say that V is a *submanifold* of E at a if it possesses a non-degenerate system of local equations at a ; it is a submanifold of E if it is a submanifold at every one of its points, that is if it possesses a nondegenerate system of local equations at each point, or if there is a family of open sets in E which cover V and such that V possesses a nondegenerate system of equations in each of the open sets.

Clearly, if V has dimension d at a then it also has dimension d at nearby points. Thus the conditions $\dim_a V = 0, 1, \ldots$ define a partition of V into open subsets. Therefore if V is connected it has the same dimension at each of its points. We say that V is *purely of dimension d* (and of codimension $n - d$) if its dimension is d at each point. More generally we let

$$\dim(V) = \sup_{a \in V} \dim_a V, \qquad \mathrm{codim}(V) = \inf_{a \in V} \mathrm{codim}_a V.$$

We then have $\dim(V) + \mathrm{codim}(V) = \dim(E)$. Note explicitly that the above definitions imply that the *empty set* is a submanifold which is purely of dimension d for every d between 0 and n inclusive.

2.2.3. According to what we saw in the last chapter, condition b) means also that the ϕ_i form part of a system of local coordinates at a. We can therefore rewrite the definition as follows:

For V to be a submanifold of E at a it is necessary and sufficient that there exist a system of local coordinates $(\phi_1, \ldots, \phi_n)$ in E centred at a such that in a neighbourhood of a the set V is the set of common zeros of $\phi_1, \ldots, \phi_m$ with $m \leq n$.

Equivalently, to say that V is a submanifold of E at a means that there exists an open set U of E containing a and a diffeomorphism[19] f from U onto an open set in $\mathbf{R}^n = \mathbf{R}^m \times \mathbf{R}^d$ such that $f(U \cap V) = f(U) \cap (\{0\} \times \mathbf{R}^d)$. Thus a submanifold of codimension m in E is simply a subset of E which becomes

---

[19]To avoid unnecessary complications we shall in general assume that the maps we consider have maximal regularity. Thus in future we shall say "diffeomorphism" to mean "diffeomorphism of class C^∞". Most theorems stated in the C^∞ setting remain true with weaker hypotheses.

a 'linear' submanifold of codimension m in adapted sytems of curvilinear coordinates (see Fig. 2.1.).

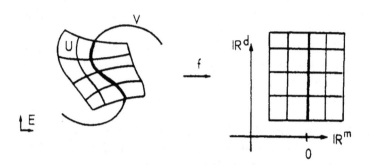

Fig. 2.1. A submanifold becomes linear in adapted coordinates

Let us consider the extreme cases. For $m = 0$ (so $d = n$) the condition a) is equivalent to $V \cap U = U$, that is $U \subset V$ and condition b) is vacuous: *the submanifolds of codimension 0 are precisely the open sets* in E. Note in passing that every open subset of a submanifold is a submanifold. At the opposite extreme, for $m = n$ (so $d = 0$) the description in terms of diffeomorphisms gives an open set U with $V \cap U = \{a\}$: *the submanifolds of dimension 0 are precisely the discrete subsets* of E. When d is 1 or 2 we speak about *curves* or *surfaces*; when $m = 1$ we refer to *hypersurfaces*.

We now turn to the fact accepted above without proof: the integer m of Definition 2.1 does not depend on the choice of nondegenerate system of equations (ϕ_i). First we prove an auxiliary result. We retain the notation above, with in particular $(\phi_1, \ldots, \phi_n)$ a system of local coordinates centred at a such that $(\phi_1, \ldots, \phi_m)$ is a local system of equations for V at a. As above, we say that C^∞ functions are *functionally independent* at a if their differentials at a are linearly independent.

Lemma 2.2.4. *Let* $\psi_1, \ldots, \psi_m$ *be* m *functions of class* C^∞ *defined on a neighbourhood of* a *in* E *which vanish on* V *and are functionally independent at* a. *Then* $(\psi_1, \ldots, \psi_m)$ *is a nondegenerate system of local equations for* V *at* a.

Write each of the functions ψ_i in the system of local coordinates:

$$\psi_i = F_i(\phi_1, \ldots, \phi_n).$$

By assumption we have

$$F_i(0, \ldots, 0, x_{m+1}, \ldots, x_n) \equiv 0$$

in a neighbourhood of 0, and in particular $\partial \psi_i / \partial x_j(0) = 0$ for $i = 1, \ldots, m$ and $j = m + 1, \ldots, n$. This implies

$$d\psi_i(a) = \sum_{j=1}^{m} \frac{\partial F_i}{\partial x_j}(0)d\phi_j(a).$$

The square matrix which thus appears has to be invertible because of the assumption of independence of the ψ_i, and the Local Inversion Theorem implies that $(\psi_1, \ldots, \psi_m, \phi_{m+1}, \ldots, \phi_n)$ is a system of local coordinates centred at a. It remains to prove that if at a point x in E close to a the functions $\psi_1, \ldots, \psi_m$ vanish then x lies in V. However, if y denotes the point in V defined by $\phi_i(y) = 0$ for $i \leq m$ and $\phi_i(y) = \phi_i(x)$ for $i > m$ the two points x and y have the same coordinates in the second system and therefore they coincide.   □

Now suppose $(\phi_1, \ldots, \phi_m)$ and $(\psi_1, \ldots, \psi_p)$ are two nondegenerate systems of local equations for V at a, with $p \geq m$. By the above lemma, $(\psi_1, \ldots, \psi_m)$ is also a nondegenerate system of local equations at a. But if $p > m$ and $\psi_1, \ldots, \psi_m$ vanish then so must ψ_{m+1}, which is absurd. Hence $p = m$.   □

2.2.5. There are several elementary properties resulting from this lemma which are quite analogous to different forms of the *exchange theorem* from linear algebra. We shall give two of these, leaving the rest as exercises. First of all we paraphrase the lemma and the several lines of proof that follow it:

a) A family of functions which are functionally independent and which vanish on V in a neighbourhood of a contains at most $\mathrm{codim}_a V$ elements. If it contains exactly this number then it is a nondegenerate system of local equations for V at a.

b) Let V' be a second submanifold of E containing V. We then have $\dim_a V \leq \dim_a V'$. If V and V' have the same dimension at a then V is open in V': there is an open subset U of E containing a such that $V \cap U = V' \cap U$.

It is enough to apply a) to a nondegenerate system of local equations for V' at a.

2.2.6. It is perhaps worth commenting on something which often seems to jar on first encounter, namely that when thinking of a curve, surface etc. we naturally imagine a *closed* set, while the definition that we have given allows us to call a set which is not closed[20] a submanifold. This choice is convenient from a technical point of view; in fact many constructions (parametrizations, for example) are 'local' and it is nice that open subsets of submanifolds, which

---

[20]In fact this definition implies that a submanifold is always a 'locally closed' subset (see 3.2.2): it is an open subset of its closure, or equivalently the intersection of an open set and a closed set. Note that the definition says nothing about the appearance of a submanifold in the neighbourhood of a point that does not belong to it (see the example at the end of Section 2.10).

we certainly have to deal with, are themselves submanifolds. But there are more fundamental reasons. Theoretical results have to be able to be applied to objects as simple as a plane curve with double points or a cone, and we have *a priori* a choice between two approaches: throw out all singular points – the solution chosen here – or broaden the definition to admit any subset defined by equations (of class C^∞, but also why not of class C^r). Unfortunately (or fortunately for students) this second solution is totally impracticable. For example, it can be shown that (*horresco referens!*) every closed subset of $\mathbf{R}^n$ is the set of zeros of a suitably-chosen C^∞ real-valued function (this is a theorem of Whitney[21]: see for example [BL], p.24). However, the choice of this type of definition will not prevent us from studying certain singularities, in particular those which occur 'inevitably' as limits of regular situations. Neither will it keep us from an inevitable abuse of language in giving the names "curve", "surface", "hypersurface", ... to objects which have singularities and therefore do not, according to our conventions, merit the name "submanifold".

The situation is different in algebraic geometry since subsets defined by arbitrary polynomial equations are much tamer. In the French language this is reflected by a major divergence in terminology: in algebraic geometry a *variété* may be singular, and we say *variété non singulière* for the analogues of the 'variétés' (namely, manifolds) in differential geometry. In English the situation is better since the two different terms *variety* and *manifold* are used.

Before going further into the general theory, let us look at some examples.

2.3 First Examples

1) A vector (or affine) subspace of E is a submanifold.

2) If V is a submanifold of E and if f is a diffeomorphism from an open set U of E containing V onto an open set U' of a space E' then $f(V)$ is a submanifold of E'. This is verified using any of the forms of the definition.

In fact we could say that the notion of submanifold is essentially defined by the examples 1) and 2). A little more precisely, the definition given in 2.1 can be paraphrased as follows :
  a) the fact that V is a submanifold may be verified 'locally' on V,
  b) the property of being a submanifold is invariant under diffeomorphisms,
  c) a vector subspace is a submanifold.

3) *Graphs.* Suppose E is given as a product $E' \times E''$ and let $f : E' \to E''$ be a C^∞ map from E' to E'' (we could just as well suppose f to be

---

[21]Hassler WHITNEY, 1907-1989, taught at Harvard from 1930 to 1952 and then at the Institute for Advanced Study from 1952 to 1977. He was one of the founders of the theory of differentiable maps. Many fine and beautiful theorems are due to him, and we shall meet some of them in this text.

defined on an open subset of E'). Then the graph Γ of f is a submanifold of E of dimension $\dim(E')$. In fact we may identify E'' with $\mathbf{R}^m$ and f with a sequence $(\phi_1, \ldots, \phi_m)$ of C^∞ functions on E'; then Γ is the set of $(x, y_1, \ldots, y_m) \in E' \times \mathbf{R}^m$ on which the functions $(x, y_1, \ldots, y_m) \mapsto y_i - \phi_i(x)$ vanish. The differentials $dy_i - d\phi_i(x)$ of these equations are linearly independent at every point.

4) *Products.* If V' is a submanifold of E' and V'' is a submanifold of E'' then $V' \times V''$ is a submanifold of $E' \times E''$. (The verification of this is an excellent exercise to see if you have understood the definition ; it is simply a question of "combining the equations".)

5) *Level hypersurfaces.* Let ϕ be a C^∞ function on an open subset U of E. The *critical points* of ϕ are the points x of U for which $d\phi(x)$ is zero (if we had chosen a basis for E this would mean that all the first partial derivatives of ϕ vanish at x). The images under ϕ of the critical points are called the *critical values* of ϕ. For $\alpha \in \mathbf{R}$ let $V_\alpha = \phi^{-1}(\alpha)$ (a 'level set' of ϕ) ; through each point x there passes one and only one V_α, namely $V_{\phi(x)}$. Then, if x is not a critical point, $V_{\phi(x)}$ is a submanifold at x since it is defined by the equation $\alpha - \phi(x) = 0$ which has nonzero derivative. Hence for every $\alpha \in \mathbf{R}$ which is not a critical value[22] of ϕ, the set V_α is a submanifold of E. Indeed, as we shall see later (3.5.1, Sard's Theorem), 'most' of the V_α are submanifolds. See Fig. 2.2.

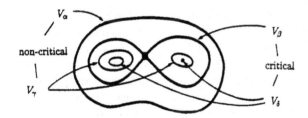

Fig. 2.2. Some level sets

In particular, let $\phi : \mathbf{R}^n \setminus \{0\} \to \mathbf{R}$ be a C^∞ function which is homogeneous of degree $k \neq 0$. The Euler identity

$$k\phi(x) = \sum_{i=1}^n x_i \frac{\partial \phi}{\partial x_i}(x),$$

which can be written in intrinsic form as

$$k\phi(x) = \langle x, d\phi(x) \rangle,$$

and is obtained by differentiating the identity $t^k \phi(x) = \phi(tx)$ at $t = 1$, shows that the only possible critical value for ϕ is 0. Hence $\phi^{-1}(\alpha)$ is a

---

[22]Take care not to confuse "not being a critical value" with "being a non-critical value", cf. 3.6.6.

submanifold of E for every real nonzero α. This applies, for example, to the sphere $\mathbf{S}^{n-1}$ of dimension $n-1$ whose equation is $\sum x_i^2 = 1$.

6) Note that it is possible for a subset of E that is 'defined by equations' to be a submanifold without the fact being evident from the given equations. For example, consider the intersection V_α of the sphere $\mathbf{S}^{n-1}$ with the hyperplane $x_n = \alpha$. For $|\alpha| < 1$ the set V_α is a submanifold of codimension 2 since the two equations have independent differentials; for $|\alpha| > 1$ we see that V_α is empty and therefore a submanifold, while for $|\alpha| = 1$ the set V_α is reduced to one point and is therefore a submanifold, but the two given 'equations' are not suitable as their differentials are proportional (and in any case there is the wrong number of them).

7) Now we give a non-example. Let $f(x,y) = y^2 - x^2 + x^4$ and consider the subset (*lemniscate*) of $\mathbf{R}^2$ consisting of pairs (x,y) with $f(x,y) = 0$. Since the partial derivatives of f are $2y$ and $4x^3 - 2x$ we verify immediately that the only critical point of f on L is the origin. Hence $L - \{0\}$ is a submanifold of dimension 1 (Example 5). On the other hand, L is not a submanifold since it is not a submanifold at the point 0. It is never easy to prove that something is *not* a submanifold. Here, for example, the fact that f has a critical point at 0 is not enough (the equation $x^3 = 0$ does define a submanifold, as does $x^2 + y^4 = 0$). To verify that L is not a submanifold at 0, one way is to consider the two curves $L_\pm$ which are the images of the segment $(-1, 1)$ under the two maps $x \mapsto (x, \pm x\sqrt{1 - x^2})$. These are submanifolds according to Example 3, they are contained in L and they have no point in common except 0. If L were a submanifold at 0 we would be able to apply 2.5 b) to L and L_+ and obtain a contradiction. We could also use a fact which will be proved in the next section on the tangent space: if L were a submanifold at 0 it would have a 1-dimensional tangent space there, but the two arcs described above which are in L have distinct tangents at 0. In fact (see 4.7.5) the locus L possesses an 'ordinary double point'.

8) Here is an example from linear algebra. Let E and F be two finite-dimensional vector spaces and consider the space $L(E; F)$ of linear maps from E to F. For every $v \in L(E; F)$ the *rank* of v is the integer $\mathrm{rk}(v)$ defined by

$$\mathrm{rk}(v) = \dim(\mathrm{Im}(v)) = \mathrm{codim}(\mathrm{Ker}(v)).$$

We have $\mathrm{rk}(v) \le \inf(\dim(E), \dim(F))$; the equality $\mathrm{rk}(v) = \dim(E)$ means that v is injective, while $\mathrm{rk}(v) = \dim(F)$ means that v is surjective. For each integer r the set of v with $\mathrm{rk}(v) \le r$ is closed; likewise the set of v with $\mathrm{rk}(v) \ge r$ is open (its complement is given by $\mathrm{rk}(v) \le r - 1$). We shall prove this below, in passing.

Proposition 2.3.1. *Let r be an integer with $0 \le r \le \inf(\dim(E), \dim(F))$. In the vector space $L(E; F)$, the set L_r of those v with $\mathrm{rk}(v) = r$ is a submanifold with codimension equal to $(\dim(E) - r)(\dim(F) - r)$.*

Let v_0 be an element of L_r; by choosing suitable bases in E and F we may suppose that v_0 is represented by the matrix $v_0 = \left(\begin{smallmatrix} I & 0 \\ 0 & 0 \end{smallmatrix}\right)$ where I is the unit matrix of order r. Represent an arbitrary element v of $L(E; F)$ by a matrix $v = \left(\begin{smallmatrix} A & B \\ C & D \end{smallmatrix}\right)$. The condition $\det(A) \neq 0$ defines an open set U containing v_0; we shall prove that $U \cap L_r$ is the graph of a C^∞ map $(A, B, C) \mapsto D$, namely $(A, B, C) \mapsto D = CA^{-1}B$, which will imply the proposition.

Suppose therefore that A is invertible and let us find the rank of v. Consider the unit matrix J of order $\dim(E) - r$; we have $AA^{-1} = I$, $BJ = B$ and $DJ = D$. From these we immediately deduce

$$\begin{pmatrix} A & B \\ 0 & J \end{pmatrix} \begin{pmatrix} A^{-1} & -A^{-1}B \\ 0 & J \end{pmatrix} = \begin{pmatrix} I & 0 \\ 0 & J \end{pmatrix},$$

$$\begin{pmatrix} A & B \\ C & D \end{pmatrix} \begin{pmatrix} A^{-1} & -A^{-1}B \\ 0 & J \end{pmatrix} = \begin{pmatrix} I & 0 \\ CA^{-1} & D - CA^{-1}B \end{pmatrix}.$$

The first equation shows that the square matrix $\left(\begin{smallmatrix} A^{-1} & -A^{-1}B \\ 0 & J \end{smallmatrix}\right)$ is invertible and therefore corresponds to an automorphism w of E. Since the linear maps v and wv have the same image, and hence the same rank, the second equation implies that the rank of v is the same as that of the block triangular matrix $\left(\begin{smallmatrix} I & 0 \\ CA^{-1} & D-CA^{-1}B \end{smallmatrix}\right)$. Therefore we have

$$\mathrm{rk}(v) = \mathrm{rk}(I) + \mathrm{rk}(D - CA^{-1}B) = r + \mathrm{rk}(D - CA^{-1}B).$$

From this we deduce the inequality $\mathrm{rk}(v) \geq r = \mathrm{rk}(v_0)$, as promised, as well as the equivalence of the condition $v \in L_r$ with $\mathrm{rk}(D - C^{-1}AB) = 0$, that is to say $D = CA^{-1}B$. This exhibits $U \cap L_r$ as a graph, and completes the proof. □

Let $r_0 = \inf(\dim(E), \dim(F))$. For each integer r with $0 \leq r \leq r_0$ the submanifold L_r is nonempty and its closure (as is easily verified) is the union of the L_s for $s \leq r$. We thus obtain what is called a *stratification*: the 'singular' object under consideration, which is here the closure of one of the L_r, has the form of a (finite) union of *strata* each of which is a submanifold, with the closure of each of these strata being in turn composed of the stratum itself and strata of lower dimensions. This is the general situation in algebraic geometry. Note that the conditions $\mathrm{rk}(v) \leq r$ are polynomials: all the minors of order $> r$ (or of order $r + 1$) of v have to vanish, and they are polynomials in the coefficients of v.

2.4 Tangent Spaces of a Submanifold

2.4.1. We retain the notation above: E is a vector space of finite dimension $\dim(E) = n$ with V a submanifold of E and a a point of V. Consider parametrized arcs traced on V and passing through a, or in other words C^1 maps

$$\gamma : (-\epsilon, \epsilon) \to E, \quad \epsilon > 0,$$

with

$$\gamma((-\epsilon, \epsilon)) \subset V \quad \text{and} \quad \gamma(0) = a.$$

To each of these arcs we associate its velocity vector $\dot\gamma(0) \in E$; the set of vectors of E so obtained is called the *tangent space* to V at a and is denoted by $T_a V$. See Fig. 2.3.

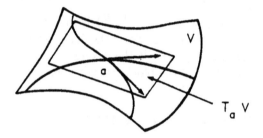

Fig. 2.3. The tangent space $T_a V$ to the submanifold V at the point a

Proposition 2.4.2. *The tangent space to V at a is a vector subspace of E with dimension $\dim_a V$. It is also the intersection of the kernels of the linear forms $d\phi(a)$ where ϕ runs through the set of C^∞ functions defined on a neighbourhood of a and vanishing on V.*

Let $(\phi_1, \ldots, \phi_n)$ be a nondegenerate system of local equations for V at a, and let $T = T_a(V)$, $T'' = \bigcap \mathrm{Ker}(d\phi_i(a))$ and let T' be the intersection as in the statement of the Proposition. Clearly $T' \subset T''$. Also, if γ is one of the arcs considered and ϕ is one of the functions as stated then $\phi(\gamma(t)) = 0$ for t small enough; differentiation yields $\langle \dot\gamma(0), d\phi(a) \rangle = 0$ which gives $T \subset T'$. Since T'' is a vector subspace of dimension $n - m = \dim_a V$, we see by the very definition of a nondegenerate system of local equations that the proof will be completed once we show $T'' \subset T'$. Thus we must show that every vector in T'' is the velocity vector of a suitably-chosen arc traced on V. Now, as we have seen, the given system of equations may be completed to a system of local coordinates $(\phi_1, \ldots, \phi_n)$ for E at a ; clearly we may assume that the additional functions, like the others, vanish at a. For every n-tuple $(\alpha_1, \ldots, \alpha_n)$ of real numbers there exists an arc $t \mapsto \gamma(t)$ in E with $\gamma(0) = a$ and $\phi_i(\gamma(t)) = \alpha_i t$. Differentiation gives $\langle \dot\gamma(0), d\phi_i \rangle = \alpha_i$; taking

$\alpha_1 = \cdots = \alpha_m = 0$ we obtain arcs traced on V whose velocity vectors fill out the whole of T'', and this completes the proof. $\qquad\qquad\square$

It is often more natural geometrically to consider the *affine tangent space* to V, which is the translate $a + T_a V$.

2.4.3. To calculate $T_a V$ we may use either of the two descriptions given above. If we use the description "by equations" it is often convenient to proceed as follows. First note that if ϕ is a C^∞ function defined on a neighbourhood of a and v is a vector in E then we have a partial expansion

$$\phi(a + tv) = \phi(a) + t\langle v, d\phi(a)\rangle + o(t).$$

Consequently in order to write simultaneously that a belongs to V and v to $T_a V$ it suffices to write $\phi_i(a + tv) = o(t)$ for a family of functions that vanish on V and contain a nondegenerate system of local equations. On the other hand, if we use the description "by parametrized arcs" it is enough to find arcs γ_j passing through a and traced on V, equal to $\dim_a V$ in number and such that the $\dot{\gamma}_j(0)$ are linearly independent; the subspace generated by these vectors will indeed be contained in $T_a V$ and will have the correct dimension and will therefore coincide with $T_a V$.

To illustrate these remarks, let us take Examples 1) to 5) from the previous section, keeping the same notation.

1) An affine subspace of E is its own affine tangent space at every point.
2) Let $a \in V$. Since f is a diffeomorphism the linear map $f'(a) : E \to F$ is bijective. The tangent space $T_{f(a)} f(V)$ is the image of $T_a V$ under $f'(a)$; indeed, if γ is an arc on V passing through a then $f \circ \gamma$ is an arc on $f(V)$ passing through $f(a)$ and we have $(f \circ \gamma)'(0) = f'(a) \cdot \dot{\gamma}(0)$.
3) The tangent space to the graph Γ of f at the point $(a, f(a))$ is the graph of the linear map $f'(a) : E \to F$. If γ is an arc in E passing through a then $t \mapsto (\gamma(t), f(\gamma(t)))$ is an arc in Γ passing through $(a, f(a))$ with velocity vector $(\dot{\gamma}(0), f'(a) \cdot \dot{\gamma}(0))$.
4) The tangent space at (a', a'') to $V' \times V''$ is $T_{a'} V' \times T_{a''} V''$.
5) The tangent space to the level hypersurface $\phi^{-1}(\phi(a))$ at a noncritical point a is the hyperplane that is the kernel of the nonzero linear form $d\phi(a)$. If E is euclidean space it is also the hyperplane orthogonal to the vector $\mathrm{grad}(\phi)(a)$.

We now assume again the notation of Example 8 from the last section.

Proposition 2.4.4. *The tangent space at the point v to the submanifold L_r of $L(E; F)$ consists of the linear maps $w : E \to F$ such that $w(\mathrm{Ker}(v)) \subset \mathrm{Im}(v)$.*

We choose suitable bases in E and F such that v is represented by the block matrix $\left(\begin{smallmatrix} I & 0 \\ 0 & 0 \end{smallmatrix}\right)$ where I is the unit matrix of order r. We then represent w by the matrix $\left(\begin{smallmatrix} A & C \\ B & D \end{smallmatrix}\right)$ so that $v + tw$ is represented by $\left(\begin{smallmatrix} I+tA & tC \\ tB & tD \end{smallmatrix}\right)$. Consider an element d_i^j of D. The $(r+1) \times (r+1)$ minor of the previous matrix that has principal diagonal consisting of the diagonal of I and the element d_i^j is of the form $td_i^j + o(t)$. If $w \in T_v L_r$ we must then have $d_i^j = 0$ for every (i,j), so $D = 0$. But this condition defines a subspace of the correct dimension, which therefore must be the space we want. Moreover, the condition $D = 0$ means that w maps $\mathrm{Ker}(v)$ into $\mathrm{Im}(v)$. $\qquad\qquad\square$

The different tangent spaces $T_a V$ organize themselves into a manifold in the following way. If $a \in V$ and $v \in T_a V$, let us call the pair (a, v) a *contact element* of V.

Proposition 2.4.5. *The contact elements of V form a submanifold TV of $E \times E$. We have $\dim_{(a,v)} TV = 2\dim_a V$.*

Let U be an open set in E in which V is described by a nondegenerate system of equations $\phi_1, \ldots, \phi_m$. In order for the point (x,y) of the open subset $U \times E$ of $E \times E$ to belong to TV it is necessary and sufficient that

$$\phi_1(x) = 0, \ \ldots, \ \phi_m(x) = 0, \qquad \phi_1'(x) \cdot y = 0, \ \ldots, \ \phi_m'(x) \cdot y = 0.$$

It is enough to show that the derivatives of these $2m$ equations are linearly independent at every point. However, the images of a vector $(\xi, \eta) \in E \times E$ under the derivatives of the maps $(x,y) \mapsto \phi_i(x)$ and $(x,y) \mapsto \phi_i'(x) \cdot y$ are $\phi_i'(x) \cdot \xi$ and $(\cdots) \cdot \xi + \phi_i'(x) \cdot \eta$ respectively. The required result follows immediately. $\qquad\qquad\square$

The submanifold TV is called the *tangent bundle* of V.

2.4.6. Let F be another vector space and for $r \in [1, +\infty]$ let f be a map of class C^r from an open set in E containing V into F. Let W be a submanifold of F, suppose that f maps V into W, and denote by $g : V \to W$ the map induced by f. Then the linear map $f'(a) : E \to F$ maps $T_a V$ into $T_{g(a)} W$ and the map $u : T_a V \to T_{g(a)} W$ that it induces depends only on g : if γ is a parametrized arc traced on V with $\gamma(0) = a$ then $f'(a) \cdot \dot\gamma(0) = (g \circ \gamma)'(0)$. The linear map u is denoted by $T_a g$ and it is called the *tangent map* of (or to) g at the point a. The map

$$(a, \xi) \mapsto (f(a), f'(a) \cdot \xi) = (g(a), (T_a g) \cdot \xi)$$

from TV into $F \times F$ is of class C^{r-1}; as we have just seen it maps TV into TW. The induced map is denoted by $Tg : TV \to TW$ and it is called the *tangent map* of g.

2.5 Transversality: Intersections

2.5.1. First, a little linear algebra. Let E be a finite-dimensional vector space, with T and T' two vector subspaces of E. Recall that $T + T'$ denotes the subspace formed by the sums $x + x'$ where x runs through T and x' runs through T', and we have the relation

$$\dim(T + T') + \dim(T \cap T') = \dim(T) + \dim(T'),$$

which can equally well be written as

$$\operatorname{codim}(T + T') + \operatorname{codim}(T \cap T') = \operatorname{codim}(T) + \operatorname{codim}(T')$$

(this follows from the usual relation between the dimensions of the kernel and image of a linear map, applied to the map $(x, x') \mapsto x + x'$ from $T \times T'$ into E). We say the subspaces T and T' are *transverse* (to each other) if the following conditions (equivalent, by the above) are satisfied:

(i) $T + T' = E$;
(ii) $\operatorname{codim}(T \cap T') = \operatorname{codim}(T) + \operatorname{codim}(T')$;
(iii) $\dim(T \cap T') = \dim(T) + \dim(T') - \dim(E)$.

Note that if $\dim(T) + \dim(T') < \dim(E)$ then T and T' cannot be transverse, and in the case when $\dim(T) + \dim(T') = \dim(E)$ to say that T and T' are transverse means that they are mutual complements.

2.5.2. More generally we shall say that a family $T_1, \ldots, T_m$ of subspaces is transverse if

$$\operatorname{codim}(T_1 \cap \ldots \cap T_m) = \operatorname{codim}(T_1) + \cdots + \operatorname{codim}(T_m).$$

2.5.3. We can express the transversality condition in another more concrete way. First recall the following elementary properties. Let $\phi_1, \ldots, \phi_d$ be linear forms (d of them) on the space E. Let T denote the subspace defined by the equations $\phi_i(x) = 0$, $i = 1, \ldots, d$, that is to say the intersections of the kernels of the ϕ_i. We then have $\operatorname{codim}(T) \leq d$ and the following conditions are equivalent:

a) $\operatorname{codim}(T) = d$;
b)   the ϕ_i are linearly independent in E^*;
c)   the linear map from E into $\mathbf{R}^d$ whose components are the ϕ_i is surjective.

We now return to the matter in hand: consider m subspaces $T_1, \ldots, T_m$ of E that have codimensions $d_1, \ldots, d_m$ respectively. Describe each T_i by means of d_i linear equations $\phi_{i,1} = 0, \ldots, \phi_{i,d_i} = 0$; the left hand sides of these d_i equations are linearly independent linear forms. Therefore the intersection of the T_i is defined by the union of all these equations, of which there are

$d = d_1 + \cdots d_m$. To say that the family of the T_i is transverse is therefore equivalent to saying that these d equations form a linearly independent set. With this version of the definition it becomes clear that if the given family is transverse then every family extracted from it is also transverse.

This criterion can be expressed without equations, by means of the quotient vector spaces E/T_i : there is a natural linear map from E into the product of the E/T_i and transversality of the family is equivalent to surjectivity of this map.

2.5.4. From the above we deduce a simple geometric criterion for transversality: in order for the family $T_1, \ldots, T_m$ of m vector subspaces to be transverse it is necessary and sufficient that every family $A_1, \ldots, A_m$ of affine subspaces which are parallel to $T_1, \ldots, T_m$ respectively should have non-empty intersection. Indeed, with the notation above, each A_i is described by equations $\phi_{i,1} = a_{i,1}, \ldots, \phi_{i,d_i} = a_{i,d_i}$ and we recover the criterion c) of independence given above.

Definition 2.5.5. Let V and V' be two submanifolds of E. We say that V and V' are *transverse* if, for every point a of $V \cap V'$, the subspaces $T_a V$ and $T_a V'$ of E are transverse. See Fig. 2.4.

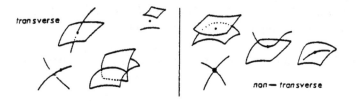

Fig. 2.4. Transverse and non-transverse intersections of submanifolds

In particular, if we have $\dim(V) + \dim(V') < \dim(E)$ this means that V and V' do not meet. Now let us look at the situation in the usual 3-dimensional space. To say that two curves are transverse is to say that they do not meet; to say that a curve C and a surface S are transverse means that at each point of intersection the tangent to C is not contained in the tangent space to S, or in other words C is nowhere tangent to S; to say that two surfaces are transverse means that at their common points their tangent planes are distinct; to say that a point is transverse to a surface is to say that it does not belong to it.

Note that the notion of transversality is relative to the space E in which V and V' are embedded. Thus two lines in a plane P in a 3-dimensional space E which meet at one point are transverse as submanifolds of P but not as submanifolds of E. We can grasp this difference by remarking that the notion of transversality is invariant under (small) deformations of the objects being considered; thus it is impossible to separate by small deformations two plane curves which cut without being tangent, while in 3-space it is possible.

The definition above can be generalized to an arbitrary number of submanifolds: we demand that at each common point the tangent spaces form a transverse family. The very definition of submanifold says that it is (locally) a transverse intersection of hypersurfaces. This leads naturally to the following proposition:

Proposition 2.5.6. *A transverse intersection of submanifolds is a submanifold.*

To simplify notation we deal with the case of two submanifolds V and V'. Let $a \in V \cap V'$. Take an open set U in E containing a in which V and V' are defined by non-degenerate systems of equations $(\phi_1, \ldots, \phi_d)$ and $(\psi_1, \ldots, \psi_{d'})$ respectively. The $d\phi_i(a)$ are independent equations for $T_a V$ and the $d\psi_j(a)$ are independent equations for $T_a V'$; the transversality hypothesis means precisely that the $d\phi_i(a)$ and te $d\psi_j(a)$ are independent. Consequently $\phi_1, \ldots, \phi_d, \psi_1, \ldots, \psi_{d'}$ form a non-degenerate system of local equations for $V \cap V'$ at a.   □

Note that this proof also yields the following statement:

Proposition 2.5.7. *If V and V' meet transversely at a then $T_a(V \cap V') = T_a V \cap T_a V'$ and $\operatorname{codim}_a(V \cap V') = \operatorname{codim}_a V + \operatorname{codim}_a V'$.*   □

Using the vocabulary of transversality, the notion of submanifold can be characterized by the following three properties:

a) the fact that V is a submanifold is verified 'locally' on V,
b) a level hypersurface is a submanifold at every non-critical point,
c) a transverse intersection of submanifolds is a submanifold.

2.6 Transversality: Inverse Images

2.6.1. We can generalize the statements in the preceding section in a very useful way. First take the linear case and consider two finite-dimensional vector spaces E and E', a linear map $u : E \to E'$, a vector subspace T of E and a vector subspace T' of E'. Recall that $u.T$ denotes the image of T under u ; we write $u^{-1}T'$ for the inverse image of T' under u. We then have the relation

$$\operatorname{codim}(u.T + T') + \operatorname{codim}(T \cap u^{-1}T') = \operatorname{codim}(T) + \operatorname{codim}(T'),$$

where, naturally, the codimensions are calculated relative to E or E' as appropriate. This can be proved just as in the special case $E = E'$, $u = Id_E$ dealt with previously, by considering the linear map $(x, x') \mapsto u \cdot x + x'$ from $T \times T'$ into E'. In particular, the following three conditions are equivalent:

(i) $E' = u.T + T'$; in other words the subspaces $u.T$ and T' of E' are transverse;

(ii) $\operatorname{codim}(T \cap u^{-1}T') = \operatorname{codim}(T) + \operatorname{codim}(T')$;

(iii) $\dim(T \cap u^{-1}T') = \dim(T) - \operatorname{codim}(T')$.

This being the case, Proposition 5.7 generalizes as follows:

Proposition 2.6.2. *Let E and E' be two finite-dimensional vector spaces, with U an open subset of E and f a C^∞ map from U into E'. Let V be a submanifold of U and let W be a submanifold of E'. Let a be a point of V such that $f(a) \in W$. Suppose that the subspaces $f'(a) \cdot T_a V$ and $T_{f(a)} W$ of E' are transverse. Then $V \cap f^{-1}(W)$ is a submanifold of E and we have*

$$T_a(V \cap f^{-1}W) = T_a V \cap f'(a)^{-1}\bigl(T_{f(a)} W\bigr),$$

$$\operatorname{codim}_a(V \cap f^{-1}W) = \operatorname{codim}_a V + \operatorname{codim}_{f(a)} W.$$

The proof of 6.2 is modelled on that for the case $E = E'$. Consider a nondegenerate system of local equations $(\phi_1, \ldots, \phi_d)$ for V at a and a nondegenerate system of local equations $(\psi_1, \ldots, \psi_{d'})$ for W at $f(a)$. The $d\phi_i(a)$ are linearly independent equations for $T_a V$ and the $d\psi_j(f(a))$ are linearly independent equations for $T_{f(a)} W$; the transversality assumption means precisely that the $d\phi_i(a)$ and the $d\psi_j(f(a)) \circ f'(a) = d(\psi_j \circ f)(a)$ are linearly independent. Hence $\phi_1, \ldots, \phi_d, \psi_1 \circ f, \ldots, \psi_{d'} \circ f$ form a nondegenerate system of local equations for $V \cap f^{-1}W$ at a ; the proposition follows immediately. $\square$

When $E = E'$ and $f = Id_E$ we recover the statements of the previous section. The other extreme case is when $V = E$; we then obtain the following criterion:

Corollary 2.6.3. *If $f(a)$ belongs to W and the image of $f'(a)$ is transverse to $T_{f(a)} W$ then the inverse image of W is a submanifold of E at a with codimension equal to that of W at $f(a)$ and with tangent space equal to the inverse image of $T_{f(a)} W$ under $f'(a)$.* $\square$

This applies in particular if W is reduced to a point, when we obtain the following corollary generalizing the example of level hypersurfaces:

Corollary 2.6.4. *If $f'(a)$ is surjective then $f^{-1}(f(a))$ is a submanifold at a with codimension equal to $\dim E'$ and with tangent space $\operatorname{Ker}(f'(a))$.* $\square$

Note that the very definition of submanifolds is in some sense a converse to the above corollary: suppose that V is a submanifold of E at a ; we saw in 2.3 that there then exists an open set U in E containing a and a diffeomorphism h of U onto an open subset of a product $E' \times E''$ such that $V \cap U$ is the inverse image of $\{0\} \times E''$; if we say that f is the map from U into E' consisting of h followed by the projection of $E' \times E''$ onto E' then $f(a) = 0$, the map $f'(a) : E \to E'$ is surjective, and we have $V \cap U = f^{-1}(0)$.

Let us now take W to be the point $f(a)$ and suppose that $f'(a)$ maps T_aV bijectively onto E' ; then the proposition implies that $V \cap f^{-1}(f(a))$ is a submanifold at a with tangent space reduced to zero, or in other words that a is an isolated point of $V \cap f^{-1}(f(a))$. This almost provides us with an inverse map for the restriction of f to V ; the situation is singularly reminiscent of the Local Inversion Theorem, and it is clearly not just by chance! It takes us to the subject of the next section: the Implicit Function Theorem.

2.7 The Implicit Function Theorem

Recall the situation above: f is a C^∞ map from E to E' and a is a point of a submanifold V of E. Consider the map $T_aV \to E'$ which is the restriction of the linear map $f'(a)$ to T_aV. Note immediately the following special case which will be essential in what follows: f is linear and so then $f'(a) = f$.

Theorem 2.7.1. *Suppose the linear map $f'(a)$ induces a bijection from T_aV to E'. Then there exists an open set U in E containing a , an open set U' in E' containing $f(a)$ and a C^∞ map $g : U' \to U$ such that for $x \in U$ and $x' \in U'$ the condition ($x \in V$ and $f(x) = x'$) is equivalent to ($x = g(x')$).*

In particular we have $g(f(x)) = x$ for every $x \in U \cap V$ and $f(g(x')) = x'$ for every $x' \in U'$. By differentiation, these relations imply that $g'(f(a))$ maps E' into T_aV and is the inverse of the restriction of $f'(a)$; the condition given in the statement is therefore necessary.

Now for the proof. Let

$$d = \operatorname{codim}_a V = \operatorname{codim}(T_aV) = \dim(E) - \dim(E'),$$

and let $(\phi_1, \ldots, \phi_d)$ be a nondegenerate system of local equations for V at the point a. Let u denote the map $x \mapsto (f(x), \phi_1(x), \ldots, \phi_d(x)) \in E' \times \mathbf{R}^d$, defined for x sufficiently close to a in E. The tangent map of u at a is the linear map $u'(a) : E \to E' \times \mathbf{R}^d$ such that

$$u'(a) \cdot \xi = (f'(a) \cdot \xi, \, d\phi_1(a) \cdot \xi, \, \ldots, \, d\phi_d(a) \cdot \xi).$$

This map is *bijective*: since $\dim(E) = \dim(E' \times \mathbf{R}^d)$ it is enough to show that it is injective. Now, if $\xi \in \operatorname{Ker}(u'(a))$ we have first of all that $d\phi_j(a) \cdot \xi = 0$ for each j, hence $\xi \in T_a V$ so $f'(a) \cdot \xi = 0$ and therefore $\xi = 0$ since the restriction of $f'(a)$ to $T_a V$ is injective. We may therefore apply the Local Inversion Theorem to u, and see that there is an open subset U of E containing a, where U may be taken small enough so that the ϕ_j form a system of equations for $V \cap U$, and open subsets U' of E' and U'' of $\mathbf{R}^d$ containing $g(a)$ and $(0, \ldots, 0)$ respectively, such that u induces a diffeomorphism from U onto $U' \times U''$. Let $v : U' \times U'' \to U$ be the inverse diffeomorphism; then for $x \in U$, $x' \in U'$ and $(\alpha_1, \ldots, \alpha_d) \in U''$ we have the equivalence

$$\left(x = v(x', \alpha_1, \ldots, \alpha_d) \right) \iff \left(f(x) = x', \ \phi_1(x) = \alpha_1, \ \ldots, \phi_d(x) = \alpha_d \right).$$

The map $g : U' \to U$ defined by $g(x') = v(x', 0, \ldots, 0)$ therefore has the required properties. □

An important special case is when E is a product $E' \times E''$ and we take for f the projection of E onto E' parallel to E''. Then since $f(g(x')) = x'$, we see that g has the form $x' \mapsto (x', h(x'))$. Thus we obtain :

Corollary 2.7.2. (Implicit Function Theorem, geometric form.) *Let V be a submanifold of $E' \times E''$ and let $a = (a', a'')$ be a point of V. Suppose the projection of $E' \times E''$ onto E' parallel to E'' induces a bijection $T_a V \to E'$ (i.e. $T_a V$ and E'' are complementary). Then there exists an open set U' in E' containing a', an open set U'' in E'' containing a'', and a C^∞ map $h : U' \to U''$ such that $V \cap (U' \times U'')$ is the graph $\{(x', h(x))\}$ of h.* □

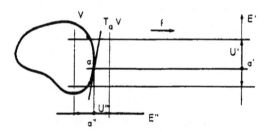

Fig. 2.5. Locally V is the graph of a map $h : U' \to U''$

In particular we have $h(a') = a''$. In classical terminology we have solved the 'system of implicit equations' $(x', x'') \in V$ for the variable x'' in a neighbourhood of (a', a''). See Fig. 2.5.

More precisely, we obtain the classical form of the Implicit Function Theorem by combining the previous corollary with Corollary 6.4 :

Corollary 2.7.3. (Implicit Function Theorem, analytic form.) *Let E', E'' and F be three finite-dimensional vector spaces. Let $\Phi : (x, y) \mapsto \Phi(x, y)$ be a C^∞ map from an open subset U of $E' \times E''$ into F; let (a', a'') be a point of U such that $\Phi(a', a'') = 0$ and such that the partial derivative of Φ along E''*

$$\Phi'_y(a', a'') : E'' \to F$$

(which by definition is the derivative of $y \mapsto \Phi(a', y)$) is bijective. Then there is an open set U' in E' containing a', an open set U'' in E'' containing a'', and a C^∞ map $h : U' \to U''$ such that for $(x, y) \in U' \times U''$ the relations $\Phi(x, y) = 0$ and $y = h(x)$ are equivalent. Moreover, the derivative $h'(a') \in L(E'; E'')$ is equal to $-\Phi'_y(a', a'')^{-1} \circ \Phi'_x(a', a'')$.

Denote by V the set of points (x, y) in U such that $\Phi(x, y) = 0$ and let $a = (a', a'') \in V$. Since $\Phi'_y(a)$ is bijective, the 'total' derivative $\Phi'(a)$: $E' \times E'' \to F$ is surjective and V is a submanifold at a by 6.4 whose tangent space is the subspace $T \subset E' \times E''$ that has equation $\Phi'_x(a) \cdot \xi + \Phi'_y(a) \cdot \eta = 0$. We then apply 7.2 and conclude by observing that the graph of $h'(a')$ is equal to T. $\qquad\square$

2.7.4. *Every submanifold may be described locally as a graph.* More precisely, let V be a submanifold of the vector space E and let (e_i), $i \in [1, n]$, be a basis for E. Let $a \in V$; set $d = \dim_a V = \dim(T_a V)$; by the theorem on completing a basis we can choose $n - d$ of the e_i to generate a complementary subspace to $T_a V$. Permuting the e_i if necessary, we can suppose that these are $e_{d+1}, \ldots, e_n$. Then by Corollary 7.2 we can describe V in a neighbourhood of a by equations of the form

$$x_{d+1} = h_{d+1}(x_1, \ldots, x_d), \quad \ldots, \quad x_n = h_n(x_1, \ldots, x_d),$$

where $h_{d+1}, \ldots, h_n$ are $n - d$ functions of class C^∞ in the d variables $x_1, \ldots, x_d$.

It is here that in algebraic geometry an essential difference appears. Namely, the fact that we do not have a Local Inversion Theorem for polynomial functions implies that a variety which is defined by polynomial equations may very well fail to have local parametrizations by polynomial functions. There is a well known and very simple example (to which we shall return in 3.6.6): take the submanifold in $\mathbf{R} \times \mathbf{R}^n$ given by the equation

$$x^n + a_1 x^{n-1} + \cdots + a_n = 0.$$

If for $a_i = \alpha_i$ this equation has a simple root ξ then the previous corollary implies that there exists a C^∞ function h such that $h(\alpha_1, \ldots, \alpha_n) = \xi$ and such that

$$x = h(a_1, \ldots, a_n)$$

is a root of the equation for (a_i) close to (α_i). But such a function is not polynomial when $n > 1$; it is not even (as has been known since the time of Abel and Galois[23]) expressible by means of radicals for $n > 4$.

---

[23]The Norwegian Niels ABEL (1802–1829) and the young Evariste GALOIS (1811–1832) born at Bourg-La-Reine near Paris had almost parallel lives and mathematical work. It was Abel who proved the insolubility by radicals of the general equation of the fifth degree; at the time of his death he had almost arrived at the general results that Galois was to obtain three years later. The latter in particular extended Abel's theorem to every degree ≥ 5.

This construction allows us to prove the following lemma making explicit the fact that the tangent space approximates the submanifold in a neighbourhood of the point under consideration, with the analogous fact for maps.

Lemma 2.7.5. *Let V be a submanifold of the finite-dimensional vector space E and let a be a point of V. Give E a norm, and let $\epsilon > 0$.*

a) *There exists $\theta > 0$ such that: for every $x \in V$ with $\|x - a\| \le \theta$ there exists $z \in T_a V$ such that $\|x - a - z\| \le \epsilon\|x - a\|$.*

b) *Let g be a C^∞ map from V into a finite-dimensional normed vector space F. There exists $\theta > 0$ such that: for every $x \in V$ with $\|x - a\| \le \theta$ there exists $z \in T_a V$ such that $\|x - a - z\| \le \epsilon\|x - a\|$ and $\|g(x) - g(a) - T_a g \cdot z\| \le \epsilon\|x - a\|$.*

With notation as before, let E'' be the subspace generated by the vectors $e_{d+1}, \ldots, e_n$. Applying 7.2 with $E' = T_a V$ we see that there exists a neighbourhood U of a in E, a neighbourhood U' of 0 in $T_a V$ and a C^∞ map $h : U' \to E$ such that $h(0) = 0$ and $V \cap U$ is the set of points $a + z + h(z)$ for z in U'. Moreover, the derivative $h'(0)$ is zero; hence for small z the size of $h(z)$ is negligible relative to z, which implies a). We deduce b) from this by applying it to the graph of g, which is a submanifold of $E \times F$ whose tangent space at the point $(a, g(a))$ is the graph of $T_a g$. $\qquad\square$

2.8 Diffeomorphisms of Submanifolds

Let E and F be two finite-dimensional vector spaces, let V be a submanifold of E and let $g : V \to F$ be a map. We shall say that g is of class C^r, $r \in [0, +\infty]$, if there exists an open set U in E containing V and a map $f : U \to F$ of class C^r whose restriction to V is equal to g. Analogously, we shall say that g is of class C^r at a point a of V if there exists an open set U in E containing a and a map f of class C^r from U into F such that f and g have the same restriction to $U \cap V$. In the case when V is an open set in E these are simply the usual notions, and to say that g is of class C^r means that it is C^r at every point of V. This is also true in the general case:

Proposition 2.8.1. *For g to be of class C^r it is necessary and sufficient that it be of class C^r at every point of V.*

The condition is clearly necessary. Suppose conversely that g is of class C^r at every point of V. Then for every point $a \in V$ we have a map $f_a : U_a \to F$, defined and of class C^r on an open set U_a in E containing a, and such that $f_a(x) = g(x)$ for all $x \in U_a \cap V$. Let U denote the union of the U_a as a runs through V ; it is an open subset of E containing V. We shall construct a C^r map f from U into F that extends g. To do this, we make use of the existence[24] of a *partition of unity* of class C^∞ which is *subordinate to the family* $(U_a)_{a \in V}$ *of open sets.* To be precise, we can find a subset A of V and for each point $a \in A$ a real function ϕ_a of class C^∞ on E such that:

a) for each $a \in A$ the function ϕ_a vanishes outside U_a ;
b) for each point $x \in U$ there exists an open set Ω_x in E containing x and a finite subset A_x of A such that for every $y \in \Omega_x$ and every $a \in A$ which is not in A_x we have $\phi_a(y) = 0$;
c) for each $x \in U$ the sum of the $\phi_a(x)$ (which by b) has only finitely many nonzero terms, namely at most those whose index a belongs to A_x) is equal to 1.

This being the case, define for every $y \in U$ an element $f(y)$ of F by

$$f(y) = \sum_{a \in A} \phi_a(y) f_a(y).$$

This expression does make sense because only finitely many of the indices $a \in A$ give coefficients $\phi_a(y)$ which are nonzero (Condition b)) and for these indices $f_a(y)$ is defined (Condition a)). More precisely, fix $x \in U$ and use the notation of b). Consider the intersection of Ω_x and the U_a for $a \in A_x$; it is an open set U containing x . Restricting to this open set, the formula above defines f as a finite sum of products of a real C^∞ function by a C^r map with values in F. Thus f is a C^r map from U into F and it remains to verify that f extends g . But for $y \in V$ all the $f_a(y)$ are equal to $g(y)$ by construction, which implies $f(y) = g(y)$ by Condition c). □

2.8.2. Naturally, if we are also given a submanifold W of F, a map g from V into W is said to be *of class* C^r if it is of class C^r as a map from V into F . Recall that for all $r \geq 1$ we have the tangent linear maps $T_a g : T_a V \to T_{g(a)} W$ (see 4.6). All the definitions and constructions made for open subsets of vector spaces may be transported to this more general case. Thus for example the Implicit Function Theorem 7.2 is just a special case of the Local Inversion Theorem for submanifolds. However, we shall for present purposes content ourselves with the following definition:

Definition 2.8.3. Let E and F be two finite-dimensional vector spaces, with V and W submanifolds of E and F respectively. We say that a map $g : V \to W$ is a *diffeomorphism* if it is bijective and if g and g^{-1} are of class C^∞.

In particular, g and g^{-1} are then continuous and g is a homeomorphism. Moreover, just as in the case for vector spaces, we see that for every $a \in V$ the tangent linear map $T_a g$ is a bijection from $T_a V$ onto $T_{g(a)} W$; in particular we have $\dim_{g(a)} W = \dim_a V$ (it is perhaps worth noting that E and F do not necessarily have the same dimension).

---

[24]See for example [BL], p.25.

2.8.4. We give an immediate example. Let Ω be an open set in E and f a C^∞ map from Ω to F. Let $\Gamma \subset E \times F$ be the graph of f ; it is a submanifold of $E \times F$ (3, Example 3). Then the bijection $g : x \mapsto (x, f(x))$ from Ω into Γ is of class C^∞ and the inverse map is induced by the projection of $E \times F$ onto E and is therefore of class C^∞. Thus g is a diffeomorphism from Ω to Γ . The fact that the tangent linear maps are bijective gives the example 3) of 4.3 again.

2.9 Parametrizations, Immersions and Embeddings

We begin with a simple case. Let E be a vector space of dimension d, with Ω an open set in E and $u : \Omega \to F$ a C^∞ map. Consider a point b in Ω and the derivative $u'(b) : E \to F$. Recall that if we identify E with $\mathbf{R}^d$ then $u'(b)$ sends the basis vectors of $\mathbf{R}^d$ to the d partial derivative vectors of u at the point b ; to say that $u'(b)$ is injective then means in this case that the partial derivatives are linearly independent.

Proposition 2.9.1. *If the derivative $u'(b)$ is injective, there exists an open subset Ω' of Ω, containing b, such that $u(\Omega')$ is a submanifold of F of dimension d and such that u induces a diffeomorphism from Ω' onto $u(\Omega')$.*

Let $e_1, \ldots, e_{n-d}$ be vectors of F which generate a subspace complementary to the image of $u'(b)$. Consider the map v from $\mathbf{R}^{n-d} \times \Omega$ to F which maps (y, z) to $y_1 e_1 + \cdots + y_{n-d} e_{n-d} + u(z)$. Its derivative map at the point $(0, b)$ is bijective by assumption. We may therefore apply the Local Inversion Theorem : there exists an open subset U of F and a diffeomorphism of U onto an open subset of $\mathbf{R}^{n-d} \times E$ of the form $U'' \times \Omega'$, where U'' is an open set in $\mathbf{R}^{n-d}$ containing 0 and Ω' is an open set in E containing b, this diffeomorphism being inverse to v. Write the diffeomorphism in the form $x \mapsto (\phi_1(x), \ldots, \phi_{n-d}(x), w(x))$. Then u maps Ω' bijectively onto the submanifold V of U given by equations $\phi_1, \ldots, \phi_{n-d}$. Moreover, the map $V \to \Omega'$ inverse to u is induced by the map w which is of class C^∞ on U ; it is therefore of class C^∞ and u does indeed induce a diffeomorphism from Ω' onto V. $\qquad\square$

2.9.2. The local description of submanifolds as graphs which we saw in the previous section shows in particular that for every point a of a submanifold V of F a suitable open neighbourhood of a in V can be described as the image of a map $u : \Omega' \to F$ of the type studied in the above proposition.

Everything therefore seems to be in good shape. However, there is one serious error which must not be committed: we demonstrate this now by a simple example. Consider the open interval $I = (-\pi, \pi)$ of $\mathbf{R}$ and the parametrized arc $u : I \to \mathbf{R}^2$ defined by $u(t) = (\sin t, \frac{1}{2} \sin 2t)$. Its velocity vector at the point t is $(\cos t, \cos 2t)$, and so is never zero. See Fig. 2.6.

Moreover, u is injective and we may well have the impression from the above proposition that $u(I)$ is a submanifold of $\mathbf{R}^2$. This is incorrect, though, because $u(I)$ is the lemniscate with equation $y^2 = x^2 - x^4$ and is not a submanifold at 0 (see 3, example 7).

Where does the error arise? The Proposition says that for every $t \in I$ there exists an open subset I' with $t \in I' \subset I$ such that $u(I')$ is a submanifold, and that is quite true; in fact $u((a, b))$ is a submanifold for every pair (a, b) with $-\pi < a < b < \pi$. On the other hand, to say that $u(I)$ is a submanifold at 0 would mean that there exists an open subset U of $\mathbf{R}^2$ containing 0 such that $u(I) \cap U$ has a certain property; but $u(I) \cap U$ is the image under u of $u^{-1}(U)$ which, however small u may be, contains the intervals $(-\pi, a)$ and (b, π). Here we have a question of pure topology: the bijection u from I to $u(I)$ is certainly continuous, but the inverse map is not. The image under u of an open set $(-\pi, a)$ with $a < 0$ *is not open in* $u(I)$. The error is precisely that of confusing two distinct topologies on $u(I)$: the one which comes from I and for which $u(I)$ is an honest open interval, and the one which comes from $\mathbf{R}^2$ for which $u(I)$ is a 'figure eight'.

This leads us to introduce a whole new range of terminology.

Let V be a submanifold of a finite-dimensional vector space E, and let g be a C^∞ map from V into another finite-dimensional vector space F.

Definition 2.9.3. We say that g is an *immersion at a* if $T_a g$ is injective. We say that g is an *immersion* if $T_a g$ is injective for every $a \in V$. We say that g is an *embedding* if $g(V)$ is a submanifold of F and if g induces a diffeomorphism from V onto $g(V)$.

2.9.4. Let us give some examples. By a *local parametrization* of the submanifold W of F in a neighbourhood of a point $p \in W$ we mean an embedding u of an open set Ω of $\mathbf{R}^d$ into F whose image is an open subset of W that contains p. As already discussed, *every submanifold posesses local parametrizations at every point.*

In particular, suppose W is an open subset of F. Then the map inverse to u is a diffeomorphism from the open set $u(\Omega)$ onto the open set Ω in $\mathbf{R}^d$, that is to say a system of curvilinear coordinates on $u(\Omega)$. For open subsets of vector spaces, local parametrizations are simply the inverse maps to systems of local coordinates.

This justifies our calling a *system of local coordinates* on a submanifold W of F any sequence $(x_1, \ldots, x_d)$ of functions defined on an open subset W' of W such that the map from W' to $\mathbf{R}^d$ they define is inverse to a local parametrization u. Each point p of W' is the image under u of the point $(x_1(p), \ldots, x_d(p))$ of $\mathbf{R}^d$.

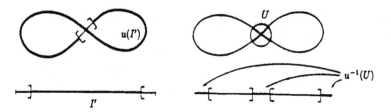

Fig. 2.6. An injective image of an open interval that is not a submanifold of $\mathbf{R}^2$

2.9.5. Let $u : \mathbf{R} \to \mathbf{R}^2$ be the map $t \mapsto (\sin t, \cos t)$. This is an immersion whose image is the circle $\mathbf{S}^1$. For every open interval I of $\mathbf{R}$ of length $< 2\pi$, or more generally for every open subset I of $\mathbf{R}$ such that the restriction v of u to I is injective but not surjective, we have that v is a local parametrization of $\mathbf{S}^1$. Conversely, on every open subset V of $\mathbf{S}^1$ whose complement is nonempty we can construct a local coordinate $x : V \to \mathbf{R}$ with $u(x(p)) = p$ for every $p \in V$.

Now let us return to the situation of Definition 9.3.

Proposition 2.9.6.

(a) *If g is an immersion at a there exists an open set U in E containing the point a such that the restriction of g to $V \cap U$ is an embedding.*

(b) *If g is an injective immersion and the map from $g(V)$ to V that is inverse to g is continuous, then g is an embedding.*

Let $a \in V$ be such that g is an immersion at a. Let $d = \dim_a V$. Choose an open set U of E that contains a, an open subset Ω of $\mathbf{R}^d$ and an embedding h of Ω into E with image $V \cap U$. Let $b \in \Omega$ be the point that corresponds to a. Then the derivative $(g \circ h)'(b)$ is injective and we may apply Proposition 9.1 to $g \circ h : \Omega \to F$. Shrinking U and Ω if necessary, we can suppose that $g \circ h$ is an embedding. This implies that the restriction of g to $h(\Omega) = V \cap U$ is an embedding, which gives (a), and also that $g(V \cap U) = (f \circ h)(\Omega)$ is a submanifold of F. Suppose now that g satisfies the hypotheses of (b). First we prove that $g(V)$ is a submanifold of F. Let $b = g(a)$ be a point of $g(V)$ and let U be the open subset of E constructed above. Since the inverse bijection to g is continuous by assumption (this is the key point) there exists an open set U' in F such that $g(V \cap U) = g(V) \cap U'$, and $g(V)$ is indeed a submanifold at b. However, since the map of $g(V)$ to V that is inverse to g coincides locally with a diffeomorphism, it is of class C^∞ by Proposition 8.1 and g is an embedding. $\qquad\square$

Here is a simple application of this proposition:

Corollary 2.9.7. *Suppose that g is an injective immersion, that $g(V)$ is a submanifold of F, and we have $\dim_a V = \dim_{g(a)} g(V)$ for all $a \in V$. Then g is an embedding.*

We shall see later (Corollary 3.2.9) that this dimension condition is superfluous if V has the same dimension everywhere.

From (b), the question is one of showing that the map $h : g(V) \to V$ inverse to g is a continuous map. Let a be a point of V. By (a) there exists an open set U in E containing a such that g induces an embedding of $V \cap U$ into F. Let $h_U : g(V \cap U) \to V \cap U$ be the inverse map, which is continuous at the point $g(a)$ by hypothesis. Since $g(V \cap U)$ and $g(V)$ are submanifolds of F having the same dimension at $g(a)$, we deduce from 2.5 b) that there exists an open set W in F containing $g(a)$ with $W \cap g(V) = W \cap g(V \cap U)$. The restriction of h to the open subset $W \cap g(V)$ of $g(V)$ then coincides with the restriction of h_U and is therefore continuous, which is what we wished to prove. (We could also conclude directly as in (b), starting from the existence of W.) □

Some authors use the term *immersed manifold* for the image of an immersion, and of course *embedded manifold* for a submanifold. An immersed manifold $g(V)$ that is the image of a non-injective immersion g has multiple points; at such a point p which is the image of several different points a_i of V there is an intersection of "branches" $(d = 1)$, "sheets" $(d = 2)$, ... which are the images of embeddings of small neighbourhoods of the a_i. In fact it is usually easier and more convenient to consider directly the 'true' manifold V and the immersion g rather than the complicated object $g(V)$.

2.10 Proper Maps; Proper Embeddings

To make good use of the results in the previous section it is worth having at our disposal an elementary notion that usually does not feature in basic courses in topology, namely that of a *proper* map.

2.10.1. First of all recall that a subset K of a finite-dimensional vector space is said to be *compact* if it has the following equivalent properties :

(i) K is bounded and closed in E,
(ii) the 'Borel-Lebesgue property' : from every cover of K by open sets it is possible to extract a finite subcover,
(iii) the 'Bolzano-Weierstrass property' : from every infinite sequence of elements of K it is possible to extract a sequence that has a limit in K.

In the property (i) the assumption of finite dimensionality is essential. In general topology the condition (ii) is taken as the definition of compactness. In metrizable spaces the condition (iii) is equivalent to it. In infinite dimensions compactness is a very strong condition.

Let $f : A \to B$ be a continuous map between subsets A and B of finite-dimensional vector spaces E and F. The image $f(K)$ of every compact subset

K of A is compact (and in particular closed in F and therefore in B); recall that almost by definition the inverse image under f of an open or closed subset of B is open or closed in A respectively. We say that f is *proper* if the inverse image under f of every compact subset of B is compact.

2.10.2. It should be noted immediately that, in contrast to continuity, the fact of being proper depends on the choice of the target B of the map. For example, take A to be the interval $(-1, 1)$ and f, g to be the map $x \mapsto x$ from A into A or into $\mathbf{R}$ respectively; then f is proper (obviously) but g is not since the inverse image of the compact subset $[-2, 2]$ of $\mathbf{R}$ under g is equal to A which is not compact. Here are some other examples :

1) If A is compact then every continuous map $f : A \to B$ is proper. Indeed, for every compact subset K of B the set $f^{-1}(K)$ is closed in A and therefore compact.

2) Let $\|\xi\|$ be a norm on E. The map $\xi \mapsto \|\xi\|$ from E to $\mathbf{R}$ is proper. Likewise, for every map $u : \mathbf{R} \to \mathbf{R}$ with $\lim_{t \to \infty} u(t) = \infty$ the map $\xi \mapsto u(\|\xi\|)$ is proper.

   Here too the assumption of finite dimensionality is essential.

3) A linear map is proper only if it is injective.

4) Let $\gamma : [a, b) \to B$ be a continuous map. To say that γ is proper is by definition to say that for every compact subset K of B there exists $b' \in [a, b)$ with $\gamma^{-1}(K) \subset [a, b']$, that is to say $\gamma((b', b)) \cap K = \emptyset$. This means also that there is no sequence (t_i) of points of $[a, b)$ tending to b such that the $\gamma(t_i)$ tend to a point of B. This is loosely translated by saying that $\gamma(t)$ "leaves every compact set in B" as t tends to b. For example, if $B = F$ and if $\|\eta\|$ is a norm on F it means that $\|\gamma(t)\|$ increases indefinitely.

5) Let $A \subset B$ be two subsets of E and let i be the canonical injection of A into B ; then i is proper if and only if A is closed in B. This is in fact a special case of Proposition 10.4 below.

Lemma 2.10.3. *Let $f : A \to B$ be a map which is continuous and proper. Then for every closed subset A' of A the set $f(A')$ is a closed subset of B.*

Suppose $f(A')$ is not closed. Then there exists a sequence (a_i) of elements of A' such that the sequence of the $f(a_i)$ converges to an element b of B that does not belong to $f(A')$. Let L' denote the set of all the $f(a_i)$ and let L be the inverse image of L'. Since f is proper and $L' \cup \{b\}$ is compact its inverse image $f^{-1}(L' \cup \{b\}) = L \cup f^{-1}(b)$ is also compact. Since A' is closed in A and $A' \cap (L \cup f^{-1}(b)) = A' \cap L$, this latter intersection is compact. Therefore it is possible to extract from the sequence (a_i), which consists of elements of $A' \cap L$, a sub-sequence which converges to an element a of $A' \cap L$.

Since f is continuous, the corresponding $f(a_i)$ converge to $f(a)$; this gives $b = f(a) \in f(A')$ which contradicts the assumption.   □

Proposition 2.10.4. *Let* $f : A \to B$ *be a map which is continuous and injective, and let* $h : f(A) \to A$ *be the inverse map. The following conditions are equivalent :*

(*i*)   *f is proper,*
(*ii*)   *the image under f of every closed subset of A is a closed subset of B,*
(*iii*)   *$f(A)$ is a closed subset of B and the map h is continuous.*

(i)⇒(ii) : this results from the previous lemma.
(ii)⇒(iii) : $f(A)$ is closed in B as the image of A ; moreover, h is continuous since for every closed subset A' of A the set $h^{-1}(A') = f(A')$ is closed in $f(A)$.
(iii)⇒(i) : if K is a compact set in B then $f^{-1}(K)$ is compact because it is $h(K \cap f(A))$.   □

Now we return to the situation in the previous section. From Propositions 9.6 b) and 10.4 we immediately have :

Proposition 2.10.5. *Let* V *be a submanifold of* E *and let* g *be a* C^∞ *map of* V *into an open set* U *of* F. *The following conditions are equivalent :*

(*i*)   *g is an injective and proper immersion,*
(*ii*)   *g is a proper embedding,*
(*iii*)   *g is an immersion and $g(V)$ is closed in U.*   □

In particular, an injective immersion of a compact submanifold is a proper embedding. More generally, if g is an injective immersion of V into F, if K is a compact subset of V and if V' is an open set in V containing K, then the restriction of g to V' is an embedding and hence $g(V')$ is a submanifold of F. Note also that it is clear directly that the bijective immersion of an open interval onto a lemniscate constructed in 9.2 is not proper even though its image is closed : as t tends to one of the ends of the interval, $\gamma(t)$ has a limit, in contradiction to the example 4) above.

2.10.6. In this connection, let us return for a moment to the very definition of submanifold. Let V be a submanifold of the vector space E. By definition, for each point a of V there exists an open subset U_a of E such that $V \cap U_a$ is defined in U_a by equations and is in particular a closed subset of U_a; in other words the complement $U_a \setminus V \cap U_a$ is open. Let U denote the union of the U_a. This is an open set containing V and the complement $U \setminus V$ is the union of the $U_a \setminus V \cap U_a$ and is therefore open. Hence V is a *closed submanifold of the open set* U. If $\overline{V}$ denotes the closure of V in E then V is the intersection of the open set U and the closed set $\overline{V}$. Now let b be a point of $\overline{V} \setminus V$; then in every open set U_b in F that contains b there will be points of V, and the sets $U_b \cap V$ and $U_b \cap \overline{V}$ will be unlikely to have a simple description.

Matters can be very complicated, as the following rudimentary example shows. Consider a diffeomorphism u from $\mathbf{R}$ onto the interval $(0, 1)$, such as

$$t \mapsto \frac{e^t}{e^t + 1},$$

and consider the map $\gamma : \mathbf{R} \to \mathbf{R}^2$ given by $\gamma(t) = (u(t) \cos t, u(t) \sin t)$. Then γ is an embedding; the spiral C which is the image of γ is closed in the open set $0 < x^2 + y^2 < 1$, and $\overline{C} - C$ is the union of the origin and the circle $x^2 + y^2 = 1$. See Fig. 2.7.

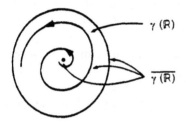

Fig. 2.7. The closure of an embedded submanifold may be complicated

2.11 From Submanifolds to Manifolds

For reasons explained in the Introduction, a deliberate choice has been taken to work within the framework of *submanifolds* throughout this text. However, in order not to give too false a perspective we now explain via a simple example (curves) how a more 'intrinsic' approach might be justified.

Obviously we know that there are at least two distinct examples of connected curves: the straight line $\mathbf{R}$ and the circle $\mathbf{S}^1$ represented, say, by the submanifold with the equation $x^2 + y^2 = 1$ in $\mathbf{R}^2$. There are essentially no others:

Proposition 2.11.1. *Let E be a finite-dimensional vector space and let C be a connected curve (1-dimensional submanifold) in E. Then:*

(a) *if C is not compact there is an embedding $\mathbf{R} \to E$ whose image is C;*
(b) *if C is compact there is an embedding $\mathbf{S}^1 \to E$ whose image is C.*

This proposition is a good example of a phenomenon that is common in differential topology: the most intuitive facts are not ones that are easy to prove, and we count ourselves lucky if we can find an argument which is more or less elementary. We here give a detailed outline of such a proof and leave it as an exercise for the reader to fill in the the details.

a)   We may take $E = \mathbf{R}^n$ equipped with the usual euclidean norm. For
     each point x of C there are local parametrizations $\gamma : (-\epsilon, +\epsilon) \to C$
     with $\gamma(0) = x$. We have $\dot{\gamma}(t) \neq 0$ for all t; replacing γ by $\gamma \circ \phi$ for a
     suitable function ϕ we can impose the condition $\|\dot{\gamma}(t)\| = 1$. Two such
     parametrizations coincide up to sign in a neighbourhood of 0.

b)   Now fix $x \in C$ and consider (C^∞) maps $\gamma : (a, b) \to E$ whose image is
     contained in C, with $a \in (-\infty, 0)$, $b \in (0, +\infty)$, $\gamma((a, b)) \subset C$, $\gamma(0) = x$
     and $\|\dot{\gamma}(t)\| = 1$. These do exist, by a). Take a minimal and b maximal
     (possibly infinite): we now prove that $\gamma((a, b)) = C$. As C is assumed
     to be connected and the image of γ is open in C, it is enough to prove
     that this image is closed in C. Suppose this is false, and let y be a point
     of C which does not belong to the image of γ and is a limit point of
     this image. There is a sequence of points (t_i) in (a, b) such that $\gamma(t_i)$
     converges to y. From the sequence (t_i) we can extract a subsequence
     which converges either to a point of (a, b) or to a or to b. In the first
     case, y would belong to the image of γ, which it does not. Hence the
     sequence converges either to a or to b; to fix the ideas let us suppose it is
     b. There is a local parametrization $\delta : (-\epsilon, +\epsilon) \to C$ with $\delta(0) = y$ and
     $\|\dot{\delta}(0)\| = 1$; since the $\gamma(t_i)$ tend to y there is an i such that $\gamma(t_i) = \delta(u)$
     with $u \in (-\epsilon, +\epsilon)$. By 'glueing together' γ and $t \mapsto \delta(u \pm (t - t_i))$
     we obtain a parametrization that extends γ beyond y, which gives a
     contradiction.

c)   We have therefore found an immersion γ of an open interval I of $\mathbf{R}$
     whose image is C. If γ is injective it is an embedding by Corollary 9.7.
     Composing with a diffeomorphism of $\mathbf{R}$ onto I we obtain an embedding
     $\mathbf{R} \to E$ whose image is C.

d)   It remains to deal with the case when γ is not injective. Here there exist
     t_0 and t_1 in I with $t_0 < t_1$ and $\gamma(t_0) = \gamma(t_1)$. Up to translation in t we
     can suppose $t_0 = 0$; then let $a = t_1 > 0$ so that $\gamma(a) = \gamma(0)$. Then the
     arc $t \mapsto \gamma(a + t)$ parametrizes the same curve as γ with the same speed
     and starting at the same point $\gamma(0)$. Therefore we have $\gamma(t) = \gamma(a + t)$
     or $\gamma(t) = \gamma(a - t)$ according to whether the two arcs leave in the same
     or opposite dirctions.

e)   The latter case is impossible, however, since it would imply $\dot{\gamma}(a/2) =$
     $-\dot{\gamma}(a/2)$ and γ would not be an immersion. Thus we have $\gamma(t) = \gamma(a+t)$
     and in view of the maximal character of I we have $I = \mathbf{R}$ and γ is
     periodic. This implies in particular that C is compact.

f)   Let a denote the period of γ. The above argument proves that if
     $\gamma(t + b) = \gamma(t)$ for one value of t then it is true for all t and hence
     b is a multiple of a. We define a C^∞ map f from $\mathbf{R}^2 \setminus \{0\}$ into E by
     putting $f(\rho \cos\theta, \rho \sin\theta) = \gamma(a\theta/2\pi)$; then the restriction of f to $\mathbf{S}^1$ is
     an injective immersion and therefore an embedding (Proposition 10.5)
     with image C, which (at last!) finishes the proof. $\square$

2.11.2. This shows at least the conceptual interest in bringing out the intrinsic notion of a circle without reference to a given embedding. Thus from the intrinsic point of view there exist only two connected curves, the straight line and the circle, that may be embedded or immersed in many ways. For example, consider the following embedding (the *trefoil knot*) of S^1 in $\mathbf{R}^3$:

$$f(\cos t, \sin t) = ((2 + \cos 3t) \cos 2t, (2 + \cos 3t) \sin 2t, \sin 3t).$$

Fig. 2.8. A trefoil knot

This circle is said to be *knotted* because we cannot extend f to an embedding of any disc $\{x^2 + y^2 < a\}$ with $a > 1$: see Fig. 2.8. Therefore we see here too the interest in not instinctively associating to S^1 the disc which is its interior in the usual embedding.

2.11.3. In any case, either of the two approaches introduces the same objects, for it can be proved that every 'abstract' manifold has a proper embedding into a space $\mathbf{R}^N$ for a suitable N. A quite simple projection method then enables us to prove that $N = 2d + 1$ works for every manifold of dimension d (Whitney's Embedding Theorem: see Proposition 3.4.5); it is then possible to go from $2d + 1$ to $2d$, which is optimal for $d = 1$ and $d = 2$ since the circle does not embed in $\mathbf{R}$ nor the Klein bottle in $\mathbf{R}^3$. It has only recently been discovered how to set about finding the best possible value of N for each value of d; we remark on this because the result is quite surprising: it involves the number of bits equal to 1 in the base 2 repesentation of d.

We do not therefore 'lose' any objects in the elementary approach, nor any maps. In fact for every C^r map $f : V \to V'$ of one 'abstract' manifold into another it is possible to find a proper embedding $p : V \to \mathbf{R}^n$, a proper emebedding $p' : V' \to \mathbf{R}^{n'}$ and a C^r map $g : \mathbf{R}^n \to \mathbf{R}^{n'}$ such that $g \circ p = p' \circ f$.

2.12 Some History

It is difficult to trace the history of the material treated in this chapter since it mainly has to do with definitions. We shall just give a few signposts.

The existence of the tangent plane to a surface was shown explicitly by Dupin in 1813. It was Gauss[25] who first used local parametrizations and curvilinear coordinates for surfaces in a systematic way. At the beginning of his *Disquisitiones Generales Supra Superficies Curvas*, presented to the Royal Scientific Society at Göttingen on 8 October 1827, he mentions in passing the three classical ways to describe a piece of curved surface in $\mathbf{R}^3$ and the three corresponding descriptions of the tangent plane: by an equation $W(x, y, z) = 0$, by a parametrization $(p, q) \mapsto (x, y, z)$, or, as a special case of the other two, as the graph of a map $(x, y) \mapsto z$. We quote the following extracts from No.3 of the Disquisitiones:

'...Duae habentur methodi generales ad exhibendam indolem superficiei curvæ. Methodus *prima* utitur æquatione inter coordinatas x, y, z, quam reductam esse supponemus ad formam $W = 0$ ubi W erit functio indeterminarum x, y, z ...

Methodus *secunda* sistit coordinatas in forma fonctionum duarum variabilium p, q ...

His duabus methodis generalibus accedit *tertia*, ubi una coordinatarum, e.g. z exhibetur in forma functionis reliquarum x, y: hæ c methodus manifesto nihil aliud est, nisi casus specialis vel methodi primæ, vel secundæ...'

Gauss does not say explicitly how to go from one to the other, nor that they define the same objects; these matters appear to be accepted intuitively in the texts of the time. It would be necessary here to have the Implicit Function Theorem.

The fact that it is possible to define a "many times extended multiplicity" (we would now say *differentiable manifold*) without reference to a particular embedding in an auxiliary space $\mathbf{R}^n$ can be found in the Inaugural Lecture of Riemann[26] in 1854, *On the Assumptions that Lie at the Foundations of Geometry*.

Roughly speaking, the basic concepts of general topology (such as connectedness, compactness, ...), of differential topology (manifolds, exterior differential calculus, ...), and of algebraic topology (universal covering, Betti numbers, ...) were formulated and refined in the period 1875–1925, and in any case they were firmly established in their present context by 1940. The study of the best bound for the Embedding Theorem is more recent and has not yet been fully completed; see [GI] for a technical survey. The problem for immersions was solved by R.L.Cohen [CO].

---

[25] Carl Friedrich GAUSS (1777–1855), one of the giants in the history of science and called "the prince of mathematicians": he was certainly one of the greatest, with Archimedes, Euler, Riemann and a few others including some living today. The (bilingual) text of his *Disquisitiones Generales* – not to be confused with the *Disquisitiones Arithmeticæ*, which is equally famous and important – accompanied by a historical study by Peter Dombrowski can be found in [DO] which is strongly recommended.

[26] Bernhard RIEMANN (1826–1866) was a German mathematician and Professor at Göttingen from 1854 until his death. He contributed decisively to all areas of mathematics. The subject for this inaugural lecture had been assigned to him by Gauss.

3 Transversality Theorems

3.1 Introduction

In this chapter we look into a topic which is the modern version of an old idea, namely 'general position', and we shall try to explain why it is important.

According to the philosophy sketched out in the Introduction, a simple (and Utopian) aim of geometry could be to apply suitable transformations to convert every geometric situation to a 'reduced form' taken from a fairly restricted list of standard types, and the study of these types would then be sufficient for us to be able to solve every problem. In the analysis of a given situation, this programme usually starts by considering the existence of *discrete invariants* (number and dimensions of the manifolds in question, dimensions of their intersections, ranks of certain maps ...). Once the list of discrete invariants has been determined – not always an easy task – we hope that for every set of values assigned to these invariants there will remain a *continuous* classification described by a few or perhaps many numerical invariants. This scheme of things works quite well in linear algebra where many of the results do have this form (linear independence, classification of matrices and of quadratic forms, solution of linear systems, ...). For the subject occupying us now, however, it is a vain hope. For example, every closed subset of $\mathbf{R}^n$ is the set of zeros of some suitable C^∞ function, and so a general classification of C^∞ functions of two variables would imply a classification of closed subsets of the plane, which is out of the question.

Nevertheless, in many problems we are able to observe the appearance of 'general positions' which are more common and easier to study than the others. For example, consider four straight lines in the plane. In general they form a complete quadrilateral, that is they meet in pairs with the six points of intersection all distinct. A quick analysis of similar examples shows that the idea of general position in fact embraces two separate notions: *density* and *openness*. Let E be the set of quadruples of straight lines in the plane (equipped with a suitable topology which models the natural idea of two straight lines being close to each other) and let A be the subset of E consisting of those quadruples which form a complete quadrilateral. The two ideas above are then made explicit as follows. First, every quadruple in E can be deformed into an element of A by a modification which is as small as we like; in the jargon of mathematicians A is *dense* in E, every element of E being a limit of points of A. Secondly, a complete quadrilateral remains a complete quadrilateral under small deformations; A is *open* in E, every point of E sufficiently close to a point of A is itself a point of A.

A less fanciful hope would be to find, within each of the continuous families identified by particular values of the discrete invariants for the problem, an open and dense subset which is sufficiently well-behaved for the programmme to be carried out for the general positions that it defines. As we shall see later, this still too much to hope for, but the idea is a fruitful one and there is much that can be learned by studying the extent to which it is valid.

Already, the demand that the general positions form an open set is for techincal reasons a little too strong and we often have to be content with a slightly weaker notion. Rather than open dense sets we find countable intersections of open dense sets, called *residual* sets for short. For example, the rational numbers form a countable subset of $\mathbf{R}$, so that a number taken at random will be irrational, but the set of irrational numbers is not an open subset of $\mathbf{R}$: it is a residual subset, being the intersection of the countable family of complements of the rational points. The expression *at random* was used intentionally, since probabilities do provide another way of mathematizing the idea of general position: describing properties which are almost certain, that is which hold everywhere except in a set of measure zero (we say more simply a *negligible* set). The connection between residual sets and sets with negligible complement is in general rather weak. However, the sets that arise will often have the two properties at the same time, and this is in any case true in what follows.

The central theme of this chapter is the study of the relative position of two submanifolds, and more generally of a submanifold and a map (the first case being deduced from the second by considering the embedding map of the second submanifold). In this setting, the useful notion of general position is that of *transversality* which we met in Chapt. 2.

The results are deduced from two basic theorems: *Baire's Theorem* asserts that in the most common spaces residual subsets are dense, while *Sard's Theorem* asserts that certain sets of exceptional points are negligible. The most convenient form of the end result is *Thom's Transversality Theorem*. The ideal setting for this type of statement is the space $C^\infty(V, W)$ of all the C^∞ maps from a manifold V into a manifold W, equipped with a suitable topology. Unfortunately, this topology is not defined by a norm, and we have to depart from the elementary setting which has been sufficient for us up to now. We shall do this in the precise case of Thom's Theorem, which loses much of its effectiveness when the context is simplified too much. For other statements we shall use a subterfuge that consists of studying finite-dimensional families of maps of V into W, defined simply to be maps $V \times \Lambda \to W$ where Λ is a parameter-manifold, and we leave the general statement as an exercise (which boils down to considering the infinite-dimensional family of *all* the maps).

The structure of the chapter is as follows. Sects. 2 and 3 contain the basic minimum of general topology and measure theory that will be needed later. In Sect. 4 we deal separately with a special case that has an elementary proof, namely the case when the dimension conditions imply that general position for the manifolds in question means that they are disjoint. Two forms of Sard's Theorem are given in Sects. 5 and 6. The Transversality Theorem is the subject of Sect. 7 (weak form) and Sect. 9 (more elaborate form). In Sect. 8 we quickly introduce the notion of *jet* which is used in the statement of Thom's Theorem. Finally, Sect. 10 is a very brief historical survey.

3.2 Countability Properties in Topology

3.2.1. In this subsection we assemble a few statements relating density and countability. Recall that a subset A of a space E (having any topology) is said to be *dense* if it has the following equivalent properties: every point of E is a limit of points in A; any closed set containing A is equal to E; every nonempty open set meets A. We shall say that a subset A of E is *locally closed* (in E) if it has the following equivalent properties: A is the intersection of an open set and a closed set; A is an open subset of its closure; for every $x \in A$ there exists an open set U in E containing x such that $A \cap U$ is closed in U. Every open or closed set is locally closed; we saw in 2.10.6 that a submanifold in a vector space is locally closed.

We use the adjective *countable* to mean "in 1:1 correspondence with a subset of $\mathbf{N}$". A finite set is therefore countable. The union of a countable family of countable sets is countable; the product of a finite family of countable sets is countable (diagonal construction due to Cantor).

3.2.2. First we state a few elementary properties. Let E denote a finite-dimensional vector space and let A be a subset of E. A family $\{U_\alpha\}$ of open sets in E is said to *cover* A if the union of the U_α contains A, which is the same as saying that the $U_\alpha \cap A$ form an open cover of A.

1) *There exists a countable dense subset of A.*

Take a countable dense subset D of E (for example, the points which have rational coordinates with respect to a given basis of E) and fix a norm on E. For each point $x \in D$ and each integer $n > 0$ consider the open ball $U_{x,n}$ with centre x and radius $1/n$. For each pair (x, n) such that $U_{x,n}$ meets A take a point $a_{x,n}$ in $U_{x,n} \cap A$. These points then form a dense subset of A.

2) *From every family $(U_i)_{i \in I}$ of open sets that cover A it is possible to extract a countable family that covers A.*

To simplify matters let $(V_j)_{j \in J}$ denote the family of all the open sets $U_{x,n}$ considered in 1). This is a countable cover of E. Let J' be the subset of J consisting of those j for which there exists an $i \in I$ with $V_j \subset U_i$. Then the

V_j cover A : for every $a \in A$ there exists an i with $a \in U_i$, and then a j with $a \in V_j \subset U_i$. For each $j \in J'$ choose an element $i(j) \in I$ with $V_j \in U_{i(j)}$. Then the $U_{i(j)}$ form a countable family of open sets whose union contains all the V_j for $j \in J'$ and therefore contains A.

3) *Every open subset of A is the union of a countable family of closed subsets of A ; every closed subset of A if the intersection of a countable family of open subsets of A.*

It is enough to prove the second assertion. Let F be a closed subset of A. For each integer $n > 0$ let U_n be the union of the intersections with A of the open balls of radius $1/n$ centred at all the points of F. Then each U_n is open in A and F is the intersection of the U_n.

4) *Every locally closed subset of E is the union of a countable family of compact subsets.*

Let U be an open subset and F a closed subset of E; we shall prove that $U \cap F$ is a countable union of compact sets. By 3) above, U is the union of a sequence $F_1, F_2, \ldots$ of closed subsets of E. Moreover, E is itself the union of an increasing sequence of compact subsets $K_1, K_2, \ldots$ (for example, the closed balls with fixed centre and radii $1, 2, \ldots$). Then $U \cap F$ is the union of the compact sets $F_n \cap F \cap K_n$.

5) *Let A be a locally closed subset of E and let $g : A \to X$ be a continuous map (we suppose X is Hausdorff). Then $g(A)$ is the union of a countable family of closed subsets.*

This is a result of the preceding property and the fact that for every compact set K in A the image $g(K)$ is compact and therefore closed in X. Recall that we saw in 2.10.3 that $g(A)$ is closed when g is proper.

Of course, all these properties are true in a much more general setting. Thus 3) holds for every metric space, 1) and 2) are true in every metric space that has a countable dense subset, and 4) is true in every locally compact metric space that has a countable dense subset.

3.2.3. One way to say that a property is "almost always true" would be to say that its domain of validity is an open dense subset. This condition is in practice a little too strong, in particular because it is not stable under countable intersections. In a general topological space it could happen that a countable intersection of open dense sets is empty (take for example the complements to the individual points in $\mathbf{Q}$). Fortunately this does not happen in the more common spaces in view of *Baire's*[27] *Theorem.* We say by definition that a topological space A has the Baire property, or is a *Baire space*, if every

---

[27]René BAIRE (1874–1932), a French mathematician, was Professor at the University of Dijon from 1905 until 1914, when illness forced him to give up research and teaching.

countable intersection of open dense sets in A is itself dense in A. The space $\mathbf{R}^n$ has the Baire property, as do all submanifolds. More generally:

Theorem 3.2.4. (Baire's Theorem.) *Let A be a locally closed subset of a finite-dimensional vector space. Every countable intersection of open dense subsets of A is dense in A.*

Write A in the form $U \cap F$ where U is an open subset and F a closed subset of the finite-dimensional vector space E. We have to prove that if $(U_n)_{n>0}$ is a sequence of open sets in E such that the $A \cap U_n$ are dense in A, and if V is an open set in E such that $A \cap V$ is non-empty, then $A \cap V \cap \bigcap_n U_n$ is non-empty. Let $V_0 = U \cap V$, $V_n = U \cap V \cap U_n$; then $F \cap V_n$ is dense in $F \cap V_0$ for every n, and we must prove that $F \cap \bigcap_n V_n$ is non-empty. Choose an open ball B_0 of radius < 1 centred at a point x_0 of $F \cap V_0$ and contained in V_0; then a ball B_1 of radius $< 1/2$ centred at a point x_1 of $F \cap B_0$ and contained in $B_0 \cap V_1$; and so on. In this way we obtain a nested sequence of open balls in F whose radii tend to 0. The intersection of this family is a single point (this is one version of Cauchy's criterion); this point belongs to all the balls B_n and hence to all the V_n. Moreover, since the balls are centred at points of F this point is a limit of points of F, and therefore belongs to F since F is closed. The intersection that we are considering is therefore non-empty, which is what we wished to prove.   □

This proof generalizes directly and shows that complete metric spaces and locally compact spaces are Baire spaces. Notice the connection between the fact that $\mathbf{R}$ is complete and the fact (established by Cantor) that it is uncountable: the family of complements of single points has empty intersection, which would contradict Baire's Theorem if it were a countable family.

We sometimes say that a subset of a topological space is *rare* if the complement of its closure is dense, *i.e.* it is contained in the complement of an open dense set, and it is *meagre* ('of first category', in the terminology of Baire) if it is the union of a countable family of rare sets. In this language the Baire property can be expressed by saying that the complement of every meagre subset is dense.

Let A be a Baire space (for example, a submanifold of a finite-dimensional vector space). It is useful to make the following definition:

Definition 3.2.5. A subset of A is called *residual* if it contains the intersection of a countable family of open dense sets.

Note the following properties of Baire spaces:

a)   every open dense subset is residual;
b)   every subset that contains a residual subset is residual;
c)   a countable intersection of residual subsets is residual;
d)   every residual subset is dense;
e)   every residual subset is a Baire space.

Indeed, a),b) and c) follow immediately from the definition of residual subsets, and d) from the fact that A is assumed to be a Baire space. We now prove e): let B be a residual subset of the Baire space A; every residual subset of B is residual in A and is therefore dense in A and *a fortiori* dense in B.

3.2.6. In a Baire space it is convenient to say that a property is *generic* if it is true in a residual (and therefore dense) subset, and d) above can then be stated as: "a countable conjunction of generic properties is generic". We use the words 'generic','generically',... with the same abuses as the words 'almost everywhere','almost surely',... in measure theory. Thus instead of saying "the set of elements x in A such that property P holds is a residual subset", we say "every generic element x of A has property P" or even "generically, every element x of A has property P".

An important example of residual subset, which we shall generalize later on, is the following:

Proposition 3.2.7. *Let E be a finite-dimensional vector space, with V and W two submanifolds of E. Suppose that at every point $a \in V \cap W$ we have $\dim_a W < \dim_a V$ (as is the case, for example, if V is purely of dimension p and W is purely of dimension q with $q < p$). The set V' of those $a \in V$ which do not belong to W is then a residual subset of V.*

By the property 4) of 2.2, V is the union of a countable family of compact sets (K_i). Then V' is the intersection of the countable family of the open subsets of V that are complementary to the $K_i \cap W$. It remains to prove that V' is dense in V, that is to say that every non-empty open subset of V meets V', or equivalently is not contained in W. Replacing V by this open set, we are reduced to showing that if V is contained in W then V is empty. But now this is quite clear: if it were not the case, then taking $a \in V \subset W$ we would have $\dim_a V \leq \dim_a W$ (by 2.2.5 b)), contrary to hypothesis.   □

Corollary 3.2.8. *If a submanifold V of E is the union of a countable family of submanifolds W_i, all being purely of dimension p, then V itself is of dimension p.*

Indeed, if V is not purely of dimension p there exists a point $a \in V$ with $\dim_a V > p$ (again by 2.2.5 b)). Consider then the non-empty open subset U of V consisting of those points where V has the same dimension as at a. For each i let U_i' be the set of points in U which do not belong to W_i. Then each U_i' is a residual subset of U by the proposition above, but their intersection is empty which contradicts Baire's Theorem.   □

We can now improve Corollary 2.9.7:

Corollary 3.2.9. *Let E and F be two finite-dimensional vector spaces, with V a submanifold of E and $g : V \mapsto F$ a C^∞ map. Suppose that V has the same dimension everywhere, that g is an injective immersion, and that $g(V)$ is a submanifold of F. Then g is an embedding.*

By the property 1) of 2.2 and 2.9.6(a), there exists a countable family of open sets U_i in U such that each $g(U_i)$ is a submanifold of F. The above Corollary then implies that $g(V)$, which is the union of the $g(U_i)$, is purely of dimension $\dim(V)$ and we apply 2.9.7 directly. $\qquad\square$

3.3 Negligible Subsets

3.3.1. Let us recall what is meant by a negligible subset (we also say a set of measure zero) in a finite-dimensional vector space.

In the space $\mathbf{R}^n$ the cube with centre a and side r, that is the set of x with $\sup_i |x_i - a_i| \leq r/2$, has volume r^n. We say that a subset A of $\mathbf{R}^n$ is *negligible* or *of (n-dimensional) measure zero* if for every $\varepsilon > 0$ there exists a sequence $K_1, K_2, \ldots$ of cubes with

$$A \subset \bigcup_i K_i \quad \text{and} \quad \sum_i \text{vol}(K_i) < \varepsilon.$$

Let $u \in L(\mathbf{R}^n; \mathbf{R}^n)$ be a linear map. Then the image under u of the unit cube is contained in a cube with side k, say. By dilation and translation we conclude that the image of an arbitrary cube of side r is contained in a cube of side kr. Therefore for every cube K we have $\text{vol}(u(K)) \leq k^n \text{vol}(K)$ and it follows that $u(A)$ is negligible for every negligible subset A.

If E is an n-dimensional vector space we say that a subset A of E is *negligible in E* if the same is true for the image of A by an isomorphism $v : E \to \mathbf{R}^n$ of vector spaces; from what we have just seen, this is independent of the choice of isomorphism v.

3.3.2. Here are some elementary properties:

1) If $A \subset B$ and B is negligible then A is negligible.

2) A countable union of negligible subsets is negligible. Indeed, if $A_1, A_2, \ldots$ are negligible we can enclose A_1 in a union of cubes of total volume $\leq \varepsilon/2$, then A_2 in a union of cubes of total volume $\leq \varepsilon/4$ etc., and thus enclose the union of the A_i in a union of cubes of total volume $\leq \varepsilon$. In particular, every countable subset is negligible.

3) Since a negligible subset of E cannot contain any nonempty open set in E, its complement is dense in E.

We have two distinct and quite natural ways of translating into mathematics the statement that something is "almost always true": in a topological context we demand that the exceptions form a rare or meagre subset; in a measure-theoretical context we demand that they form a negligible subset. Contrary to intuition, these two notions are very strongly distinct. Thus in the interval $[0, 1]$ there exist:

a)   residual subsets which are negligible,

b)   subsets of measure 1 which are not residual.

We might hope that this is hair-splitting. Unfortunately, situations of this type – where the property that we are considering is almost always true in one sense but almost always false in another – are actually encountered in certain applications.

4) If (U_α) is a family of open sets in E which cover A and if $A \cap U_\alpha$ is negligible for every α then A is negligible: by the property 2) of 2.2 we can cover A by countably many of the U_α and then apply 2) above.

5) Every affine subspace of E other than E itself, and hence every countable union of such subspaces, is negligible.

Proposition 3.3.3. *Let E and F be two vector spaces of the same finite dimension, let U be an open set in E and let $f : U \to F$ be a C^1 map. Then the image under f of every negligible subset of U is negligible in F.*

Clearly we may take $E = F = \mathbf{R}^n$. Let $a \in U$. There exists $r > 0$ such that the cube K with centre a and side $2r$ is contained in U. Let K' denote the cube with centre a and side r. Since we can cover U by a countable family of such K' it is enough (in view of 2) above) to prove that $f(A)$ is negligible for every negligible subset A of K'. Let M be an upper bound for $\|f'(x)\|$ for $x \in K$. Then every cube L of sufficiently small side ε (say $\varepsilon < r/4$) which meets K' is contained in K and hence by the Mean Value Theorem $f(L)$ is contained in a cube of side $M\varepsilon\sqrt{n}$ and hence of volume $\leq M^n\sqrt{n}^n\mathrm{vol}(L)$. The proposition follows immediately.   □

The fact of being negligible is therefore preserved by diffeomorphisms; since it is also *local* (property 4)) we are able to define what is meant by a negligible subset of an 'abstract' manifold, and the results of this chapter may be transported to that context without difficulty.

3.4 The Complement of the Image of a Submanifold

The following proposition is a particular case of Sard's Theorem that we shall see later on. Although it is much more elementary than the general case it already has numerous applications that justify giving it separate treatment.

Let E and F be two finite-dimensional vector spaces, let V be a submanifold of E and let g be a map from V into F. Recall (Section 2.8) that g is said to be of class C^r, $r \in [1, \infty]$, if there exists an open subset U of E containing V and a C^r map f from U into F which induces g.

Proposition 3.4.1. *Let E and F be two finite-dimensional vector spaces, let V be a submanifold of E and let $g : V \to F$ be a C^1 map. Suppose $\dim(V) < \dim(F)$. Then $g(V)$ is a negligible subset of F.*

Let a be a point of V. By 2.9.4. there exists a local parametrization of V in a neighbourhood of a : we can find an open set Ω in $\mathbf{R}^d$ with $d = \dim_a V < \dim(F)$ and an embedding $h : \Omega \to F$ whose image $h(\Omega)$ is an open set in V containing a. Since we can cover V by a countable family of such open sets it suffices to prove that $g(h(\Omega))$ is negligible. Let $m = \dim(F) - d > 0$ and consider the map $u : (x, y) \mapsto g(h(x))$ from $\Omega \times \mathbf{R}^m$ into F. Then the subset $\Omega \times \{0\}$ of $\Omega \times \mathbf{R}^m$ is negligible (by 1) and 4) of 3.2) and $g(h(\Omega))$, which is its image under u, is negligible in F (Proposition 3.3). $\qquad\square$

Corollary 3.4.2.

a) *The complement of $g(V)$ is a dense residual subset of F; it is the countable intersection of open (dense) sets.*

b) *If g is proper (for example, if V is compact) then the complement of $g(V)$ is an open dense set in F.*

Indeed, we know by 4) in 2.2 that V is the union of a countable family $\{K_i\}$ of compact subsets. Therefore $g(V)$ is the union of the countable family of compact subsets $g(K_i)$. If g is proper then $g(V)$ is closed (Lemma 2.10.3). Moreover, the complement of $g(V)$ is dense by Proposition 4.1. $\qquad\square$

Note that the image $g(V)$ of V may itself be dense: for example, there exist parametrized arcs $\mathbf{R} \to \mathbf{R}^n$ whose image is dense ("scribbles").

Often we use Corollary 4.2 via the next result as intermediary:

Proposition 3.4.3. *Let E and F be two finite-dimensional vector spaces with V a submanifold of E and $g : V \to F$ a C^1 map. For each $b \in F$ let $g + b$ denote the map $x \mapsto g(x) + b$ from V into F. Let (W_i) be a countable family of submanifolds of F with $\dim(V) < \mathrm{codim}(W_i)$. The subset of F consisting of those b for which $(g + b)(V)$ does not meet any of the W_i is residual and therefore dense.*

Since the countable intersection of residual subsets is residual, it is enough to treat the case where the family $\{W_i\}$ consists of just one element W. Consider now the map $(x, y) \mapsto g(x) - y$ from $V \times W$ into F. Its image is the set of those b such that $(g + b)(V) \cap W \neq \emptyset$. Since $\dim(V \times W) = \dim(V) + \dim(W) < \dim(F)$ we have only to apply the above Corollary. $\qquad\square$

Here is a typical application of this Proposition:

Corollary 3.4.4. *Let E and F be two finite-dimensional vector spaces with $\dim(F) \geq 2\dim(E)$, let U be an open set in E and let $f : U \to F$ be a C^2 map.*

The subset of $L(E; F)$ consisting of linear maps u for which $f + u : U \to F$ is an immersion (Definition 2.9.3) is residual and therefore dense.

To see this, consider the map $x \mapsto f'(x)$ from U into $L(E; F)$. To say that f is an immersion at x means that $f'(x)$ belongs to none of the submanifolds L_r of Proposition 2.3.1 for $r < \dim(E)$. But for these values of r we have

$$\mathrm{codim}(L_r) = (\dim(F) - r)(\dim(E) - r) \geq \dim(E) + 1,$$

and we simply have to apply the Proposition.   □

Proposition 4.1 also enables us to prove a weak form of *Whitney's Embedding Theorem* (see the subsequent remark for the full statement).

Proposition 3.4.5. *Let V be a compact n-dimensional submanifold of a finite-dimensional vector space E. There exists an embedding of V into $\mathbf{R}^{2n+1}$.*

This is immediate if $\dim(E) \leq 2n + 1$. Moreover, since V is compact the injective immersions of V are embeddings (Proposition 2.10.5). Therefore by induction it suffices to prove that if V is a submanifold of $E = \mathbf{R}^N$ with $N > 2n + 1$ then V admits an injective immersion g in $\mathbf{R}^{N-1}$. We shall construct g as the restriction of a surjective linear map f from E to $\mathbf{R}^{N-1}$. Let k denote a nonzero vector in the kernel of f, so that $\mathrm{Ker}(f) = \mathbf{R}k$. To say that g is an immersion means that k does not belong to any of the spaces $T_a V$; to say that g is injective means that k cannot be written in the form $\lambda(x - y)$ with $\lambda \in \mathbf{R}$, $x \in V$, $y \in V$. It is now clear what we have to do. On the one hand we consider the tangent bundle $TV \subset E \times E$ (2.4.5) and the map $h_1 : (a, v) \mapsto v$ of TV into E, and on the other hand the submanifold $V \times V \times \mathbf{R} \subset E \times E \times \mathbf{R}$ and the map $h_2 : (x, y, \lambda) \mapsto \lambda(x - y)$ of $V \times V \times \mathbf{R}$ into E. Since $N > 2n = \dim(TV)$ the image of h_1 is negligible; since $N > 2n + 1 = \dim(V \times V \times \mathbf{R})$ the image of h_2 is negligible. Therefore we can find an element k having the desired properties, and then choose an arbitrary element of $L(E; \mathbf{R}^{N-1})$ that has kernel $\mathbf{R}k$.   □

The full statement of Whitney's Theorem is that for every n-dimensional manifold V the set of immersions or embeddings is a residual subset (and therefore dense and in particular non-empty, which implies the above weak form of the Theorem) of the topological vector space $C^\infty(V, \mathbf{R}^{2n})$ or $C^\infty(V, \mathbf{R}^{2n+1})$, respectively. For example, take the circle $\mathbf{S}^1$. Every C^∞ map of $\mathbf{S}^1$ into $\mathbf{R}^2$ or $\mathbf{R}^3$ can be approximated by an immersion or an embedding, respectively. Note that $\mathbf{S}^1$ can be embedded in $\mathbf{R}^2$, but an immersion of $\mathbf{S}^1$ onto a planar "figure eight" cannot be approximated by embeddings since an extra dimension is needed in order to uncross the two branches.

Let $m = \mathrm{codim}(V) = \dim(F) - \dim(V) > 0$. We can also generalize Proposition 4.1 by proving not only that a 'generic' point of F is not in

$g(V)$, but that a 'generic' affine subspace of F of dimension $m-1$ does not meet $g(V)$. Recall that the *affine subspace generated* by the elements $\xi_1, \ldots, \xi_m$ of F is the affine subspace which is parallel to the vector subspace generated by the differences $\xi_i - \xi_j$ and which passes through the ξ_i. It may also be constructed as the set of linear combinations $\alpha_1\xi_1 + \cdots + \alpha_m\xi_m$ where $\alpha_1 + \cdots + \alpha_m = 1$.

Proposition 3.4.6. *With the notation of Proposition 4.1, let $m = \dim(F) - \dim(V) > 0$. In F^m the set S of sequences $(\xi_1, \ldots, \xi_m)$ generating an affine subspace not meeting $g(V)$ is a dense subset.*

We have to show that for every sequence $U_1, \ldots, U_m$ of non-empty open sets in F we can find an element $(\xi_1, \ldots, \xi_m)$ of S with $\xi_i \in U_i$ for $i \in [1, m]$. By Proposition 4.1 there exists an element ξ_1 of U_1 which does not belong to $g(V)$. Consider the map $(x, \lambda_1) \mapsto (1 - \lambda_1)g(x) + \lambda_1\xi_1$ from $V \times \mathbf{R}$ into F; if $m > 1$ its image is negligible and we can find $\xi_2 \in U_2$ which does not belong to this image. Now consider the map $(x, \lambda_1, \lambda_2) \mapsto (1 - \lambda_1 - \lambda_2)g(x) + \lambda_1\xi_1 + \lambda_2\xi_2$ from $V \times \mathbf{R} \times \mathbf{R}$ into F, and so on. We show that the sequence $(\xi_1, \ldots, \xi_m)$ constructed in this way serves our purpose. If that were not the case then we could find an element x of V and scalars $\alpha_1, \ldots, \alpha_m$ with

$$g(x) = \alpha_1\xi_1 + \cdots + \alpha_m\xi_m, \quad \alpha_1 + \cdots + \alpha_m = 1.$$

Let p be the largest index such that $\alpha_p \neq 0$. Dividing the relation above by α_p we would obtain

$$\xi_p = \lambda_0 g(x) + \lambda_1\xi_1 + \cdots + \lambda_{p-1}\xi_{p-1}$$

with $\lambda_1 = -\alpha_1/\alpha_p, \ldots, \lambda_{p-1} = -\alpha_{p-1}/\alpha_p$ and

$$\lambda_0 = \frac{1}{\alpha_p} = \frac{\alpha_1}{\alpha_p} + \cdots + \frac{\alpha_p}{\alpha_p} = -\lambda_1 \cdots - \lambda_{p-1} + 1,$$

which would contradict the construction of ξ_p. $\qquad\square$

Let G be the set of affine $(m-1)$-dimensional subspaces of F (an 'affine grassmannian'[28]) and let H be the set of those which do not meet $g(V)$. The grassmannian G can be given a natural topology and the proposition above implies that H is a dense subset of G. It can be shown that H is a residual subset of G (open when $g(V)$ is closed). Note that the above result provides a purely topological definition of codimension.

---

[28] A word coined from the name of the German mathematician Hermann GRASSMANN (1809-1877), one of the founders of multilinear algebra.

3.5 Sard's Theorem

Let E and F be two finite-dimensional vector spaces with U an open set in E and $f : U \to F$ a C^∞ map. A point a of U is called a *critical point* for f if the derived map $f'(a) \in L(E; F)$ is not surjective. We say that $y \in F$ is a *critical value* of f if it is the image under f of a critical point of f.

Theorem 3.5.1. (Sard's Theorem.) *Let E and F be two finite-dimensional vector spaces with U an open set in E and $f : U \to F$ a C^∞ map. The set of critical values of f is a negligible subset of f.*

We shall not give the proof for the general case, but be content with a few remarks.

1) If $\dim(E) < \dim(F)$ all the points of U are critical and the critical values of f are the points of $f(U)$. In this case the theorem follows from Proposition 4.1, and we may even weaken the assumptions on regularity of f from C^∞ to C^1. The serious things happen when $\dim(E) \geq \dim(F)$.

2) We give the proof in the particular case when $E = F = \mathbf{R}$, the case $\dim(E) = \dim(F)$ being quite analogous. Let C be the set of critical points of f. It suffices to show that for every compact interval $[a, b]$ contained in U the set $f(C \cap [a, b])$ is negligible. Let M be an upper bound for f'' on $[a, b]$, let N be an integer, and let $h = (b - a)/N$. Consider each of the N intervals $I_n = [a + nh, a + (n + 1)h]$, $n = 0, \ldots, N - 1$, and let A be the set of those n for which $C \cap I_n \neq \emptyset$, so that $f(C \cap [a, b])$ is the union of the $f(C \cap I_n)$ for $n \in A$. Let $n \in A$. By assumption there exists a point in I_n where f' vanishes; therefore by the Mean Value Theorem applied to f' we have $|f'(x)| \leq Mh$ for $x \in I_n$. Applying the Mean Value Theorem once again, but this time to f, we deduce that $f(I_n)$ is contained in an interval of length $\leq L = Mh^2$. Thus $f(C \cap [a, b])$ is contained in the union of a family of at most N intervals of length at most L. The sum of the lengths of these intervals is bounded above by

$$NL = NMh^2 = \frac{M}{N}(b - a)^2,$$

which is arbitrarily small.

3) The proof in the general case is by induction on the dimension of E, the case $\dim(E) = 1$ being a result of 1) and 2) above. See [BL] for example. □

The conclusion of the theorem remains valid if we suppose only that f is of class C^r with $r \geq 1$ and $r \geq \dim(E) - \dim(F) + 1$, but the proof is much more difficult. This minimal value for r cannot be improved upon. There exists the following horrible example (due to Whitney, as are many others). Let $I = [0, 1]$; it is possible to construct a C^1 function $f(x, y)$ on $I \times I$ and with values in I such that the set of critical points contains a connected 'curve' C, which is the image of a continuous injective map $\gamma : I \to I \times I$ and along which f varies from 0 to 1. Hence every value of I is a critical value of f. At first sight there is something impossible going on here: the derivative f' of f vanishes identically along C and yet f is not constant! What then has happened to the Mean Value Theorem? The answer is that it doesn't apply: γ is not differentiable, nor is $f \circ \gamma$, and the derivative $(f \circ \gamma)'$ – which would be identically zero if it existed – doesn't exist. In fact, we start by constructing γ, whose image is the curve C (this curve is of 'Peano' type having infinite length between any two of its points) and the continuous surjective function $f \circ \gamma : I \to I$; then we prove that, thanks to Whitney's Extension Theorem, the function defined in this way on C can be extended to a C^1 function on $I \times I$.

Here is an instructive example that conveys the full power of this theorem. Let $E = \mathbf{R}^n$ and let A be a closed subset of E. There exists a C^∞ function f on E such that $A = f^{-1}(0)$ (Whitney; see for example [BL], p.24). Then, however complicated A and hence also f may happen to be, there exist arbitrarily small ε (better still: the set of unsuitable ε is negligible) such that $f^{-1}(\varepsilon)$ is a submanifold.

3.6 Critical Points, Submersions and the Geometrical Form of Sard's Theorem

Let E and F be two finite-dimensional vector spaces, with V a submanifold of E and g a C^∞ map from V into F. Let $a \in V$ and let $T_a g : T_a V \to F$ be the tangent linear map to g at a (2.4.6).

3.6.1. The rank of the linear map $T_a g$ is called the *rank* of g at the point a and is denoted by $\mathrm{rk}_a(g)$. For example, to say that g is an immersion at a (Definition 2.9.3) means that $\mathrm{rk}_a(g) = \dim_a V$. Combining the definitions in the previous section, we say that a point a of V is *critical* for g, or is a *critical point* of g, if the linear map $T_a g : T_a V \to F$ is *not* surjective, that is $\mathrm{rk}_a(g) < \dim(F)$. The set of critical points of g is called the *critical locus* of the map g. The critical locus of g is closed. More generally:

Lemma 3.6.2. *Let r be an integer. The set of $a \in V$ such that $\mathrm{rk}_a(g) \leq r$ is closed in V.*

Let A denote this set. We have to show that if a point a of V is a limit point of points of A then it belongs to A. Let Ω be an open set in a vector space E' with $h : \Omega \to V$ a parametrization of an open set in V containing a. For every $x \in \Omega$ the linear map $h'(x)$ is a bijection from E' onto $T_a V$ and hence $\mathrm{rk}_x(g \circ h) = \mathrm{rk}_{h(x)}(g)$. But the set of x in Ω such that $\mathrm{rk}_x(g \circ h) \leq r$ is closed (Example 8 of Section 2.3) and so since h is a homeomorphism from Ω onto $h(\Omega)$ we deduce the result. $\qquad\square$

We have given the details of this proof so that it may serve as a model. In a 'normal' mathematical text we might have found something of the following kind:

"it is immediate, since the fact of being closed in V is local in V and the assertion is true for an open set in a vector space". If necessary we shall make use of this telegraphic style. As Arnol'd has rightly remarked, there are some proofs which are easier to do than to read.

Definition 3.6.3. We say that g is a *submersion* at a if $T_a g$ is surjective, or in other words if a is not a critical point of g. We say that g is a *submersion* if it is a submersion at every point, that is to say its critical locus is empty.

The origin of the word 'submersion' is simple: it is derived from *immersion*, following the *injective/surjective* model.

Proposition 3.6.4. *Suppose that g is a submersion at a, and let $b = g(a)$. Then*

a) *at the point a the set $g^{-1}(b)$ is a submanifold of dimension $\dim_a V - \dim(F)$ with tangent space $\operatorname{Ker}(T_a g)$;*
b) *there exists an open set Ω of F containing b and a C^∞ map $h : \Omega \to E$ with $h(b) = a$ and such that $h(y) \in V$ and $g(h(y)) = y$ for every $y \in \Omega$.*

Part a) is the particular case $W = \{b\}$ of Proposition 2.6.2. We shall prove b). Fix a basis for F and let $\phi_1, \ldots, \phi_p$ be the components of the map $f : U \to F$ that induces g. Let $\psi_1, \ldots, \psi_m$ be local equations for V at a. Then since the common kernel of the $p + m$ linear forms $d\phi_i$ and $d\psi_j$ is of codimension $p + m$ (because it is $\operatorname{Ker} T_a g$) we may complete the ϕ_i and the ψ_j by some θ_k in order to obtain a local system of coordinates on E at a. It is then enough to take as h the map which associates to $(y_1, \ldots, y_p)$ the point x such that $\phi_i(x) = y_i$, $\psi_j(x) = 0$ and $\theta_k(x) = \theta_k(a)$. $\square$

Note that b) is roughly half of the Implicit Function Theorem (2.7.1): assuming that g is a submersion at a we have shown that in a neighbourhood of a there exists locally a solution $z = h(y)$ to the implicit equation $g(z) = y$, for $z \in V$. By 2.9.6 a) such a solution is (locally) unique if g is an immersion at a, and putting these two statements together gives 2.7.1 exactly.

3.6.5. Now we give an example which will be useful in what follows and which we could have discussed much earlier. Fix an integer n, and let E denote the space $M_n(\mathbf{R})$ of square $n \times n$ matrices and F the subspace of symmetric matrices. Take $Q \in F$ with $\det(Q) \neq 0$ and consider the map $f : E \to F$ which takes $A \in E$ to ${}^t A Q A$. Its derivative at the point $I = I_n$ is $a \mapsto {}^t a Q + Q a$; it is surjective because for every $q \in F$ we can solve $Q a = q$ and then we have ${}^t a Q = {}^t(Q a) = {}^t q = q$ and therefore $f'(I) \cdot (a/2) = q$. Consequently f is a submersion at I and we can apply b) above: there exists an open set Ω in F containing Q and a C^∞ map $h : \Omega \to F$ with $h(Q) = I$ and ${}^t h(q).Q \cdot h(q) = q$. In other words we have established that every symmetric matrix q sufficiently close to Q can be converted into Q by a coordinate change that *depends in a C^∞ way on q.*

3.6.6. We say that $y \in F$ is a *critical value* of g if it is the image under g of a critical point of g. The points of F which are not critical values of g are called *regular values* of g. Beware that with this terminology a regular value of g need not be a value of g at all: every point of F which is not in $g(V)$ is a regular value of g. Let b be a point of F; by the above Proposition, $g^{-1}(b)$ is a submanifold of V of dimension $\dim(V) - \dim(F)$ at every non-critical point and is thus a submanifold everywhere since b is a regular value of g.

Recall an example from 2.7.4. For $a = (a_1, \ldots, a_n) \in F = \mathbf{R}^n$, consider the polynomial in one variable

$$P(a; T) = T^n + a_1 T^{n-1} + \cdots + a_n;$$

take V to be the subset of $E = F \times \mathbf{R}$ consisting of those (a, x) with $P(a, x) = 0$ and take f to be the projection $(a, x) \mapsto a$. Then V is a submanifold (it is the graph of a function, since the given equation can be solved for a_n) and $T_{(a,x)}V$ is the hyperplane with equation

$$\frac{\partial P}{\partial a_1}(a; x)da_1 + \cdots + da_n + \frac{\partial P}{\partial T}(a; x)dT = 0.$$

In order for the projection (which is its own derivative) to map $T_{(a,x)}V$ surjectively onto F it is necessary and sufficient that $\frac{\partial P}{\partial T}(a; x)$ be nonzero, that is x be a *simple* root of the polynomial $P(a; T)$. Hence the critical locus of g consists of those (a, x) such that x is a multiple root of the polynomial $P(a; T)$ and the critical values of g are those a for which the polynomial has a multiple root. If we were working in the field of complex numbers rather that $\mathbf{R}$ these critical values would be precisely the roots of the discriminant of the polynomial $P(a; T)$. It is because of this example that the set of critical values of g is often called the *discriminant locus* (or simply the discriminant) of g.

The geometric form of Sard's Theorem is the following:

Theorem 3.6.7. *Let E and F be two finite-dimensional vector spaces with V a submanifold of E and $g : V \to F$ a C^∞ map.*

a) *The critical values of g in F form a negligible subset, while the regular values form a residual dense subset which is a countable intersection of open sets.*

b) *If g is proper (for example, if V is compact) the critical values of g form a closed negligible subset, while the regular values form an open dense subset.*

Let A denote the critical locus of g, which is closed in V by Lemma 6.2. If g is proper then $g(A)$ is closed in F. In any case, A is locally closed in E since V is, and therefore it is a countable union of compact sets and $g(A)$ is a countable union of closed sets. Thus it is enough to prove that $g(A)$ is negligible, which will imply that its complement is dense. Now, for each $a \in V$ there exists a parametrization $h : \Omega \to V$ whose image is an open set in V containing a. Since V is the union of a countable number of such open sets it suffices to prove that each $g(A \cap h(\Omega))$ is negligible; but, as we saw above, $g(A \cap h(\Omega))$ is also the set of critical values of $g \circ h : \Omega \to F$, and so we are back at Theorem 5.1.   □

It should be noted explicitly that this theorem *says nothing about the critical locus itself*. For example, if g is constant (and $\dim(F) > 0$) then the critical locus of g is the whole of V and the set of critical values reduces to a single point. Less drastically, we can manufacture a Whitney-type example as follows. For every closed subset A of $E = \mathbf{R}^n$ it is possible to construct (exercise) a C^∞ map $g : E \to E$ whose critical locus is A. Then Sard's Theorem says simply that for almost every value $y \in E$ the set $g^{-1}(y)$ is a discrete subset of E; indeed for every regular value it is a submanifold of dimension 0 (Theorem 6.4a). Note finally that if $\dim(V) < \dim(F)$ then all the points of V are critical, the critical values of g are the points of $g(V)$ and we recover the particular case handled in Proposition 4.1.

3.7 The Transversality Theorem: Weak Form

Let E and F be two finite-dimensional vector spaces, let V and W be submanifolds of E and F respectively, and let g be a C^∞ map from V to F.

Definition 3.7.1.

a) Let a be a point of V. We say that g is *transverse* to W at a if either $g(a)$ does not belong to W or $g(a)$ does belong to W and the image of $T_a g$ is transverse (see 2.5.1) to the subspace $T_{g(a)} W$.

b) We say that g is *transverse* to W if it is transverse at every point of V. We say that g is transverse to W on a subset A of V if it is transverse at every point of A.

Let $a \in V$ with $g(a) \in W$. Consider local equations $\phi_1, \ldots, \phi_m$ for W on a neighbourhood of $g(a)$. The $d\phi_i(g(a))$ are linear forms on F and their compositions with $T_a g$, that is the $T_a(\phi_i \circ g)$, are linear forms on $T_a V$. Then we have the following useful criterion: *in order that g be transverse to W at a it is necessary and sufficient that the linear forms $T_a(\phi_i \circ g)$ be linearly independent on $T_a V$.* Explicitly this means that

$$\dim\left((T_a g)^{-1}(T_{g(a)} W)\right) = \dim T_a V - m = \dim T_a V - \operatorname{codim} T_{g(a)} W,$$

which is one of the definitions of transversality (2.6.1).

Here are a few examples. The map g is always transverse to F. At the other extreme, to say that g is transverse to the submanifold $\{y\}$ of F means that y is a regular value of g. If g is a submersion, that is if $T_a g$ is surjective for every $a \in V$, then g is transverse to every submanifold of F. If g is an embedding, to say that g is transverse to W means that the submanifold $g(V)$ is transverse to W. If $\dim(V) < \mathrm{codim}(W)$, to say that g is transverse to W means that $g(V)$ does not intersect W.

Proposition 2.6.2 may be rewritten as follows:

Proposition 3.7.2. *Let* $a \in g^{-1}(W)$. *If* g *is transverse to* W *at* a *then* $g^{-1}(W)$ *is a submanifold at* a *and we have*

$$\dim_a(g^{-1}(W)) = \dim_a V - \mathrm{codim}_{g(a)} W, \quad T_a(g^{-1}(W)) = (T_a g)^{-1}(T_{g(a)} W).$$

$\square$

Corollary 3.7.3. *Suppose* V *is purely of dimension* n *and* W *is purely of codimension* m *in* F, *and suppose* g *is transverse to* W. *Then* $g^{-1}(W)$ *is a (possibly empty) submanifold of dimension* $n - m$. $\square$

The Weak Transversality Theorem is a theorem that applies to a family (f_λ) of maps of a manifold V (a submanifold of an auxiliary space E) into a space F, parametrized by the elements λ of a parameter-space Λ (a submanifold, or more simply an open subset of another auxiliary space L). This comes down to considering the map $f : (x, \lambda) \mapsto f_\lambda(x)$ from $V \times \Lambda$ into F. To give a regularity condition on f is the same as giving a regularity condition on each f_λ and a regularity condition for the dependence of f_λ on λ. Therefore we are considering three finite-dimensional vector spaces E, F and L, a submanifold V of E, an open subset Λ of L and a C^∞ map $f : V \times \Lambda \to F$. For each $\lambda \in \Lambda$ let $f_\lambda : V \to F$ denote the map $x \mapsto f(x, \lambda)$.

Theorem 3.7.4. (Weak Transversality Theorem.) *Let* W *be a submanifold of* F, *and suppose* f *is transverse to* W. *Then the set* Λ_1 *of those* $\lambda \in \Lambda$ *such that* f_λ *is transverse to* W *is a dense residual subset of* Λ, *a countable intersection of open sets and with negligible complement. If* V *is compact and if* W *is closed in* F *then* Λ_1 *is an open dense subset of* Λ.

Since we assume that f is transverse to W it follows that $f^{-1}(W)$ is a submanifold of $V \times L$; let $p : f^{-1}(W) \to L$ be the projection onto the second factor. We prove the following lemma:

Lemma 3.7.5. *Those* λ *such that* f_λ *is not transverse to* W *are the critical values of* p. *More precisely, let* a *be a point of* V *and* λ *a point of* Λ; *then in order for* f_λ *to be transverse to* W *at* a *it is necessary and sufficient that* p *be a submersion at* (a, λ).

This is pure linear algebra. Let $(a, \lambda) \in V \times \Lambda$ and let us prove that in order for f_λ to be non-transverse to W at a (which requires that (a, λ) belong to $f^{-1}(W)$) it is necessary and sufficient that (a, λ) be a citical point of p. We simplify notation by letting $S = T_a V \subset E$, $T = T_{f(a,\lambda)} W \subset F$, and writing the map $T_{(a,\lambda)} f : S \times L \to F$ in the form $(\xi, \eta) \mapsto u \cdot \xi + v \cdot \eta$, where $u = T_a f_\lambda \in L(S; F)$ and $v \in L(L; F)$. The tangent space to $f^{-1}(W)$ at (a, λ) is the set R of those $(\xi, \eta) \in S \times L$ with $u \cdot \xi + v \cdot \eta \in T$. To say that p is a submersion at the point (a, λ) is to say that the projection $R \to L$ onto the second factor is surjective, or again, in view of the definition of R, to say that for every $\eta \in L$ we can find $\xi \in S$ and $\theta \in T$ with $u \cdot \xi + v \cdot \eta = \theta$. This means also that $\text{Im}(v)$ is contained in $\text{Im}(u) + T$. But the transversality assumption can be written $\text{Im}(u) + \text{Im}(v) + T = F$ and our condition means that $\text{Im}(u) + T$ is equal to F, or that f_λ is indeed transverse to W at a. $\square$

We return now to the proof of the Theorem. According to the Lemma, the set of those λ such that f_λ is not transverse to W is the set of critical values of the map $p : f^{-1}(W) \to L$. The first assertion is therefore a direct result of Sard's Theorem (Theorem 6.7). To prove the second, it is enough to establish the following proposition:

Proposition 3.7.6. (Openness of transversality.) *Suppose W is closed in F.*

a) *The set of pairs $(x, \lambda) \in V \times \Lambda$ such that f_λ is transverse to W at x is an open subset of $V \times \Lambda$.*

b) *Let K be a compact subset of V. The set of $\lambda \in \Lambda$ such that f_λ is transverse to W on K is an open subset of Λ.*

Part a) results from the Lemma above: the set in question is the complement of the critical locus of p, which is closed (see 6.1) in $f^{-1}(W)$, which is itself closed in $V \times \Lambda$. We now prove b). Let $\mu \in \Lambda$ be such that f_μ is transverse to W on K. By a), for each $x \in K$ there exists an open subset V_x of V containing x and an open subset Λ_x of Λ containing μ such that f_λ is transverse to W on U_x for all $\lambda \in \Lambda_x$. We may extract a finite covering from the covering $\{U_x\}$ of K; the intersection of the corresponding Λ_x is an open set S containing μ, and all the f_λ for λ in S are transverse to W on K. This completes the proof of the Proposition and also that of the Theorem. $\square$

The second part of the Proposition above is not at all specific. What we have proved in fact is that if we have any open subset U of a product $K \times L$ where K is compact, then the set of $y \in L$ such that $K \times \{y\}$ is contained in U is open in L.

We often use the theorem above by taking f to be a submersion, since f is automatically transverse to W. This is the case in the following Corollary, which generalizes Proposition 4.3:

Corollary 3.7.7. *Let E and F be two finite-dimensional vector spaces, with V a submanifold of E and $g : V \to F$ a C^∞ map, and let W be a submanifold* *of F. For every $b \in F$ let $g + b$ denote the map $x \mapsto g(x) + b$ from V to F. Then the set of b for which $g + b$ is transverse to W is a dense residual subset of F with negligible complement, and is open when V is compact and W is closed.*

This is an immediate consequence of the Weak Transversality Theorem since the map $(x, b) \mapsto g(x) + b$ from $V \times F$ to F is a submersion.   □

In particular take $W = \{0\}$. Then to say that $g + b$ is transverse to W means that 0 is a regular value of $g + b$, that is $-b$ is a regular value of g. We thus see that Sard's Theorem is a special case of the Weak Transversality Theorem.

3.8 Jet Spaces

3.8.1. In practice it is often difficult to apply Sard's Theorem (or the Weak Transversality Theorem) directly. We give an example. First consider a system of three equations in three variables of the form

$$f(x, y, z) = 0, \quad g(x, y, z) = 0, \quad h(x, y, z) = 0,$$

where f, g and h are of class C^∞. It follows immediately from the previous discussion that in general such a system defines a discrete subset. This is the case when 0 is a regular value of the map from $\mathbf{R}^3$ to $\mathbf{R}^3$ whose components are (f, g, h); if (thanks to Sard) we take a regular value (α, β, γ) of this map then 0 will be a regular value of the map whose components are $(f - \alpha, g - \beta, h - \gamma)$. In other words, perturbing the left hand sides of the equations by adding generic constants (which may be taken to be as small as we wish) we are led back to the desired situation. By the usual technology of function spaces we could deduce from this an impressive statement to the effect that in a suitably chosen function space the set of acceptable left hand sides does indeed form a residual dense subset. So far, so good.

3.8.2. Now suppose that we are interested in a system of equations of the form

$$f_x(x, y, z) = 0, \quad f_y(x, y, z) = 0, \quad f_z(x, y, z) = 0$$

(finding the critical points of a function f), or even

$$f(x, y, z) = 0, \quad f_z(x, y, z) = 0, \quad f_{zz}(x, y, z) = 0$$

(finding the cusp points of an apparent contour). There are now 'automatic' relations between the left hand sides, and an assumption of genericity on the function $f(x, y, z)$ does not imply that the system of three equations is itself generic. Of course, in the first case is suffices to adapt slightly the technique already used: we replace $f(x, y, z)$ by $f(x, y, z) - \alpha x - \beta y - \gamma z$ with (α, β, γ) generic. In the second case it is a little less immediate: if we replace $f(x, y, z)$ by $f(x, y, z) - \alpha - \beta z - \frac{1}{2}\gamma z^2$ the system becomes

$$f(x, y, z) = \alpha + \beta z + \frac{1}{2}\gamma z^2, \quad f_z(x, y, z) = \beta + \gamma z, \quad f_{zz}(x, y, z) = \gamma,$$

or

$$f - z f_z + \frac{1}{2}z^2 f_{zz} = \alpha, \quad f_z - z f_{zz} = \beta, \quad f_{zz} = \gamma.$$

Therefore it is enough to take (α, β, γ) to be a regular value of this newly created map. Thom's Theorem generalizes this procedure, by formalizing the vague notion of 'conditions applying to the derivatives' as *submanifold of jet space*. It is this notion that we shall now explain.

3.8.3. Let E and F be two finite-dimensional vector spaces and let $r \geq 0$ be an integer. Let $P^r(E; F)$ be the vector space of polynomial maps of degree $\leq r$ from E to F. For every open set U in E we write

$$J^r(U; F) = U \times P^r(E; F),$$

and call this the *space of jets of order r of maps from E to F*. Thus we have

$$J^0(U; F) = U \times F, \quad J^1(U; F) = U \times F \times L(E; F), \quad \cdots$$

Let $f : U \to F$ be a C^∞ map and let $a \in U$. Write the Taylor expansion for f at a in the form

$$f(a + h) = (j_a^r f)(h) + (R_a^r f)(h) = \big(f(a) + f'(a) \cdot h + \cdots\big) + (R_a^r f)(h),$$

with $j_a^r f \in P^r(E; F)$ and $R_a^r(h) = o(\|h\|^r)$. We say that $j_a^r f$ is the *jet of order r* or r-*jet* of f at a. The map

$$j^r f : U \to J^r(U; F) \qquad (j^r f)(a) = (a, j_a^r f)$$

is called the r-*jet* of f. In particular we have $j^0 f(a) = (a, f(a))$, $j^1 f(a) = (a, f(a), f'(a))$.

3.8.4. To explain the use of these constructions, take the particular case when U, E and F are equal to $\mathbf{R}$. The space $J^r(U; F)$ is then identified with $\mathbf{R}^{r+2}$, and the jet $j^r f$ of a map $f : \mathbf{R} \to \mathbf{R}$ is the map which associates to x in $\mathbf{R}$ the vector $(x, f(x), f'(x), \cdots, f^{(r)}(x))$. Note that the maps $j^r f$ do not describe all the maps from U into $J^r(U; F)$. To specify a subset of $J^r(U; F)$, for example (for $r = 1$) the set consisting of those (x, y, p) with $y = px$, is to regard a condition such as $f(x) = f'(x)x$ (the tangent passes through the origin) as a 'pointwise' condition on the triple $(x, f(x), f'(x))$ regardless of the fact that $f(x)$ is the value of a function and $f'(x)$ is the value of its derivative.

3.9 The Thom Transversality Theorem

Let E and F be two finite-dimensional vector spaces with U an open set in E and $r \geq 0$ an integer. Let W be a submanifold of $J^r(U; F)$ of codimension c, and let f be a C^∞ map from U to F. If $j^r f$ is transverse to W then the set of those a in U such that $(a, j_a^r f)$ belongs to W is a submanifold of U of codimension c (Proposition 7.2 or 2.6.2) – and in particular it is empty when $c > \dim(U)$. The theorem of Thom says that this is the situation for generic f. First we give a 'concrete' version:

Proposition 3.9.1. *Let W be a submanifold of $J^r(U; F)$. The set of those $p \in P^r(E; F)$ such that $j^r(f+p)$ is transverse to W is a dense residual subset of $P^r(E; F)$ with negligible complement. If W is closed, the set of p such that $j^r(f + p)$ is transverse to W on a given compact subset K of U is an open dense subset of $P^r(E; F)$ with negligible complement.*

We have $j^r(f + p)(a) = (a, j_a^r f + j_a^r p)$. But since p is a polynomial of degree $\leq r$ we have $j_a^r p(h) = p(a + h)$ and the map which associates to (a, p) the pair $(a, j_a^r p) = (a, x \mapsto p(a + x))$ is bijective with inverse $(a, q) \mapsto (a, x \mapsto q(x - a))$. We deduce from this that the map $g : (a, p) \mapsto j_a^r(f + p)(a)$ from $U \times P^r(E; F)$ to $J^r(E; F) = U \times P^r(E; F)$ is a diffeomorphism with inverse $(a, q) \mapsto (a, x \mapsto q(x - a) - j_a^r f)$ (here we see again a calculation carried out earlier in the special case when $p(x, y, z) = \alpha + \beta z + \frac{1}{2}\gamma z^2$). In particular the map g is a submersion and we can apply Theorem 7.4 directly.   □

3.9.2. To make the most convenient use of this statement we shall introduce a topology for spaces of C^∞ maps. Let E and F be two finite-dimensional vector spaces, and let U be an open subset of E. Let $C^\infty(U; F)$ denote the space of all C^∞ maps from U to F. We shall define what it means for a sequence to be convergent in this space. We could restrict to the case $F = \mathbf{R}$ by taking a basis for F and arguing component by component. We shall work directly in the general case; the reader may take $F = \mathbf{R}$ in all that follows if he or she wishes.

We start by taking a basis in E (it will be clear that the result is independent of the choice of basis) and therefore suppose that $E = \mathbf{R}^n$. Let $f \mapsto \partial_\alpha f$ denote the operators of mixed partial differentiation, where the multi-index $\alpha = (\alpha_1, \ldots, \alpha_n)$ runs through $\mathbf{N}^n$; in particular we have $\partial_0 f = f$. A sequence of elements f_i of $C^\infty(U; F)$ is said to *converge* to an element $f \in C^\infty(U; F)$ if, for every compact set K in U and every multi-index α, the sequence of the $\partial_\alpha f_i$ converges to $\partial_\alpha f(x)$ uniformly for $x \in K$.

Having thus defined convergent sequences, we now know what are the closed sets (sets stable under the process of taking limits), open sets (complements of closed sets), dense sets (those which meet every non-empty open set), residual sets (countable intersections of open dense sets), and so on. We then have the key result:

Proposition 3.9.3. *The space $C^\infty(U; F)$ is a Baire space: every residual subset is dense.*

We sketch the proof. We shall show that $C^\infty(U; F)$ is a complete metric space, which will suffice. First we define a metric.

Fix a norm on F. Fix for the moment an integer $r \geq 0$ and a compact subset K of U. For each $f \in C^\infty(U; F)$ let

$$p_{K,r}(f) = \sup_{x \in K, |\alpha| \leq r} \|\partial_\alpha f\|,$$

where $|\alpha| = \alpha_1 + \cdots + \alpha_n$ is the length of the multi-index α. To say that $p_{K,r}(f_i - f)$ tends to 0 is to say that all the partial derivatives of order $\leq r$ of the sequence f_i tend to the corresponding derivatives of f uniformly on K.

In fact $p_{K,r}$ is almost a norm on the vector space $C^\infty(U)$ (almost, because $p_{K,r}(g) = 0$ does not mean that g is zero but only that its restriction to K is zero); we say that it is a *semi-norm*.

It amounts to the same thing to say that $\inf(1, p_{K,r}(F_i - f))$ tends to zero. Now, we can find a sequence $K_0, K_1, \ldots$ of compact subsets of U whose union is U. If for f and g in $C^\infty(U; F)$ we let

$$d(f, g) = \sum_{r=0}^{\infty} 2^{-r} \inf\left(1, p_{K,r}(g - f)\right),$$

then it is easy to see that the convergence of the f_i to f can be expressed by saying that $d(f_i, f)$ tends to zero. It is immediate that $d(f, g) = d(g, f)$; since the K_r fill all of U it is clear that $d(f, g) = 0$ implies that $f = g$; the triangle inequality is an easy consequence of the triangle inequality for the $p_{K,r}$.

The fact that $C^\infty(U; F)$ is complete under this metric is a simple translation of the classical theorem on sequences of C^∞ functions which, together with their derivatives, converge uniformly on every compact set.   □

We may therefore talk about *generic* properties in the space $C^\infty(U; F)$: these are the properties that are satisfied in a residual subset; we then have properties a) to e) of 2.5. The general form of the Transversality Theorem is as follows:

Theorem 3.9.4. (Thom's Transversality Theorem.) *Let E and F be two finite-dimensional vector spaces, with U an open set in E. Let $r \geq 0$ be an integer, and let W be a submanifold of $J^r(U; F)$. Then the set of maps $f \in C^\infty(U; F)$ such that $j^r f$ is transverse to W is a dense residual subset of $C^\infty(U; F)$.*

The proof is based on Proposition 9.1 and the fact (Weierstrass' Theorem) that polynomial maps are dense in $C^\infty(U; F)$.   □

More generally, if V and V' are two manifolds and W is a submanifold of the manifold $J^r(V; V')$ of r-jets of maps from V to V', then the set consisting of those $f \in C^\infty(V; V')$ for which the map $j^r f : V \to J^r(V; V')$ is transverse to W is a dense residual subset of $C^\infty(V; V')$, open when V is compact and W is closed. Note finally that when V is not compact we can define another topology on the space $C^\infty(V; V')$ – called the *fine topology* or the *Whitney topology* – which is in some respects nicer than the topology of compact C^∞-convergence constructed above, and for which the statements given above remain valid *ne varietur*.

The conclusion of Theorem 9.4 is often abbreviated by saying "*for generic f in $C^\infty(U; F)$ the map $j^r f$ is transverse to W*". Note that if this is the case then the set of x in U such that $j^r f(x)$ belongs to W is a submanifold of U of codimension equal to $\mathrm{codim}(W)$.

3.9.5. We give an elementary example. Let $\dim(E) = p$ and $\dim(F) = q$, and suppose $p \leq q$. We have $J^1(U; F) = U \times F \times L(E; F)$, with $j^1 f(a) = (a, f(a), f'(a))$. In $L(E; F)$ the set of maps of rank $p - r$ (with $r \leq p$) is a submanifold of codimension $c = r(q - p + r)$. For a generic map f the set A of those a in U such that $f'(a)$ has rank $p - r$ is therefore (if it is non-empty) a submanifold of dimension $p - c$. For $q \geq 2p$ and $r > 0$ we have $p - c < 0$ and A is empty; hence f is an immersion and in this way we recover Corollary 4.4. For $q = 2p - 1$ we have $p - c < 0$ for $r > 1$ and $p - c = 0$ for $r = 1$, and we deduce from this that the set of $a \in V$ where f is not an immersion is discrete, and that at these isolated points the rank of $f'(a)$ is $p - 1$.

The theorem above is quite remarkable, and we shall see some spectacular applications of it in the next chapter. Nevertheless, it has a serious limitation: it is expressed using jets at *a point* of the function and is therefore powerless when we need to take into account conditions that apply simultaneously at several points in the source space. Here is a simple example. As we have just seen, we can express the property "f is an immersion" by means of a condition on the jet of f at each point, but this will not be the case for the condition "f is injective" which implies the comparison of the values of f at two different points. This is the justification for extending Thom's Theorem to multijets, as follows.

3.9.6. We retain the previous notation, and also fix an integer $n \geq 1$ (the case $n = 1$ will give the case already dealt with). Let $\Delta_{(n)}(U)$ denote the subset of U^n consisting of sequences $(a_1, \ldots, a_n)$ of *pairwise distinct* points of U ; this is an open set in U^n. Likewise let $J^r_{(n)}(U; F)$ denote the subset of $(J^r(U; F))^n = (U \times P^r(E; F))^n$ consisting of those $((a_1, p_1), \ldots, (a_n, p_n))$ with $a_i \neq a_j$ for $i \neq j$ (the "space of n-multijets of order r of maps from U into F"). The map

$$j^r_{(n)} f : \Delta_{(n)}(U) \to J^r_{(n)}(U; F)$$

that associates to $(a_1, \ldots, a_n) \in \Delta_{(n)}(U)$ the sequence $(j^r f(a_1), \ldots, j^r f(a_n))$ is called the *n-multijet of order r* of the map $f \in C^\infty(U; F)$.

For example, for $r = 0$ and $n = 2$ the map $j^0_{(2)} f$ sends the pair (x, y) (with $x \neq y$) to the element $(x, f(x), y, f(y))$ of $U \times F \times U \times F$. To say that f is injective is to say that the image of $j^0_{(2)} f$ does not meet the submanifold W of $J^0_{(2)}(U; F) \subset U \times F \times U \times F$, of codimension $\dim(F)$, consisting of quadruples (x, u, y, v) with $x \neq y$ and $u = v$. More generally, the inverse image of W by $j^0_{(2)} f$ is the set of pairs $(x, y) \in U \times U$ with $x \neq y$ and $f(x) = f(y)$, which is often called the set of *double points* of f. Hence if $j^0_{(2)} f$ is transverse to W, the set of double points of f is a submanifold of $U \times U$ of codimension $\dim(F)$, and in particular f is injective if $\dim(F) \geq 2 \dim(E)$. This shows in this particular case the interest in the extension of Thom's Transversality Theorem which we now state.

Theorem 3.9.7. (Multijet Transversality Theorem.) *Let E and F be two finite-dimensional vector spaces and let U be an open subset of E. Let $r \geq 0$ and $n \geq 1$ be two integers, and let W be a submanifold of $J^r_{(n)}(U; F)$. The set of $f \in C^\infty(U; F)$ such that $j^r_{(n)} f$ is transverse to W is a dense residual subset of $C^\infty(U; F)$.* $\square$

The proof uses no new ideas. Indeed, Theorem 9.7 is deduced from Proposition 9.1 by a standard approximation technique (exercise). Taking $n = 1$ in Theorem 9.7 recovers Theorem 9.4. Taking $r = 0$ gives a sequence of statements valid for a generic map f, of which the first two are:

a) The set of double points of f is a submanifold of U^2 of dimension $2 \dim(E) - \dim(F)$. In particular, f is injective when $\dim(F) > 2 \dim(E)$.
b) The set of triple points of f is a submanifold of U^3 of dimension $3 \dim(E) - 2 \dim(F)$. In particular, f has no triple points when $\dim(F) > \frac{3}{2} \dim(E)$.

3.10 Some History

This essential notion of transversality has a very short history. It dates in fact from the 1950s. The theorem of A.Sard was published in 1942; a particular case of it was proved in 1935 by A.B.Brown. Thom's Theorem dates from 1954-1956. From that moment on, transversality took its place among the fundamental concepts of differential topology. See, for example, the Introduction and Chapt. 4 of [AR].

4 Classification of Differentiable Functions

4.1 Introduction

In this chapter we make a start on the programme sketched out in Sect. 3.1 by giving the first steps in the classification of differentiable functions. We follow the method suggested by the Transversality Theorem in going from 'generic' situations to more particular ones. First of all, as the Local Inversion Theorem shows, for a generic function f at a generic point a there is nothing to say: such a function can be written as $f(a) + x$ where x is one member of a system of local coordinates centred at a.

The study of a generic function at an arbitrary point, not necessarily a generic one, is hardly more complicated. If the space has dimension n the worst that can be expected, still as a consequence of the Transversality Theorem, is that the n partial derivatives of f at a will vanish and nothing more; at such a point ('nondegenerate critical point') the principal part of $f - f(a)$ is a quadratic form H which is nondegenerate (since the degeneracy of H would impose a further condition). The key result in this situation is the *Morse Lemma*: not only can the function f be written in the given coordinates as $f(x) = f(a) + H(x) +$ 'higher order terms', but it can be written *precisely* as $f(x) = f(a) + H(x)$ for an appropriate system of curvilinear coordinates. With this result a new theme appears which will be developed in the following chapter: when does the Taylor expansion of f at a up to a certain order, which is *a priori* an approximation to f, become an exact expression when we take suitable coordinates?

If we are no longer interested in a single function but in a family of functions, say a p-parameter family, then we have to consider more complicated 'singularities' that satisfy p further conditions. First, the form H may become degenerate at certain points (at least one further condition, or to be precise $r(r+1)/2$ of them to say that its rank is $n-r$) and at these points we have to go to degree 3 or more. In fact the Morse Lemma has a parametrized version which shows that the entire problem is concentrated on the kernel of the form H, the simplest case being when the dimension of the kernel (the *corank* of H) is 1. Since $r(r+1)/2$ is equal to 3 for $r = 2$ and is equal to 6 for $r = 3$ the corank will never exceed 1 in a generic family with $p \leq 2$ parameters and will never exceed 2 in a family with $p \leq 5$ parameters.

The first two singularities of this classification – the *fold* and the *cusp* – are enough to describe the local form of generic families of $p \leq 2$ parameters. We shall consider three related examples: the apparent outlines of surfaces, maps of the plane to itself, and envelopes of plane curves. Later we shall sketch the more difficult case of caustics.

To handle these examples we shall anticipate the following chapter and use a result from deformation theory. The question is this: if the element f_0 of the family under consideration has a given singularity at the point a, what happens for nearby functions f_λ? The first case is an immediate application of the Implicit Function Theorem: if f_0 has a non-degenerate critical point at a, then for λ sufficiently small the function f_λ has a unique critical point $a(\lambda)$ close to a which is also non-degenerate and which depends C^∞ on λ. The next cases are more complicated, as 'bifurcation' phenomena appear. We introduce a key result (a particular case of the main theorem of the following chapter) which answers the above question for functions of one variable. This will be enough for us to deal with the three examples cited.

Sects. 2 and 3 are preliminary sections in which we consider different forms of 'Hadamard's Lemma' and sketch the general classification problem. In Sect. 4 we define the Hessian form H introduced above and show that for a generic function the Hessian forms at critical points are non-degenerate. Sect. 5 contains various useful versions of the Morse Lemma; in Sect. 6 we state the Deformation Theorem for functions of one variable. Sects. 7,8 and 9 contain the three stated applications. In Sect. 10 we treat the problem of caustics. In Sect. 11 we return to the general problem of structural stability.

4.2 Taylor Formulae Without Remainder

We begin with the following observation: a C^∞ function of a variable x which vanishes for $x = 0$ is divisible by x in the ring of C^∞ functions. More generally, introducing an auxiliary parameter $y = (y_1, \ldots, y_p)$ we have:

Lemma 4.2.1. *Let I be an interval in $\mathbf{R}$ containing 0, let U be an open set in $\mathbf{R}^p$, and let $f(x, y) = f(x, y_1, \ldots, y_p)$ be a C^∞ function on $I \times U$. Then there exists a unique C^∞ function $g(x, y)$ defined on $I \times U$ such that $f(x, y) - f(0, y) = xg(x, y)$. Moreover $g(0, y) = \frac{\partial f}{\partial x}(0, y)$.*

Let $(x, y) \in I \times U$. Define a function h on $[0, 1]$ by $h(t) = f(tx, y)$. We have $\dot{h}(t) = x \frac{\partial f}{\partial x}(tx, y)$ and so

$$f(x, y) - f(0, y) = h(1) - h(0) = \int_0^1 \dot{h}(t)dt = x \int_0^1 \frac{\partial f}{\partial x}(tx, y)dt = xg(x, y).$$

Differentiating under the integral sign we see that g is C^∞ and we have shown the existence of a function with the required properties. Furthermore, the given relation defines the values of the continuous function $g(x, y)$ for $x \neq 0$ and g is uniquely determined. Finally, differentiating at $x = 0$ we indeed obtain $g(0, y) = \frac{\partial f}{\partial x}(0, y)$.   □

Note the extrememly general nature of this construction. We could have taken y to belong to a manifold, or made some other assumption about the regularity of f with respect to the variable y, for instance.

Proposition 4.2.2. (Parametrized Taylor formula without remainder.) *Let I be an interval of $\mathbf{R}$ containing 0, let U be an open set in $\mathbf{R}^p$, let $f(x, y) = f(x, y_1, \ldots, y_p)$ be a C^∞ function on $I \times U$, and let $m > 0$ be an integer. Then there exists a unique C^∞ function $r_m(x, y)$ defined on $I \times U$ such that*

$$f(x, y) = \sum_{i=0}^{m-1} \frac{\partial^i f}{\partial x^i}(0, y) \frac{x^i}{i!} + x^m r_m(x, y). \tag{4.2.1}$$

We have $r_m(0, y) = \frac{1}{m!} \frac{\partial^m f}{\partial x^m}(0, y)$.

For $m = 1$ this is the lemma above. Suppose the result has been obtained up to order $m - 1$ and let us apply the lemma to r_{m-1}. We obtain a function r_m with

$$r_{m-1}(x, y) - \frac{1}{(m-1)!} \frac{\partial^i f}{\partial x^{m-1}}(0, y) = x r_m(x, y),$$

which implies (2.1). The uniqueness is proved as above. Differentiating m times at $x = 0$ we obtain the value for $r_m(0, y)$ as claimed.   □

Rather than apply the lemma again to g relative to x as above, we could reapply it with respect to the first component of y. Consider a C^∞ function $f(x_1, \ldots, x_n)$ on an open set V in $\mathbf{R}^n$ that is a product of open intervals containing 0. Functions $g_1, \ldots, g_n$ can then be found such that

$$f(x_1, x_2, \ldots, x_n) = x_1 g_1(x_1, \ldots, x_n) + f(0, x_2, \ldots, x_n),$$
$$f(0, x_2, \ldots, x_n) = x_2 g_2(x_2, \ldots, x_n) + f(0, 0, x_3, \ldots, x_n),$$
$$\cdots = \cdots$$
$$f(0, \ldots, 0, x_n) = x_n g_n(x_n) + f(0, \ldots, 0).$$

With a slight change of notation we obtain the following result, known as 'Hadamard's[29] Lemma':

Proposition 4.2.3. *There exist C^∞ functions $h_1, \ldots, h_n$ on V such that*

$$f(x_1, \ldots, x_n) - f(0, \ldots, 0) = \sum_{i=1}^{n} x_i h_i(x_1, \ldots, x_n).$$ $\square$

4.2.4. By differentiation at $0 = (0, \ldots, 0)$ we immediately have $h_i(0) = \frac{\partial f}{\partial x_i}(0)$. Hadamard's Lemma may then be applied to the h_i; rearranging the terms we arrive at an equation of the form

$$f(x_1, \ldots, x_n) = f(0) + \sum_{i=1}^{n} x_i \frac{\partial f}{\partial x_i}(0) + \sum_{i,j=1}^{n} x_i x_j h_{i,j}(x_1, \ldots, x_n). \quad (4.2.2)$$

Replacing each $h_{i,j}$ by $\frac{1}{2}(h_{i,j} + h_{j,i})$ we can make $h_{i,j} = h_{j,i}$; differentiating at 0 we then obtain

$$h_{i,j}(0) = \frac{\partial^2 f}{\partial x_i \partial x_j}(0). \quad (4.2.3)$$

Clearly we could combine these two procedures and obtain a *general Taylor formula without remainder* with p parameters, n variables and of order m that the reader will be able to imagine without difficulty.

4.2.5. Consider the following special case: let f be a C^∞ function defined on a neighbourhood of 0 in $\mathbf{R}$. Suppose its derivatives at 0 are not all zero. Let m be the smallest integer such that $\partial^m f/\partial x^m(0)$ is nonzero. Then by (2.1) we may write $f(x) = x^m r(x)$ with $r(0) \neq 0$. Let $\varepsilon = -1$ if m is even and $r(0) < 0$ and $\varepsilon = 1$ otherwise. Let $a \in \mathbf{R}$ with $a^m = \varepsilon r(0)$; then (Local Inversion Theorem) the function $s(x)$ defined by $s(0) = a$ and $s(x)^m = \varepsilon r(x)$ is of class C^∞. Therefore $z = xs(x)$ is a local coordinate on $\mathbf{R}$ in a neighbourhood of 0 for which we have $f = \varepsilon z^m$.

4.3 The Problem of Classification of Maps

Let us pose the problem in a general setting. We wish to classify C^∞ maps up to diffeomorphism. More precisely, we consider objects of type (E, F, a, b, f) where E and F are two finite-dimensional vector spaces with a a point of E and b a point of F and with f a C^∞ map defined on a neighbourhood of a in E with values in F and such that $f(a) = b$. Two such objects $(E_1, F_1, a_1, b_1, f_1)$ and $(E_2, F_2, a_2, b_2, f_2)$ will be called *equivalent* if there are local diffeomorphisms u and v such that $f_2 = v \circ f_1 \circ u$ (we suppose that u is defined on a neighbourhood of a_2 with values in E_1, that v is defined on a neighbourhood of b_1 with values in F_2, and that $u(a_2) = a_1$ and $v(b_1) = b_2$). This is clearly an equivalence relation, and the problem is to determine a list of equivalence classes if possible. We can observe straight away that

$$f_2'(a_2) = v'(b_1) \circ f_1'(a_1) \circ u'(a_2)$$

and therefore the first invariant of an equivalence class is the linear equivalence class of the derivative, that is to say its *rank* $\mathrm{rk}_a f = \mathrm{rk}(f'(a))$.

We naturally begin by posing the *linearization* problem: is the system (E, F, a, b, f) equivalent to a system $(E, F, 0, 0, g)$ where g is linear? If the answer is positive then g is linearly equivalent to $f'(a)$ and we can take $g = f'(a)$. Clearly there are cases where the answer is negative, if only those where $f'(a)$ is zero and f is non-constant on a neighbourhood of a. On the other hand we have met two cases where the answer is positive, although this was not made explicit in the above terms, namely immersions ($f'(a)$ is injective) and submersions ($f'(a)$ is surjective). The following form of the Local Inversion Theorem generalizes these two cases by characterizing those situations where linearization is possible.

Proposition 4.3.1. *For f to be equivalent to a linear map on a neighbourhood of a it is necessary and sufficient that $\mathrm{rk}_x f = \mathrm{rk}_a f$ for all x sufficiently close to a.*

Note first that in any case we have $\mathrm{rk}_x f \geq \mathrm{rk}_a f$ for x close to a (see Example (8) in 2.3, or argue directly observing that the minors of $f'(x)$ depend continuously on x). Moreover, the condition is obviously necessary since every linear map has constant rank. We state the converse in the following way resembling a well known result for linear maps:

Theorem 4.3.2. (Constant Rank Theorem.) *Let f be a C^∞ map defined on a neighbourhood of a point a in an n-dimensional vector space E and with values in an m-dimensional vector space F. Let $p = \mathrm{rk}_a f = \mathrm{rk}(f'(a))$. Suppose that $\mathrm{rk}_x f = p$ for every element x of E sufficiently close to a. Then there exists a local coordinate system $(x_1, \ldots, x_n)$ on E centred at a and a local coordinate system $(y_1, \ldots, y_m)$ on F centred at $f(a)$ with respect to which the expression for f is*

$$y_1 = x_1, \ \ldots \ , y_p = x_p, \ y_{p+1} = 0, \ \ldots \ , y_m = 0.$$

□

Clearly we may suppose that $a = 0$ and $f(a) = 0$. We begin by choosing linear coordinates $(u_1, \ldots, u_n)$ on E and $(v_1, \ldots, v_m)$ on F in which the matrix of $f'(a)$ has the desired form. In particular, the matrix $(\partial f_i / \partial u_j)(a)_{i,j \in \{1, \ldots, p\}}$ is the identity matrix and the functions

$$x_1 = f_1, \ \ldots, \ x_p = f_p, \ x_{p+1} = u_{p+1}, \ \ldots, \ x_n = u_n$$

form a system of local coordinates at a. In the system of coordinates (x) and (v) the map f is expressed by

$$v_1 = x_1, \ \ldots, \ v_p = x_p, \ v_{p+1} = \phi_{p+1}(x_1, \ldots, x_n), \ \ldots, \ v_m = \phi_m(x_1, \ldots, x_n).$$

At each point the Jacobian matrix of f therefore contains the identity matrix of order p as its leading diagonal block; since it has to remain of rank p in a neighbourhood of a this implies that all the $\partial\phi_i/\partial x_j$ are zero in a neighbourhood of a, for $i > p$ and $j > p$. Thus the ϕ_i depend only on the first p variables x in a neighbourhood of a, and f may in fact be written

$$v_1 = x_1, \ \ldots, \ v_p = x_p, \ v_{p+1} = \phi_{p+1}(x_1, \ldots, x_p), \ \ldots, \ v_m = \phi_m(x_1, \ldots, x_p).$$

If we now set
$$y_1 = v_1, \ \ldots, \ y_p = v_p,$$

$$y_{p+1} = v_{p+1} - \phi_{p+1}(v_1, \ldots, v_p), \ \ldots, \ y_n = v_n - \phi_n(v_1, \ldots, v_p),$$

we obtain a system of local coordinates on F (true because the v can be calculated immediately as functions of the y) in which f is expressed in the desired form.   □

Let us now spell out the two special cases that we have already mentioned. If f is an immersion at a we have $p = n$; then in a neighbourhood of a we have $p \le \mathrm{rk}_x f \le n = p$ and f does indeed have constant rank locally. If f is a submersion at a we have $p = m$ and argue similarly.

In these two special cases the result can be neatly formulated as follows. Take $a = 0$ and $b = 0$ to simplify the notation. If f is an immersion at 0 there exists a local diffeomorphism v of F with $v(0) = 0$, $v'(0) = Id_F$ and $f = v \circ f'(0)$. If f is a submersion at 0 there exists a local diffeomorphism u of E with $u(0) = 0, u'(0) = Id_E$ and $f = f'(0) \circ u$.

In the general case f is the composition of the submersion $(x_1, \ldots, x_n) \mapsto (x_1, \ldots, x_p)$ and the immersion $(x_1, \ldots, x_p) \mapsto (x_1, \ldots, x_p, 0, \ldots, 0)$. This is why maps with constant rank are sometimes called *subimmersions*.

4.3.3. The cases which interest us are those where the rank 'jumps'. A little later (in 4.7) we shall discuss a case in which F has dimension 2. For the moment we concentrate on the case where the dimension of F is 1; here we are concerned with classifying C^∞ scalar functions f in the neighbourhood of a point a of a vector space E. We equip the set of triplets (E, a, f) with the following equivalence relation: (E_1, a_1, f_1) and (E_2, a_2, f_2) are *equivalent* if there exists a diffeomorphism u from a neighbourhood of a_2 in E_2 to a neighbourhood of a_1 in E_1 and a constant c such that $u(a_2) = a_1$ and $f_2 = c + f_1 \circ u$. Here we automatically have $c = f_2(a_2) - f_1(a_1)$ and the above relation can also be written as $(f_2 - f_2(a_2)) = (f_1 - f_1(a_1)) \circ u$, which reduces us to classifying triplets (E, a, f) with $f(a) = 0$. Making a list of equivalence classes comes down to giving a list of 'reduced forms' for the functions in suitably chosen coordinates.

4.3.4. As an example, the case dealt with in 2.5 can be expressed as follows: if the function f of one variable is of order m at the point a, it is equivalent to $\pm(x-a)^m$. Likewise the case of submersions treated above can be translated in two ways as follows: if $df(a) \neq 0$ then (E, a, f) is equivalent to $(\mathbf{R}^n, 0, pr_1)$ where pr_1 is the projection of $\mathbf{R}^n$ onto the first coordinate $\mathbf{R}$, or, alternatively, if $df(a) \neq 0$ there exists a system of local coordinates $(x_1, \ldots, x_n)$ on E at a with $x_i(a) = 0$ and $f = f(a) + x_1$. All noncritical points (for $\dim E$ fixed) are therefore equivalent, so it is a question of classifying critical points. At a critical point $f - f(a)$ is of second order and this brings in a new invariant: the quadratic form which is the principal part of $f - f(a)$, called the *Hessian form* for f at a, and on which we shall dwell in the next section.

4.4 Critical Points: the Hessian Form

4.4.1. A *quadratic form* on a finite-dimensional vector space T is a map $q : T \to \mathbf{R}$ which can be written as a homogeneous polynomial of degree 2 when a basis for T is given. The matrix for q in a basis is the symmetric matrix Q such that $q(x) = {}^t x Q x$ where each vector x of T is represented by the column vector of its components. We have $q(x+y) = q(x) + q(y) + 2b(x, y)$ where b is a symmetric bilinear form on T with matrix Q, called the *bilinear form associated* to q; note that $q(x) = b(x, x)$. The *kernel* of q is the kernel of b, meaning the linear subspace N of T consisting of those x such that $b(x, y) = 0$ for all y, that is $q(x + y) = q(y)$ for all y. The *corank* of q is the dimension of the kernel of q. Forms defined by invertible matrices, *i.e.* with corank zero, are said to be *invertible* or *non-degenerate*. Recall (*Sylvester's Law of Inertia*) that there exist bases for T in which q can be written in the *reduced form*

$$q(x) = x_1^2 + \cdots + x_{p_+}^2 - x_{p_+ + 1}^2 - \cdots - x_{p_+ + p_-}^2, \qquad (4.4.1)$$

where the integers p_+ and p_- do not depend on the chosen basis. The rank of the matrix for q in an arbitrary basis is $p_+ + p_-$; the corank of q is $n - p_+ - p_-$. The integer p_- is often called the *index* (or index of negative inertia) of the non-degenerate quadratic form q; a non-degenerate quadratic form of index p on an n-dimensional space can therefore be written in the reduced form

$$q(x) = -(x_1^2 + \cdots + x_p^2) + (x_{p+1}^2 + \cdots x_n^2). \qquad (4.4.2)$$

4.4.2. Let V be a submanifold of a finite-dimensional vector space E, with f a real C^∞ function on V and a a point of V. By definition (see 3.6.1), to say that a is a critical point of f means that the tangent map $T_a f$ is zero.

If f is induced by a C^∞ function F on an open set U of E containing a, then $T_a f$ is the restriction of the linear form $dF(a)$ to the subspace $T_a V$. Let $\phi_1, \ldots, \phi_m$

be a nondegenerate system of local equations for V at a. Then the $d\phi_i(a)$ form a basis for the vector space of linear forms on V that are zero on T_aV. Hence, to say that a is a critical point of f means that there exist scalars $\lambda_1, \dots, \lambda_m$ (*Lagrange multipliers*) such that

$$dF(a) = \lambda_1 d\phi_1(a) + \cdots + \lambda_m d\phi_m(a).$$

Another way to say that a is a critical point of f is to say that for every parametrized arc γ traced on V with $\gamma(0) = a$ we have $(f \circ \gamma)'(0) = 0$: this is because this derivative is equal to $(T_a f) \cdot \dot\gamma(0)$. Thus all the functions $f \circ \gamma - f(a)$ are of second order; their principal parts are given by a quadratic form on T_aV :

Proposition 4.4.3. *Suppose f is of class C^2. There exists a unique quadratic form $H_f(a)$ on the space T_aV such that*

$$(f \circ \gamma)''(0) = H_f(a)(\dot\gamma(0))$$

for every arc γ on v with $\gamma(0) = a$.

The quadratic form $H_f(a)$ is called the *Hessian form* of f at the critical point a.

The uniqueness is clear since the $\dot\gamma(0)$ run through the whole of T_aV. Suppose first that V is an open set in $E = \mathbf{R}^n$. Let $\gamma : I \to V$ be a parametrized arc with $\gamma(0) = a$. We have

$$(f \circ \gamma)'(t) = \sum_{i=1}^{n} \frac{\partial f}{\partial x_i}(\gamma(t))\dot\gamma_i(t) = \langle \dot\gamma(t), df(\gamma(t)) \rangle, \tag{4.4.3}$$

and differentiation gives

$$(f \circ \gamma)''(0) = \sum_{i=1}^{n} \frac{\partial f}{\partial x_i}(a)\ddot\gamma_i(0) + \sum_{i,j=1}^{n} \frac{\partial^2 f}{\partial x_i \partial x_j}(a)\dot\gamma_i(0)\dot\gamma_j(0)$$

$$= \langle \ddot\gamma(0), df(a) \rangle + f''(a) \cdot (\dot\gamma(0), \dot\gamma(0)).$$

Since $df(a)$ is zero this gives the result stated, as well as the fact that the matrix of the form $H_f(a)$ consists of the second partial derivatives of f at a. The general case is deduced from the case of an open set in $\mathbf{R}^n$ using a local parametrization of V.   □

As we have just seen, if we choose a local parametrization $h : \Omega \to V$ with $h(b) = a$ then the coefficients of the Hessian form of f with respect to the basis of T_av formed by the $\frac{\partial h}{\partial x_i}(b)$ are the second partial derivatives of $f \circ h$ at b.

The first discrete invariants of a critical point are therefore the index and the corank of the Hessian form, also called the *index* and the *corank* of the critical point. The simplest critical points (so-called *Morse points*) are those of corank 0; we shall see below in Corollary 5.2 that they are completely determined up to equivalence by the index of their Hessian form.

Definition 4.4.4. We say that the critical point a of f is *nondegenerate* or is a *Morse point* if the Hessian form $H_f(a)$ is nondegenerate. We say that f is a *Morse function* on V if all its critical points are nondegenerate.

It will shortly become clear that nondegenerate critical points are isolated points of the critical locus. In particular, the critical locus of a Morse function is closed and discrete, and the critical locus of a Morse function on a compact manifold is finite.

There are plenty of Morse functions; indeed, *a function chosen at random will be a Morse function*. This is an easy consequence of the Transversality Theorem, as we shall now see. First take V to be an open subset of a vector space E. To each point x of V we associate the differential $df(x) \in E^*$, which gives a map $df : V \to E^*$. This map has a derivative

$$(df)'(a) \in L(E; E^*).$$

If we choose dual bases in E and E^* then

$$(df)'(a) \cdot e_i = \sum_{j=1}^{n} \frac{\partial^2 f}{\partial x_i \partial x_j}(a) e_j^*.$$

Hence we have the following description:

Lemma 4.4.5. *To say that a is a nondegenerate critical point of f means that $df(a) = 0$ and that $(df)'(a)$ is invertible. To say that f is a Morse function means that 0 is a regular value of df.* □

Thus df is a local diffeomorphism of V into E^* at every Morse point, and the Morse points are isolated critical points.

Proposition 4.4.6. *Let f be a real function of class C^2 on V and let $i : V \to F$ be an immersion. The set of linear forms $u \in F^*$ such that $f + u \circ i$ is a Morse function is a dense residual subset of F^*, and is open when V is compact.*

Let $h : \Omega \to V$ be a local parametrization for V, where Ω is an open subset of an auxiliary vector space G, and let K be a compact subset of V. By the reasoning used several times in Chapt. 3, it suffices to prove that the set A of those $u \in F^*$ such that all the critical points of $f + u \circ i$ contained in $h(K)$ are nondegenerate is an open dense set in F^*. Now consider the map $r : \Omega \times F^* \to E^*$ that takes (x, u) to $((f + u \circ i) \circ h)'(x) = d(f \circ h)(x) + u \circ ((i \circ h)'(x))$. This is a submersion; explicitly, its derivative $r'(x, u)$ takes a vector (ξ, λ) of

$G \times F^*$ to an element of E^* of the form $(\cdots) \cdot \xi + \lambda \circ ((i \circ h)'(x))$ and the map

$$\lambda \mapsto \lambda \circ ((i \circ h)'(x))$$

is already surjective since $(i \circ h)'(x)$ is injective. The result then follows directly from the Weak Transversality Theorem 3.7.4. $\square$

Corollary 4.4.7. *The set of linear forms on E whose restriction to V is a Morse function is a dense residual subset of E^*, open when V is compact.*

Take $f = 0$ and i the given embedding of V in E. $\square$

In fact the set of Morse functions is an open dense subset of $C^\infty(V, \mathbf{R})$ for the 'fine' topology indicated in 3.9.

4.5 The Morse Lemma

Proposition 4.5.1. *Let f be a C^∞ function defined on a neighbourhood of 0 in a finite-dimensional vector space E. Suppose f has a non-degenerate critical point at 0, and let H denote the Hessian form of f at 0. There exists a C^∞ map u defined on a neighbourhood of 0 in E and with values in E such that $u(0) = 0$, $u'(0) = Id_E$ and $f = f(0) + H \circ u$.*

This proposition is a particular case of Theorem 5.3 below, but we will give a direct proof nevertheless. Choose a basis for E and identify the form H with its matrix, so $H(\xi) = {}^t\xi H\xi$. By 2.4 we can find a symmetric matrix $h(x) = (h_{i,j}(x))$, depending C^∞ on x, with $f(x) - f(0) = {}^t x h(x) x$. However, by 3.6.5 there exists a C^∞ map A from the space of symmetric matrices into the space of invertible matrices such that $A(H) = Id$ and $h = {}^t A(h) H A(h)$ for all h close to H. This gives

$$f(x) - f(0) = {}^t x h(x) x = {}^t x \, {}^t A(h(x)) H A(h(x)) x = {}^t u(x) H u(x) = H(u(x)),$$

with $u(x) = A(h(x)) \cdot x$, so $u(0) = 0$ and $u'(0) = A(h(0)) = A(H) = Id$. $\square$

The conclusion of this proposition may be expressed in the language introduced in Sect. 3 by saying that $(E, 0, f)$ is equivalent to $(E, 0, H)$. We have already noted that at a non-critical point a function is equivalent to its derivative; what we have just said means that *at a non-degenerate critical point a function is equivalent to its Hessian form.*

Corollary 4.5.2. (Morse Lemma.) *Suppose f has a non-degenerate critical point at 0. Let p denote the index of the Hessian form at this point. Then* *there exists a system of local coordinates* $(x_1, \ldots, x_n)$ *on* E *centred at* 0 *in which* f *can be written as*

$$f = f(0) - (x_1^2 + \cdots + x_p^2) + (x_{p+1}^2 + \cdots + x_n^2).$$

It is enough to take as the family (x_i) the components of u in a basis for E with respect to which H is in reduced form. □

This corollary immediately gives us the local form of f in a neighbourhood of 0: the function f has a local maximum at 0 if $p = n$ and a local minimum at 0 if $p = 0$. The proposition and its proof can be generalized as follows:

Theorem 4.5.3. (Decomposition Theorem.) *Let* E *and* F *be two finite-dimensional vector spaces and let* $f(x, y)$ *be a* C^∞ *function defined on a neighbourhood of* $(0, 0)$ *in* $E \times F$. *Suppose the function* $x \mapsto f(x, 0)$ *has a nondegenerate critical point at* 0 *with Hessian form denoted by* H. *Then there exists a* C^∞ *map* u *defined on a neighbourhood of* 0 *in* $E \times F$ *and with values in* E *for which* $u(0, 0) = 0$ *and* $(x \mapsto u(x, 0))'(0) = Id_E$, *and a* C^∞ *function* g *on a neighbourhood of* 0 *in* F *such that* $f(x, y) = H(u(x, y)) + g(y)$.

Let $f_x(x, y)$ denote the derivative of the map $x \mapsto f(x, y)$. We have $f_x(0, 0) = 0$. By assumption the partial derivative map $x \mapsto f_x(x, 0)$ has invertible derivative at 0. By the Implicit Function Theorem (Corollary 2.7.3) we can find a C^∞ map z defined on a neighbourhood of 0 in F and with values in E such that $z(0) = 0$ and $f_x(z(y), y) = 0$ for all y. The function $(x, y) \mapsto f(x + z(y), y)$ therefore has a partial derivative with respect to x that is zero at every point of the form $(0, y)$. Taylor's formula of order 2 without remainder can therefore be written as

$$f(x + z(y), y) = f(z(y), y) + {}^t x h(x, y) x,$$

where $h(x, y)$ is a symmetric matrix that depends C^∞ on (x, y) and is such that $h(0, 0) = H$. Arguing as in the proof of Proposition 5.1 we write $h(x, y) = {}^t A(x, y) H A(x, y)$ and then we obtain the desired map u in the form $u(x, y) = A(x, y) \cdot (x - z(y))$. □

As in Corollary 5.2 we obtain:

Corollary 4.5.4. *Let* $(z_1, \ldots, z_m)$ *be a local coordinate system on* F *at* 0. *Set* $y_j(x, y) = z_j(y)$ *and let* $p \in [0, n]$ *denote the index of* H. *Then there exists a local coordinate system on* $E \times F$ *centred at* $(0, 0)$ *having the form* $(x_1, \ldots, x_n, y_1, \ldots, y_m)$ *in which* f *can be written*

$$f = -(x_1^2 + \cdots + x_p^2) + (x_{p+1}^2 + \cdots + x_n^2) + g(y_1, \ldots, y_m).$$

□

The following result, often known as the Splitting Lemma, is a direct generalization of Corollary 5.2:

Corollary 4.5.5. *Let E be a finite-dimensional vector space and let f be a C^∞ function defined on a neighbourhood of 0 in E with $df(0) = 0$. Let m be the corank of the Hessian form $H_f(0)$. Then there exists a local coordinate system $(x_1, \ldots, x_{n-m}, y_1, \ldots, y_m)$ on E centred at 0 in which f can be written*

$$f = f(0) - (x_1^2 + \cdots + x_p^2) + (x_{p+1}^2 + \cdots + x_{n-m}^2) + g(y_1, \ldots, y_m),$$

where the function g is of order ≥ 3 at 0.

It suffices to apply the previous corollary after having decomposed E into a direct sum of the kernel of $H_f(0)$ and a complementary subspace.   □

4.5.6. Now let us return to our initial classification problem. Consider a generic map f from $\mathbf{R}^p$ to $\mathbf{R}^q$. Clearly we can take $p, q \geq 1$. So far we have obtained or indicated the following results:

a) if $q > 2p$ then f is an embedding (Sect. 3.4);
b) if $q = 2p$ then f is an immersion (Sect. 3.4);
c) if $q = 1$ then f is a Morse function (Proposition 4.6).

This takes care of all cases where $p = 1$; for $p = 2$ there remain the two cases $q = 2$ and $q = 3$. We shall see the first of these in Sect. 7, while for the second we refer to the literature: see [GG] for example.

4.6 Bifurcations of Critical Points

The general problem is as follows. We wish to describe the behaviour of the critical points of a function as the function varies. More precisely, we suppose the function depends on auxiliary parameters and we wish to know how the critical points vary as the parameters vary. Since we are limiting ourselves to a *local* study in the neighbourhood of a given point in space and given values of the parameters, we lose no generality in restricting ourselves to the following situation. Consider two finite-dimensional vector spaces E and Λ and a C^∞ function $f = f(x, \lambda)$ defined on a neighbourhood of $(0,0)$ in $E \times \Lambda$. For $\lambda \in \Lambda$ sufficiently small we put $f_\lambda(x) = f(x, \lambda)$, and regard f as describing a family of functions (f_λ) parametrized by λ or, as it is called, a *deformation* of the function f_0. We suppose the behaviour of f_0 at 0 is known and we seek to describe the critical points of f_λ in a neighbourhood of 0. The first two cases are given as follows:

Proposition 4.6.1.

a) *If 0 is not a critical point of f_0 then f_λ has no critical points close to 0 for λ sufficiently small.*

b) *If f_0 has a nondegenerate critical point at 0 then f_λ has a critical point $a(\lambda)$ in a neighbourhood of 0, for λ sufficiently small; this critical point is nondegenerate, and the map $\lambda \mapsto a(\lambda)$ is of class C^∞.*

The statement a) is clear since the derivative of f_λ at the point (x, λ) depends continuously on (x, λ). To show b) we can either apply the Implicit Function Theorem to the equation $f_x(x, \lambda) = 0$, or refer directly to the conclusion of Theorem 5.3. $\qquad\square$

Recall from 2.5 that a C^∞ function defined on a neighbourhod of 0 in $\mathbf{R}$ and not all of whose derivatives vanish at 0 can be written as $\pm x^{m+1}$ for a suitably chosen local coordinate x on $\mathbf{R}$ in a neighbourhood of 0. Later we shall use the following theorem:

Theorem 4.6.2. *Let $f(x, \lambda)$ be a C^∞ function defined on a neighbourhood of $(0,0)$ in $\mathbf{R} \times \Lambda$ and let $m > 0$ be an integer with $f(x, 0) = \pm x^{m+1}$. Then there exists a C^∞ function $X(x, \lambda)$ defined on a neighbourhood of $(0,0)$ in $\mathbf{R} \times \Lambda$, with $X(x, 0) = x$, and C^∞ functions $a_0(\lambda), \ldots, a_{m-1}(\lambda)$ defined on a neighbourhood of 0 in Λ and vanishing at 0, such that*

$$f(x, \lambda) = \pm X(x, \lambda)^{m+1} + a_{m-1}(\lambda)X(x, \lambda)^{m-1} + \cdots$$
$$\cdots + a_i(\lambda)X(x, \lambda)^i + \cdots + a_0(\lambda).$$

$\qquad\square$

For $m = 0$, so that $f_0 = \pm x$, we obtain $f_\lambda = \pm X_\lambda$ so $\partial f_\lambda / \partial X_\lambda = \pm 1$; for $m = 1$, so $f_0 = \pm x^2$, we obtain $f_\lambda = \pm X_\lambda^2 + a_0(\lambda)$, so $\partial f_\lambda / \partial X_\lambda = \pm 2X_\lambda$. This implies (in the case $\dim(E) = 1$) the statements a) and b) above. These cases ($m = 0, 1$) are contained in results already known (Corollary 1.7.5 and Theorem 5.3). As soon as m is ≥ 2 the proof becomes vastly more difficult and we shall not give it. The material in the following sections will illustrate the power of this assertion even for $m = 2$. Intuitively, and in a way that can be made more precise (see Sect. 5.6), we may translate the result as follows: the deformation with m parameters $a_0, \ldots, a_{m-1}$

$$F_{(a)} = \pm x^{m+1} + a_{m-1}x^{m-1} + \cdots + a_0$$

of the function $\pm x^{m+1}$ is *universal* in that a given deformation can be induced from it by maps $(x, \lambda) \mapsto X(x, \lambda)$ and $\lambda \mapsto (a_0(\lambda), \ldots, a_{m-1}(\lambda))$ whose existence is claimed by the Theorem. The critical points of $F_{(a)}$ in a neighbourhood of 0 are the zeros of a polynomial of degree m

$$\pm(m + 1)x^m + (m - 1)a_{m-1}x^{m-2} + \cdots + a_1 .$$

For general values of the a_i there are therefore m of them and they are nondegenerate. There are 'bifurcation' phenomena produced at those points a where the discriminant of the polynomial vanishes (see 3.6.6) and in particular at the point $a = 0$ at which the m critical points all coalesce at 0.

4.7 Apparent Contour of a Surface in $\mathbf{R}^3$

In this section we consider a surface S in $\mathbf{R}^3$ given by $f(\xi, \eta, \zeta) = 0$ where f is a C^∞ function on $\mathbf{R}^3$, and we aim to study the projection $p : (\xi, \eta, \zeta) \mapsto (\xi, \eta)$ of S into the plane $P = \mathbf{R}^2$.

Let us define the *horizon* of S relative to the projection p to be the set C of points $a \in S$ with $(\partial f / \partial \zeta)(a) = 0$. Its projection $p(C)$ in P is called the *apparent outline* or *apparent contour*.

To study the local situation in the neighbourhood of a point a of C we shall use local coordinate systems that are compatible on $\mathbf{R}^3$ and on P. To be precise, we shall say that a system of local coordinates (x, y, z) on $\mathbf{R}^3$ centred at a is *adapted* to p if x and y depend only on the first two coordinates and are therefore of the form $x(\xi, \eta) = x \circ p$, $y(\xi, \eta) = y \circ p$; then x and y are also local coordinates on P at $p(a)$, and p is still expressed in local coordinates by $(x, y, z) \mapsto (x, y)$ and C is still given by the equation $\partial f / \partial z = 0$.

We begin by looking at three particular local models.

4.7.1. The point $a \in S$ is called *ordinary* if it does not belong to the horizon, that is if $(\partial f / \partial z)(a) \neq 0$. In a neighbourhood of a, the surface S is then the graph of a C^∞ function $\zeta = h(\xi, \eta)$ (as in Corollary 2.7.3). Similarly, the three functions $x = \xi$, $y = \eta$ and $z = f$ form an adapted system of local coordinates at a, in which we have $f = z$. It will also be convenient to consider the system of local coordinates x, y, z formed by ξ, η and $f + \eta$, in which f is then written as $z - y$. Note that all points sufficiently close to an ordinary point are also ordinary points.

4.7.2. We say that the point $a \in S$ is a *fold point* (for the projection p) if we can find an adapted system of local coordinates at a with

$$f = z^2 - y. \tag{4.7.1}$$

In a neighbourhood of such a point and in these coordinates S is the graph of the function $(x, z) \mapsto y = z^2$, the horizon C is the curve $\{y = 0, \ z = 0\}$ in S and all the points of the horizon are fold points. The projection p induces (in a neighbourhood of a) an embedding of C in P, and $p(C)$ is (in a neighbourhood of $p(a)$) the curve $\{y = 0\}$. The projection of a small open neighbourhood of a in S is an open set in the closed half-plane $\{y \geq 0\}$ bounded by the curve $\{y = 0\}$ in P that is the projection of the horizon. The points of P near $p(a)$ have either zero, one or two preimages in S in the neighbourhood of a

according to their position relative to $p(C)$: the points of $p(C)$ have a single preimage, while the points not on $p(C)$ have zero or two preimages according to the side of the curve $p(C)$ on which they lie. See Fig. 4.1.

4.7.3. We say that the point $a \in S$ is a *cusp point* (for the projection p) if we can find an adapted system of local coordinates at a with

$$f = z^3 - xz - y. \tag{4.7.2}$$

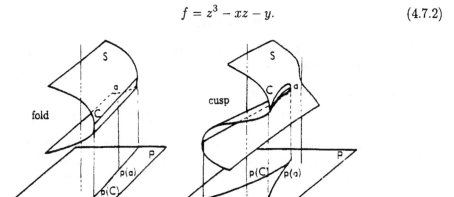

Fig. 4.1 A fold and a cusp

In a neighbourhood of such a point and in these coordinates S is the graph of the function $(x, z) \mapsto y = z^3 - xz$, the horizon C is the curve $\{x = 3z^2,\ y = -2z^3\}$ in S, and all the points of the horizon, with the exception of the point a itself, are fold points. The projection p induces (in a neighbourhood of a) an embedding of $C - \{a\}$ in P and the projection $p(C)$ of C is the parametrized arc given by the relations above, where z is the parameter, and therefore consists of those $(x, y) \in P$ with $(x/3)^3 + (y/2)^2 = 0$. The arc $p(C)$ has a cusp (see below) at $p(a)$, and is a submanifold away from this point. In a neighbourhood of a, the point $p(a)$ has a single preimage in S, the other points of $p(C)$ have two preimages, and the points of P which are not on $p(C)$ have one or three preimages according to whether they are on the 'outside' or 'inside' of $p(C)$. Again see Fig. 4.1.

These three models suffice to describe the situation when f is *generic*, as we shall now see.

Theorem 4.7.4. *A generic C^∞ function f on $\mathbf{R}^3$ has the following properties:*

a) *$S = \{f = 0\}$ is a surface (closed 2-dimensional submanifold);*
b) *its horizon $C = \{f = f_z = 0\}$ is a curve (1-dimensional submanifold, possibly empty);*
c) *all the points of C are fold points or cusp points; the cusp points form a discrete subset of C.*

The proof consists of several steps.

1) We begin by applying the Transversality Theorem. First we prove that for generic f the set S is indeed a surface, the set $C = \{f = f_z = 0\}$ is a curve and the set $D = \{f = f_z = f_{zz} = 0\}$ consists of isolated points. We can deduce this from Sard's theorem (see 3.8.2) or apply the Transversality Theorem by brute force as follows. Consider the 2-jet

$$j^2 f : \mathbf{R}^3 \to J^2(\mathbf{R}^3, \mathbf{R})$$

of f. The space $J^2(\mathbf{R}^3, \mathbf{R})$ may be identified with $\mathbf{R}^3 \times \mathbf{R} \times \mathbf{R}^3 \times \mathbf{R}^6 = \mathbf{R}^{13}$ and the jet map $j^2 f$ with the map

$$m \mapsto \big(m, f(m), (f_x(m), f_y(m), f_z(m)),$$
$$(f_{xx}(m), f_{xy}(m), f_{xz}(m), f_{yy}(m), f_{yz}(m), f_{zz}(m))\big)$$

from $\mathbf{R}^3$ into $\mathbf{R}^{13}$; we then apply the Transversality Theorem to the (linear!) submanifolds of codimensions $1, 2$ and 3 in $\mathbf{R}^{13}$ consisting of elements of the form $(m, h, (a, b, c), (p, q, r, s, t, u))$ with $h = 0$, $h = c = 0$ and $h = c = u = 0$ respectively.

2) Considering the 3-jet of f, we likewise see that the set $\{f = f_z = f_{zz} = f_{zzz} = 0\}$ is empty; at every point a of D we therefore have $f_{zzz}(a) \neq 0$. At such a point we also have $f_x(a)f_{yz}(a) - f_y(a)f_{xz}(a) \neq 0$. Using the notation above, we have to avoid the subset of $\mathbf{R}^{13}$ defined by the equations

$$h = c = u = at - br = 0.$$

But this set is the union of the linear submanifold of codimension 4 with equations $h = c = u = a = 0$ and the codimension 4 submanifold with equations

$$h = c = u = 0, \quad a \neq 0, \quad t = br/a.$$

3) We now prove that the points of $C - D$ are folds. Let $a \in C - D$; assuming $a = 0 = (0,0,0)$ we have $f(0) = 0$, $f_z = 0$ and $f_{zz}(0) \neq 0$. By Theorem 5.3 we can write f in the form $f = Z(x, y, z)^2 - Y(x, y)$ in a neighbourhood of a, where Y and Z are C^∞ functions, vanishing at zero, with $Z_z(0) \neq 0$. But since $df(0)$ is nonzero, $dY(0)$ is nonzero and we can complete Y to a system of local coordinates (X, Y) on P centred at 0. Then (X, Y, Z) is an adapted system of local coordinates such that $f = Z^2 - Y$.

4) Finally we come to the points of D and prove that they are cusp points. As above, taking 0 to be such a point, we have (as seen already) $f(0) = f_z(0) = f_{zz}(0) = 0$, $f_{zzz} \neq 0$ and $f_x(0)f_{yz}(0) - f_y(0)f_{xz}(0) \neq 0$. By Theorem 6.2, we can find C^∞ functions $Z(x, y, z)$, $X(x, y)$ and $Y(x, y)$ vanishing at 0, with $Z_z(0) \neq 0$ and such that $f = Z^3 - XZ - Y$. Differentiating twice at 0 we easily obtain $X_x(0)Y_y(0) - X_y(0)Y_x(0) \neq 0$. Thus (X, Y, Z) is an adapted system of local coordinates and f does indeed exhibit a cusp point at a. $\square$

4.7.5. Next we wish to describe the apparent contour of S, that is the projection $p(C)$ of the curve C. To state the result we employ the standard local forms as above.

Fig. 4.2 Ordinary double point and ordinary cusp

Let A be a subset of $E = \mathbf{R}^2$ and let a be a point of A. By definition, to say that A is a (regular) curve at a means that there is a system of local coordinates (x, y) on E, centred at a, such that A is defined in a neighbourhood of a by the equation $y = 0$. We say that A exhibits an *ordinary double point* or an *ordinary cusp* at a if there exists a system of local coordinates (x, y) on E, centred at a, such that A is defined in a neighbourhood of a by the equation $xy = 0$ or $x^2 = y^3$, respectively. See Fig. 4.2.

4.7.6. At a fold point, C is a curve and the restriction of p to C is an immersion. Therefore in the generic case the restriction of p to C is thus an immersion away from the discrete set of cusp points, and $p(C)$ is an immersed curve away from the image of the cusp points.

The image under p of a neighbourhood in C of a cusp point exhibits an ordinary cusp. It can happen that two distinct fold points have the same image under p, which gives a double point of $p(C)$.

In the generic case there are no other coincidences possible. To be more precise, *for a generic function the following coincidences are impossible:*

a) three distinct fold points have the same image under p, which is therefore a triple point (at least) of $p(C)$,

b) a fold point and a cusp point have the same image under p, which is therefore a triple point (at least) of $p(C)$,

c) at two fold points having the same image under p the tangents to C have the same image under the tangent map at p, thus giving a non-ordinary double point (common tangent) on $p(C)$.

4.7.7. These facts are all straightforward consequences of the Multijet Transversality Theorem 3.9.7.

For example, let us prove the first statement. With the notation of 3.9.6, consider the space $\Delta_{(3)}(\mathbf{R}^3)$ of triplets of distinct points in $\mathbf{R}^3$, and the 'tri-jet map of order 1'

$$j : \Delta_{(3)}(\mathbf{R}^3) \to J^1_{(3)}(\mathbf{R}^3; \mathbf{R}) \subset (\mathbf{R}^3 \times \mathbf{R} \times \mathbf{R}^3)^3 = \mathbf{R}^{21}.$$

Coincidence a) implies that there exists a triplet (m, m', m'') of distinct points of $\mathbf{R}^3$ for which we have

$$p(m) = p(m') = p(m''), \quad f_z(m) = f_z(m') = f_z(m'') = 0.$$

Now this means precisely that $j(m, m', m'')$ belongs to the subset A of $\mathbf{R}^{21}$ consisting of the families

$$\big((x, y, z), (x', y', z'), (z'', y'', z''), u, u', u'', (a, b, c), (a', b', c'), (a'', b'', c'')\big)$$

that satisfy the conditions

$$x = x' = x'', \ y = y' = y'', \ u = 0, \ u' = 0, \ u'' = 0, \ c = 0, \ c' = 0, \ c'' = 0.$$

But A is therefore a linear submanifold of codimension 10. Since $\Delta_{(3)}(\mathbf{R}^3)$ has dimension 9, it follows that $j^{-1}(A)$ is empty for generic f, which implies our assertion. We argue similarly in cases b) and c). To summarise:

Proposition 4.7.8. *The apparent contour of a generic surface in $\mathbf{R}^3$ is a curve having as singularities at most a discrete set of ordinary double points and ordinary cusps.* □

4.8 Maps from $\mathbf{R}^2$ into $\mathbf{R}^2$

In this section we shall determine what generic maps of the plane $\mathbf{R}^2$ into itself can look like. The results are very similar to those just obtained, although the situation is *a priori* a little more complicated since the graph of a map of $\mathbf{R}^2$ into itself is a surface in $\mathbf{R}^4$ and not in $\mathbf{R}^3$.

As in the previous case, we start by exhibiting three local models for maps of $\mathbf{R}^2$ into $\mathbf{R}^2$. Each of them is deduced from the corresponding local form in the previous section as follows: to the function f we associate the map that sends $(x, z) \in \mathbf{R}^2$ to (x, y) where y is determined by the fact that $f(x, y, z) = 0$.

4.8.1. If $a \in \mathbf{R}^2$ is not a critical point of f (in which case we say it is an *ordinary* point), then f is a local diffeomorphism at a ; hence there exist local coordinates (x, y) centred at a and local coordinates (x', y') centred at $f(a)$ such that f is expressed in a neighbourhood of a by

$$x' = x, \quad y' = y. \tag{4.8.1}$$

Every point sufficiently close to an ordinary point is ordinary.

4.8.2. We shall say that f has a *fold* at a if, in suitable systems of local coordinates centred at a as above, f is given in a neighbourhood of a by

$$x' = x, \quad y' = y^2. \tag{4.8.2}$$

In a neighbourhood of a fold point a (and of its image $f(a)$) the situation is as follows. There are two sorts of points: the point (x, y) is ordinary if $y \neq 0$ and is a fold point if $y = 0$. The fold points form a curve C (with equation $y = 0$) which is sent bijectively by f onto a curve (with equation $y' = 0$); in a neighbourhood of a the points of $f(C)$ have a single preimage, the points not on $f(C)$ have zero or two preimages according to the side of $f(C)$ on which they lie. See Fig. 4.3.

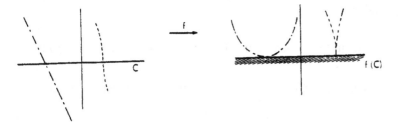

Fig. 4.3. Behaviour of f along a fold curve C

4.8.3. We shall say that f has a *cusp point* at a if, in suitable systems of local coordinates, f is given in a neighbourhood of a by

$$x' = x, \quad y' = y^3 - xy. \tag{4.8.3}$$

In a neighbourhood of the cusp point a the situation is as follows. Away from a there are ordinary points and fold points. The fold points (and a) form the curve C whose equation is $x = 3y^2$. The image $f(C)$ is the parametrized arc $x' = 3y^2$, $y' = -2y^3$ which has a cusp at $f(a)$. In a neighbourhood of a, the point $f(a)$ has a single preimage, the other points of $f(C)$ have two preimages, and the points not on $f(C)$ have one or three preimages according to whether they are on the 'outside' or 'inside' of $f(C)$. See Fig. 4.4.

As in the previous section, these three local models are sufficient to describe the generic situation:

Theorem 4.8.4. (Whitney) *A generic C^∞ map f from $\mathbf{R}^2$ to $\mathbf{R}^2$ has the following properties:*

a) *The set of critical points of f is a curve C (1-dimensional submanifold, possibly empty);*
b) *the points of C are either fold points or cusp points;*
c) *the cusp points, if they exist, are isolated: they form a discrete set.* $\qquad\square$

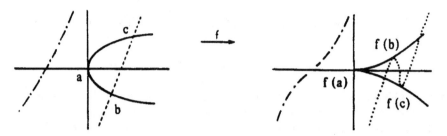

Fig. 4.4. Behaviour of f close to a cusp point

In fact it can be proved not only that these properties hold for generic f but that the f which possess them form an open and dense subset of the topological space $C^\infty(\mathbf{R}^2, \mathbf{R}^2)$.

We shall argue in two stages as in Theorem 7.4. To simplify the terminology we put $E = \mathbf{R}^3$.

1) We begin by using the Transversality Theorem in considering the 1-jet

$$j^1 f : E \to J^1(E; E).$$

We have $J^1(E; E) = E \times E \times L(E; E)$ with $j^1 f(a) = (a, f(a), f'(a))$. Since $E \times E \times \{0\}$ is a submanifold of $J^1(E; E)$ of codimension 4 and therefore greater than 2, for generic f the image of $j^1 f$ does not meet this submanifold (Theorem 3.9.4). Likewise, since the set of those (a, b, u) with $\det(u) = 0$ is a closed submanifold of codimension 1 in $E \times E \times (L(E; E) \setminus \{0\})$, for generic f the inverse image under f of this submanifold is a curve. We have already proved a), as well as the fact that at every point $a \in C$ we have $f'(a) \neq 0$.

2) Let us place ourselves at a point a of C; clearly we may assume that $a = (0, 0)$ and $f(a) = (0, 0)$. Let $u(x, y)$ and $v(x, y)$ denote the components of f. Since $f'(a)$ is nonzero we may, after permuting the coordinates if necessary, suppose that $(\partial u/\partial x)(0, 0) \neq 0$. Hence (u, y) is a system of local coordinates at a, and we are reduced to considering the case where f has the form $(x, y) \mapsto (x, v(x, y))$. Since f is of rank 1 at a, we have $(\partial v/\partial y)(0, 0) = 0$.

3) The next step is to apply the Transversality Theorem to v. To complete the proof we would have to verify that v is indeed generic when f is (the standard pitfall of which we need to beware — see below). We assume this has been done, and will make some comments about it later.

Consider then the 2-jet

$$j^2 v : E \to J^2(E; \mathbf{R}).$$

We have $J^2(E; \mathbf{R}) = E \times \mathbf{R}^6$ and

$$j^2 v(m) = \big(m, v(m), v_x(m), v_y(m), v_{xx}(m), v_{xy}(m), v_{yy}(m)\big).$$

It follows that, for generic v, the condition $v_y = 0$ defines a curve, the condition $v_y = v_{yy} = 0$ defines isolated points, and nowhere do we have $v_y = v_{xy} = v_{yy} = 0$. Consideration of the 3-jet shows analogously that for generic v we nowhere have $v_y = v_{yy} = v_{yyy} = 0$. We then use the following lemma:

Lemma 4.8.5. *Let v be a C^∞ function defined on a neighbourhood of $0 = (0,0)$ in $\mathbf{R}^2$ with $v(0) = 0$.*

a) *If $v_y(0) = 0$ and $v_{yy}(0) \neq 0$ there exist C^∞ functions $r(x,y)$ and $t(x)$ with $r(0,0) = 0$, $t(0) = 0$, $r_y(0) \neq 0$ and $v = r^2 + t$.*

b) *If $v_y(0) = v_{yy} = 0$, $v_{xy}(0) \neq 0$ and $v_{yyy}(0) \neq 0$, there exist C^∞ functions $r(x,y)$, $s(x)$ and $t(x)$ with $r(0,0) = 0$, $s(0) = 0$, $t(0) = 0$, $r_y(0,0) \neq 0$, $s_x(0) \neq 0$ and $v = r^3 + sr + t$.*

Using this lemma we can now finish the proof. In case a) we use as local coordinate systems $(x, r(x,y))$ and $(x, y - t(x))$, and then f takes (x, r) to (x, r^2). In the second case we use as coordinate systems $(-s(x), r(x,y))$ and $(-s(x), y - t(x))$ respectively, and f takes $(-s, r)$ to $(-s, r^3 + sr)$.

It remains to prove the lemma. In case a) we apply Theorem 5.3 directly (reversing the roles of x and y). In case b) we similarly apply Theorem 6.2; we obtain functions $r(x,y), s(x), t(x)$ with $r(0,0) = 0$, $r_y(0,0) \neq 0$ and $v = r^3 + sr + t$. Differentiating and bearing in mind the conditions imposed on v we obtain $s(0) = 0, t(0) = 0$ and $s_x(0) \neq 0$. □

4.8.6. Let return for a moment to those parts of the proof where we invoked the use of a rather large sledge-hammer. We could also have proceeded as follows. Returning to the situation of 2) above, this time we let $u(x,y)$ and $w(x,y)$ denote the components of f. As before, $(\partial u / \partial x)(0,0) \neq 0$ and we can take u and y as coordinates in the initial plane. Then f can be written as $(u,y) \mapsto (u, v(u,y))$ with $v(u(x,y),y) = w(x,y)$. We express the partial derivatives of v in terms of partial derivatives of u and w, and translate the conditions of Lemma 8.5 in terms of the 2-jet of f. The Transversality Theorem is then applied to the latter.

4.8.7. If we now wish to describe the set of *critical values* of f, that is the image of the curve C under f, we must be careful that we do *not* argue as follows: C is a curve, $\mathbf{R}^2$ has dimension 2 and f is generic so the restriction of f to C is an immersion. The error is that C depends on f, and the restriction of f to C is certainly not generic! In fact, reasoning as in 7.6, 7.7 we obtain:

Proposition 4.8.8 *For a generic map f from $\mathbf{R}^2$ to $\mathbf{R}^2$ the set of critical values of f is a curve having as singularities at most a finite set of ordinary double points and ordinary cusps.* □

4.9 Envelopes of Plane Curves

4.9.1. Consider a family of plane curves C_λ depending on a (1-dimensional) parameter λ. To fix the ideas, suppose that each C_λ is parametrized; thus we have a map $(t, \lambda) \mapsto f(t, \lambda) = (x(t, \lambda), y(t, \lambda))$ from an open set U of $\mathbf{R}^2$ into the plane $E = \mathbf{R}^2$ such that $\partial f/\partial t$ is everywhere nonzero.

If $a = (t, \lambda) \in U$ is an ordinary point (by which we mean as in 7.1 that f is a local diffeomorphism at the point a), we can choose a system of local coordinates at $f(a)$ in E in which $f(t, \lambda) = (t, \lambda)$. In this system the curves C_λ are segments of parallel straight lines in a neighbourhood of $f(a)$.

Suppose on the other hand that a is a critical point of f. Then since $(\partial f/\partial t)(a)$ is nonzero it follows that $(\partial f/\partial \lambda)(a)$ is proportional to it which, as we know, means that $f(a)$ belongs to the *envelope* of the C_λ. Thus this envelope is the set of critical values of f. We can also see this in a more geometric way. Consider the map $g : (t, \lambda) \mapsto (f(t, \lambda), \lambda)$ from U into $E \times \mathbf{R} = \mathbf{R}^3$. The tangent map to g at the point a is the linear map

$$g'(a) : (u, v) \mapsto (u\frac{\partial f}{\partial t}(a) + v\frac{\partial f}{\partial \lambda}(a), v),$$

which is injective. Therefore g is an immersion and (if we take U small enough) the image $g(U)$ is a surface S. To say that the tangent plane at $g(a)$ to S is vertical (relative to the projection $p : E \times \mathbf{R} \to E$) means that $p \circ g$ — namely f — has a critical point at a. Thus the set of critical values of f is the apparent contour of S. The curves C_λ in E are the projections in E of curves traced on S (images of $t \mapsto g(t, \lambda)$); therefore they are tangent to the apparent contour.

Now applying Proposition 7.8 or Proposition 8.8 we deduce:

Proposition 4.9.2. *The envelope of a generic 1-parameter family of plane curves is a curve whose singularities are no more that a finite set of ordinary double points and ordinary cusps.*   □

4.9.3. We ought not to think, however, that this exhausts the question and that we have thus obtained a classification of local forms of generic families of plane curves. With the notation used above, consider the map $(t, \lambda) \mapsto f(t, \lambda)$. Naturally we could choose from the start an arbitrary system of local coordinates on E, which amounts to replacing f by $v \circ f$ where v is an arbitrary local diffeomorphism of E. In the source, on the other hand, the situation is more subtle. We could in fact reparametrize each of the curves independently, which amounts to replacing t by $t(u, \lambda)$ where u is a new parameter; we could also relabel the family, which means replacing the parameter λ by $\lambda(\mu)$. The combination of two such modifications replaces f by $f \circ u$ where u is a local diffeomorphism in U, but we do not obtain all possible u in this way but only the 'triangular' transformations

$$(u, \mu) \mapsto (t(u, \mu), \lambda(\mu)).$$

Therefore what we need to do is to classify (generic) maps f, modulo the equivalence relation $f \equiv v \circ f \circ u$, where v is arbitrary and u is triangular, whereas in Sect. 7 both u and v were allowed to be arbitrary. We thus have a finer equivalence relation and the classification established in Theorem 8.4 (ordinary points, folds, cusps) is further subdivided. The situation is very similar to that in Sect. 7 where we classified (generic) maps from $\mathbf{R}^3$ into $\mathbf{R}$ but under an equivalence that was finer than the one given by all local diffeomorphisms of $\mathbf{R}^3$: there we were restricted to those diffeomorphisms which respected the projection of $\mathbf{R}^3$ into $\mathbf{R}^2$ ('systems with adapted coordinates').

To return to our problem, note without proof the following results. In a neighbourhood of a regular point of the envelope there are three possible local forms as given in the list below. For each of these forms the point considered is $(0, 0)$ obtained for $t = 0$ and $\lambda = 0$, and the envelope has for its local equation $y = 0$:

$$x = (t + \lambda)^2, \quad y = t;$$
$$x = t^3 + \lambda t + \lambda, \quad y = t^2;$$
$$x = (t^2 + \lambda)^2, \quad y = t.$$

In a neighbourhood of a cusp point of the envelope there is on the other hand an infinite number of local forms.

4.10 Caustics

In what follows we shall be working in euclidean space E with norm denoted $\|\xi\|$ and scalar product $(\xi \mid \eta)$. We consider a hyperspace $H \subset E$ and shall find it convenient to use a unit normal vector field n on H; this is a C^∞ map from H to E such that for every $x \in H$ the vector $n(x)$ has norm 1 and is orthogonal to the tangent hyperplane $T_x H$.

4.10.1. We fix a C^∞ function τ on H which for simplicity we assume to be given as the restriction of a C^∞ function with the same name on an open subset of E containing H. The three cases that will interest us are the following:

a) $\tau(x) = 0$ ('emission'),
b) $\tau(x) = \|x\|/\nu$ with $\nu \geq 1$ ('refraction'),
c) $\tau(x) = -\|x\|$ ('reflection').

In the last two cases we assume that H does not pass through the origin and that τ is of class C^∞ on H.

Consider the map $f : H \times E \to E \times \mathbf{R}$ given by

$$f(x, y) = (y, \tau(x) + \|y - x\|). \tag{4.10.1}$$

In the three cases above we recognize in the second component of f the path length of a light ray arriving at the point y of E and emitted by the point x of H (case a)), issuing from the origin and refracted by H at the point x with refractive index ν (case b)), or issuing from the origin and reflected by H at the point x (case c)).

4.10.2. Let us calculate the critical locus C of f. For $x \in H$, $y \in E$, $\xi \in T_x H$ and $\eta \in E$ we have

$$(T_{x,y}f)(\xi, \eta) = \left(\eta, (\operatorname{grad} \tau(x) \mid \xi) + \frac{(x - y \mid \xi - \eta)}{\|x - y\|} \right). \qquad (4.10.4)$$

For (x, y) to be a critical point of f it is necessary and sufficient that the vector $\operatorname{grad} \tau(x) + (x - y)/\|x - y\|$ be orthogonal to $T_x H$. This requires first of all that $\|\operatorname{grad} \tau(x)\|$ be ≤ 1; suppose this holds at every point x of H, and let v denote a unit vector on H such that $v - \operatorname{grad} \tau$ is normal. Then the condition above can be written $(y - x)/\|y - x\| = v(x)$. In other words, C is the image of the map $(x, t) \mapsto (x + tv(x), \tau(x) + t)$ from $H \times \mathbf{R}$ to $E \times \mathbf{R}$, and we see the family of straight lines $D_x = \{x + tv(x) \mid t \in \mathbf{R}\}$ appear.

Before continuing, let us look at the three particular cases indicated above. In case a) we have $v = n$ and D_x is the normal to H at x. In cases b) and c) the vector $v(x)$ is given by *Descartes' laws*[30]. To see this, let $u(x)$ denote the unit vector $x/\|x\|$ (the 'direction vector of the incident ray'). Then we have $\operatorname{grad} \tau = \alpha u$ with $\alpha = 1/\nu$ in case b) or $\alpha = -1$ in case c); thus $v(x)$ is defined by the fact that it is of unit length and has the form $v(x) = \alpha u(x) + \beta n(x)$. In case c) we have that $u(x) + v(x)$ is normal and $v(x)$ is indeed the direction vector of the reflected ray. In case b) we see that $v(x)$ is in the plane generated by $u(x)$ and $n(x)$; we can write

$$u(x) = (\sin i)w + (\cos i)n(x), \quad v(x) = (\sin r)w + (\cos r)n(x),$$

where w is the unit vector in the direction of the orthogonal projection of $u(x)$ into $T_x H$. Then to say that $v(x) - u(x)/\nu$ is normal gives the usual relation $\sin i = \nu \sin r$.

We have thus just verified that *Fermat's Principle* (minimization of optical path length) implies Descartes' laws.

4.10.3. The envelope of the family of straight lines D_x (the *caustic*) is obtained from the map $f : H \times E \to E \times \mathbf{R}$ by the following procedure: take the critical locus C of f, then the restriction $f|_C : C \to E \times \mathbf{R}$, then the map $C \to E$ which is the composition of $f|_C$ and the projection p from $E \times \mathbf{R}$ onto E, and finally the set of critical values of $p \circ f|_C$. If E has dimension n, then the source of f has dimension $2n - 1$ (hence 5 for $n = 3$) and its target

---

[30]Known to anglophones as Snell's Law.

has dimension $n + 1$ (hence 4 for $n = 3$). Even if we restrict ourselves to a planar problem ($n = 2$) we have $2n - 1 = 3$ and $n + 1 = 3$ and we are dealing with a map from $\mathbf{R}^3$ into $\mathbf{R}^3$ and not, as in the case of an ordinary envelope, from $\mathbf{R}^2$ into $\mathbf{R}^2$. This explains why *generic singularities of caustics are more complicated than ordinary simple double points or cusp points*. For more details, see the copiously illustrated account [BN], or see [BG] or [AG].

The above argument is open to objection on the grounds that f is not an arbitrary map from $\mathbf{R}^{2n-1}$ to $\mathbf{R}^{n+1}$ but has a particular form. In fact the objection does not stand. To see this, let f be a generic map from $\mathbf{R}^{2n-1}$ into $\mathbf{R}^{n+1}$. By Proposition 2.3.1, the set of linear maps of rank $< n$ from $\mathbf{R}^{2n-1}$ to $\mathbf{R}^{n+1}$ has codimension

$$(2n - 1 - (n - 1))(n + 1 - (n - 1)) = 2n > \dim \mathbf{R}^{2n-1};$$

likewise the set of linear maps of rank n has codimension $n - 1$. The Transversality Theorem therefore implies that f is everywhere of rank $\geq n$ and its critical locus C is a submanifold of dimension n. Consider a generic linear projection p from $\mathbf{R}^{n+1}$ into $\mathbf{R}^n$. Then $f \circ p$ is a submersion and we can locally identify $\mathbf{R}^{2n-1}$ with $\mathbf{R}^{n-1} \times \mathbf{R}^n$ so that $p \circ f(x, y) = y$. Similarly identifying $\mathbf{R}^{n+1}$ with $\mathbf{R}^n \times \mathbf{R}$ so that $p(y, z) = y$ we obtain $f(x, y) = (y, \tau(x, y))$, where τ is a generic map from $\mathbf{R}^{2n-1}$ to $\mathbf{R}$. But then we are in exactly the situation that we would have arrived at by a small perturbation of the theoretical model studied above.

4.11 Genericity and Stability

4.11.1. We now return to a theme outlined in the Introduction. Suppose for example that a mathematical model of a given physical system leads to a description in terms of a map from $\mathbf{R}^p$ to $\mathbf{R}^q$. Let E denote the set of C^∞ maps from $\mathbf{R}^p$ into $\mathbf{R}^q$, so that the system (Σ) under consideration is described by an element f_Σ of E. For nearby systems there will be corresponding nearby points of E (for a suitable topology, such as envisaged in Sect. 3.9). Moreover, there will in general be some notion of admissible change of coordinates providing an equivalence relation on E, often of the type considered in 3.3. We say that an element f of E is *stable* (relative to the chosen topology and equivalence) if every element of E sufficiently close to f is equivalent to f. By definition, the stable elements form an open set.

4.11.2. Following this idea, we soon arrive at an apparent contradiction. On the one hand, as already noted, it seems unlikely that we could in practice observe features of (Σ) which are not equally evident in nearby systems, and so it is natural to suppose that the element f_Σ is stable. On the other hand, there is no reason why the model chosen for the system (Σ) should actually provide a stable element f_Σ. In fact it is usually the opposite, since we tend to introduce simplifying assumptions such as symmetry into the model which automatically create situations that are not stable.

This contradiction is not so serious. In fact we do not have to assume that f_Σ is stable, but only that there exist stable elements as close to it as we wish. This will essentially imply that the real system (Σ') can be as close to (Σ) as we wish and effectively observable. This discussion leads naturally to the following question: is the open set of stable points of E a dense set? Or, in our usual terminology, *is a generic map from $\mathbf{R}^p$ to $\mathbf{R}^q$ stable?*

4.11.3. When p and q are small enough all goes well. Thus, the stable C^∞ maps f from $\mathbf{R}^n$ to $\mathbf{R}$ form an open dense set. They are characterized by the following two properties:

a) the critical points of f are Morse points,
b) the values of f at its critical points are pairwise distinct.

It is not hard to see that if a map f satisfies these two conditions then every map sufficiently close to f can be written as $v \circ f \circ u$ where u is a diffeomorphism of $\mathbf{R}^n$ and v is a diffeomorphism of $\mathbf{R}$. Moreover, a generic map satisfies a) and b) (this is a consequence of Sect. 4 for a) and of the Multijet Transversality Theorem for b)).

4.11.4. Likewise, the stable C^∞ maps f from $\mathbf{R}^2$ to $\mathbf{R}^2$ form an open dense set. They are characterized by the following properties:

a) the critical locus of f is a (regular) curve;
b) the critical points of f are fold points or cusp points; the cusps form a discrete set;
c) three distinct critical points never have the same image; if two critical points have the same image then they are folds, and the tangents to the critical locus at these points have different projections.

Note that the set of critical values of a stable map has the structure as described several times in previous sections (Propositions 7.8, 8.8 and 9.2).

4.11.5. These positive results (due to Morse (1931) for 11.3 and to Whitney (1955) for 11.4) suggest that the general situation is just as nice as these special cases. Unfortunately, this is not what happens. When p and q are large enough it is no longer true that the stable maps from $\mathbf{R}^p$ to $\mathbf{R}^q$ are dense. This is due in fact to two simultaneous phenomena: on the one hand the equivalence relation associated to the action of diffeomorphisms is too fine, and on the other hand the spaces considered are not compact.

Let us begin with the first difficulty. If f and g are two (C^∞) maps from a manifold X into a manifold Y, we say that f and g are *topologically conjugate* if there are homeomorphisms u and v of X and Y respectively such that $g = v \circ f \circ u$. This is clearly an equivalence relation. We say that the map f is *topologically stable* if every map g sufficiently close to f (in the C^∞ topology) is topologically conjugate to f. With this we now can hope that

a) the topologically stable maps form a dense subset (it is open by definition);

b) there is only a finite number of topological conjugacy classes.

These two hopes turn out to be too optimistic. In fact, for p and q large enough it is not true that the topologically stable maps from $\mathbf{R}^p$ to $\mathbf{R}^q$ are dense; the counterexample given by Thom in 1962 can be found in [BL], pp.85-91. For the finiteness, things are not much better: the same counterexample shows that polynomial maps of degree k from $\mathbf{R}^p$ to $\mathbf{R}^q$ form infinitely many topological conjugacy classes for k, p and q large enough; we know now that the same phenomenon occurs as soon as k, p and q are ≥ 3.

4.11.6. Fortunately, these difficulties disappear under suitable topological assumptions. If X is taken to be *compact* (or, more generally, if we restrict ourselves to *proper* maps) then the two properties a) and b) above do hold. They are consequences of a very deep theorem hoped for by Thom and whose long-awaited proof was due to Mather[31]. This proof can be found in [MR] (see also [GW]). Here are some extracts from the introduction to Mather's paper:

"...I apologize to mathematicians for the long delay between the announcement and this proof...The ideas of Thom had an enormous influence in this work...I discussed these problems a great deal with Thom when I was in France. However, Thom was never inclined to work out the details of his theory, and this is what I have done here. In addition, I have contributed several ideas of my own..."

---

[31] John MATHER, American mathematician, born 1942.

5 Catastrophe Theory

5.1 Introduction

Thom's theory of catastrophes essentially consists of the study of C^∞ *families of functions* on a manifold, and in particular the variation of their critical points. In the most common applications we are concerned with potentials depending on a finite sequence of control parameters and we study the bifurcation of their equilibrium states. For the reasons given in the Introduction, we are particularly interested in *stable* families. Moreover, what we want to do essentially is to carry out a *local* study in the neighbourhood of an equilibrium point and of a given value of the parameters. Specifically, we would like to achieve a *classification* of the simplest and most common 'catastrophes'.

This local study can in fact be made totally algebraic, thanks to the results of a sequence of difficult papers by Mather. As a start we observe that in many cases a function is determined ("up to local diffeomorphism") by its Taylor expansion to a sufficiently high order; this is what we have already seen, for example, in the case of noncritical points or nondegenerate critical points (for which, as we have also seen, there are no bifurcation phenomena).

This notion of *sufficiency* of a polynomial is of major importance for mathematical modelling. If the Taylor polynomial of a function f up to order r is 'r-sufficient' we may replace the study of the function by the study of this polynomial; in so doing we are not making a 'small error' (there may be no reason why this could not have considerable – even catastrophic – consequences), but we are making no error whatsoever! This replacement simply amounts to a (small) change of coordinates. Take the Morse Lemma, for example: the fact that $f(x, y)$ can be written as $x^2 + y^2 +$'higher order terms' in the original coordinates does not *a priori* imply very much about the curves $f^{-1}(\varepsilon)$; on the other hand, the fact that f can be written as $x^2 + y^2$ in suitable curvilinear coordinates assures us that the $f^{-1}(\varepsilon)$ are concentric circles in these coordinates. Using this language the Morse Lemma can be stated as follows: the 2-variable polynomial $x^2 + y^2$ is 2-sufficient.

What is particularly pleasant is that the sufficiency of a polynomial can be recognised by a simple algebraic criterion. This criterion involves the partial derivatives of the polynomial and, more precisely, the ideal (*Jacobian* ideal) that they generate. Analogously, we have a criterion, also expressed in terms of partial derivatives, enabling us to recognise if a *deformation* of a function (that is a family of functions containing it as one particular member) is sufficiently general for us to be able to recover, by simple modifications, all of the deformations of this function (in which case we say it is *universal*).

A first measure of the complexity of a function f at a critical point is the *Milnor number* $\mu(f)$; it is the codimension of the Jacobian ideal $J(f)$, or the number of parameters in a universal deformation. In the specialist literature the term *singularity* is often used instead of critical point, and the *codimension* of a singularity is its Milnor number minus 1. The origin of this discrepancy is the fact that in the interpretation of functions as potentials the addition of a constant is regarded as a trivial deformation. Thus the codimension of a nondegenerate critical point is zero. The *elementary catstrophes* are the singularities with codimension between 1 and 4, thus with $2 \leq \mu(f) \leq 5$. There are seven of them. For $\mu(f) = 2$ there is the *fold*; for $\mu(f) = 3$ there is the *cusp*; after that the *swallowtail*, *umbilics* and *butterfly* make an appearance.

In fact throughout the whole of this story the dimension of the original manifold, that is the number of state variables, plays practically no role at all. The reason is that the addition to a potential of the square of a new variable is an operation that is totally neutral. The Decomposition Theorem which we saw in the previous chapter (a weaker version is known in mechanics as the 'Lyapunov-Schmidt principle') allows us to get rid of redundant variables and systematically reduce the problem to critical points whose Hessian is identically zero. The elementary catastrophes therefore involve at most two state variables.

The interpretation of physical bifurcation phenomena in terms of catastrophes requires an additional piece of information concerning the mechanism whereby the equilibrium states jump from one position to another. We shall give two examples of this, for a simple mechanical system and for liquid-vapour equilibrium, both exhibiting cusps.

Sects. 2,3 and 4 contain the specialised terminology (germs, jets, sufficiency and determinacy, Jacobian ideal). The two key criteria of Mather (sufficiency and universality) are stated without proof in Sects. 5 and 6. We then come to grips with catastrophe theory proper : the principles set out in Sect. 7 are made explicit in the case of cusps in Sect. 8 and illustrated in Sects. 9 and 10 by the examples already quoted. The list of elementary catastrophes is established in Sect. 11. The chapter ends with some historical glimpses.

5.2 The Language of Germs

5.2.1. Let $n > 0$ be an integer. In what follows we shall be interested in C^∞ functions defined on a neighbourhood of a given point in $\mathbf{R}^n$. We certainly lose no generality in assuming that this point is the origin, as the general case is deduced from this by translation. If f and g are two C^∞ functions with f defined on an open set U containing 0 and g defined on an open set V containing 0 we say that f and g *have the same germ* at 0 if they coincide on an open set W with $0 \in W \subset U \cap V$. See Fig. 5.1. We denote the set of these function germs by $\mathcal{E}_n$.

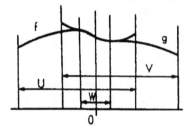

Fig. 5.1. Functions f and g with the same germ at 0

In formal terms, $\mathcal{E}_n$ is thus the quotient of the set of such functions by the given equivalence relation.

We often use the same letter to denote both the function and its germ at 0; thus x_i, $i = 1, \ldots, n$ denote the germs of the coordinate functions. It is important not to forget, however, that if a is a point of $\mathbf{R}^n$ and ϕ is a germ then the expression $\phi(a)$ has no meaning except when $a = 0$. Whenever it is necessary to distinguish between germs and functions we shall say that the function is a *representative* of the germ.

5.2.2. Under the operations derived from the sum and product of functions, $\mathcal{E}_n$ is a commutative ring whose identity element is the germ of the constant function 1. We have $fg(0) = f(0)g(0)$ and hence $f(0) \neq 0$ for every invertible element f. Conversely, every element f of $\mathcal{E}_n$ with $f(0) \neq 0$ is invertible, since a representative of f in a sufficiently small open set vanishes nowhere.

Let m_n denote the set of germs f with $f(0) = 0$. Hadamard's Lemma 4.2.3 may be stated by saying that the elements of m_n are precisely those elements of $\mathcal{E}_n$ that can be written as $\sum g_i x_i$. More generally, for every integer $r \geq 0$ we denote by m_n^{r+1} the set of germs of functions that vanish r times at 0, that is all of whose derivatives of order $\leq r$ at 0 are zero. As in 4.2.4, we prove by iterating Hadamard's Lemma that in order for a germ f to belong to m_n^{r+1} it is necessary and sufficient that it can be written as a sum

$$\sum_{\alpha} x_1^{\alpha_1} \cdots x_n^{\alpha_n} f_\alpha,$$

where α runs through all the multi-indices of total degree $r+1$ and where the f_α are suitable elements of $\mathcal{E}_n$. We thus obtain a purely algebraic description of the subset m_n^{r+1} of $\mathcal{E}_n$.

5.2.3. As in Sect. 3.8 we denote by $j^r f$ the jet of order r of f at 0, that is the polynomial of degree $\le r$ in the x_i given by the Taylor expansion of f up to order r. For example, we have $j^0 f = f(0)$ and

$$j^1 f = f(0) + f'(0) \cdot x = f(0) + \sum_{i=1}^{n} \frac{\partial f}{\partial x_i}(0) x_i.$$

If f has a critical point at 0 we have $j^2 f = f(0) + H_f(0)(x)$ (see Proposition 4.4.3), and so on.

Every germ f can be written as $f(0) + g$ with $g \in \mathrm{m}_n$. More generally, let $f \in \mathcal{E}_n$ be a germ and let $r \ge 0$ be an integer. The Taylor formula up to order r can be written

$$f = j^r f + R^r f,$$

where the remainder $R^r f$ belongs to m_n^{r+1}. In fact $j^r f$ is the unique polynomial p of degree $\le r$ such that $f - p$ belongs to m_n^{r+1}: this is immediate since 0 is the only polymomial of degree $\le r$ which belongs to m_n^{r+1}. This provides us with a purely algebraic description of Taylor's formula: in $\mathcal{E}_n$ the polynomials of degree $\le r$ form a linear subspace P_r complementary to m_n^{r+1}. In particular, the latter is of finite codimension; more precisely we have

$$\mathrm{codim}(\mathrm{m}_n^{r+1}) = \dim(P_r) = \binom{n+r}{r} = (n+1) \cdots (n+r)/r! \ .$$

Note that for $s < r$ the s-jet $j^s f$ is obtained from $j^r f$ by suppressing the terms of degree $> s$.

We could introduce the infinite order jet $j^\infty f$; it is a formal series that enables us to reconstruct all the $j^r f$. Take care to avoid confusion: being given the infinite order jet is not as much as being given the germ. In fact there exist functions (said to be *flat* at 0) whose infinite order jet at 0 is zero but which are not identically zero as functions in any neighbourhood of 0 (consider $\exp(-x^{-2})$).

5.3 r-sufficient Jets; r-determined Germs

5.3.1. Let f and g be two germs in $\mathcal{E}_n$; as in 4.3.3 we say that f and g are *equivalent* if there exists a local diffeomorphism u of $\mathbf{R}^n$ defined in a neighbourhood of 0 with $u(0) = 0$ and $g = f \circ u$, that is (still using f and g to denote suitable representatives of germs with the same names)

$$g(x_1, \ldots, x_n) = f(u_1(x_1, \ldots, x_n), \ldots, u_n(x_1, \ldots, x_n)).$$

This is indeed an equivalence relation, as may immediately be verified. It is also immediate that the following conditions are logically equivalent:

(i)   the germs f and g are equivalent,
(ii) $f(0) = g(0)$ and the germs $f - f(0)$ and $g - g(0)$ are equivalent.

In fact, we could here introduce germs of diffeomorphisms of $\mathbf{R}^n$ that fix the origin; they form a group that acts on $\mathcal{E}_n$ and the equivalence relation above is simply the relation associated to this group action.

Various examples that we have already met and which we shall recall below suggest the following definitions:

Definition 5.3.2 Let $r > 0$ be an integer.

a) We say that a polynomial P of degree $\leq r$ is *r-sufficient* if every germ g such that $j^r g = P$ is equivalent to P.
b) We say that a germ f is *r-determined* if $j^r f$ is r-sufficient.

This last definition may be expressed a little differently: to say that f is r-determined means that every germ g such that $j^r g = j^r f$ is equivalent to the germ f. Indeed, if f is r-determined then f and g are equivalent to $j^r f$ and are therefore equivalent to each other. Conversely, the condition above implies first of all that f and $j^r f$ are equivalent, and then that every germ g with $j^r g = j^r f$ is equivalent to $j^r f$. The same type of reasoning shows that if f is r-determined then it is r'-determined for $r' > r$. Also, to verify the sufficiency of a polynomial P (or the determinacy of a germ f) it is always possible to reduce to the case where $P(0) = 0$ (or $f(0) = 0$). Note finally that in the definition a) we have not required that P be of degree r; that is why the 'order of sufficiency' has to be mentioned explicitly.

For example, we shall see below that the polynomial $x^5 + y^5$ of degree 5 in two variables is 6-sufficient but is not 5-sufficient.

We have already met several particular cases of the above definition. We already noted as a corollary to the Local Inversion Theorem that every nonzero linear form is 1-sufficient (and likewise that every germ f with

$f'(0) \neq 0$ is 1-determined). In the same way, the Morse Lemma (Proposition 4.5.1) means that every nondegenerate quadratic form is 2-sufficient (or that every germ with zero derivative at 0 and nondegenerate Hessian form at 0 is 2-determined). Finally, 4.2.5 means that the 1-variable polynomial x^r is r-sufficient.

The germ f is said to be *finitely determined* if there exists an integer r such that f is r-determined.

5.3.3. The notion of equivalence can be introduced in a slightly different way. If U is an open set (or a submanifold) of a finite-dimensional vector space E, the ring A of germs of C^∞ functions at a point a of U is defined just as in the case of $\mathbf{R}^n$. For each system $(x_1, \ldots, x_n)$ of local coordinates on U centred at a, the elements of A may be expressed in these coordinates thus giving a ring isomorphism $A \to \mathcal{E}_n$. To say that two elements of $\mathcal{E}_n$ are equivalent then means that they are expressions for the same germ in two systems of local coordinates.

There is technically convenient variant of the above notions. First, in the situation of 3.1, we say that f and g are *strongly equivalent* if the local diffeomorphism u can be chosen to be tangent to the identity, that is such that $j^1(u_i) = x_i$. We then define *strongly r-sufficient polynomials* and *strongly r-determined germs* by replacing equivalence by strong equivalence in Definition 3.2. The three examples given above satisfy the strong conditions.

5.4 The Jacobian Ideal

In this section we shall see some algebraic criteria enabling us to verify if a germ is r-determined. These criteria use a fundamental construction, namely the Jacobian ideal associated to a germ.

5.4.1. We call the *Jacobian ideal* of the germ $f \in \mathcal{E}_n$ the set $J(f)$ of germs of the form $\sum g_i \partial f / \partial x_i$. It is a vector subspace of $\mathcal{E}_n$; its codimension (finite or infinite) $\mu(f)$ is called the *Milnor[32] number* of f. By definition, the Milnor number $\mu(f)$ is therefore the maximum number of germs linearly independent modulo $J(f)$: to say that an integer s is $\leq \mu(f)$ means that it is possible to find germs $f_1, \ldots, f_s$ having the following property: every relation of the form

$$\sum_{i=1}^{s} \lambda_i f_i + \sum_{j=1}^{n} g_j \frac{\partial f}{\partial x_j} = 0, \tag{5.4.1}$$

where the λ_i are real numbers and the g_i are germs, implies that the λ_i are all zero. If $\mu(f)$ is finite, we likewise introduce the notion of *basis modulo $J(f)$*;

---

[32]John MILNOR, American mathematician, born 1931, Fields medal 1962.

this means a system of $\mu(f)$ germs that are linearly independent modulo $J(f)$, or a basis for a subspace complementary to $J(f)$.

Naturally, all this can be translated immediately into the language of quotient spaces.

It will turn out as a consequence of results to come later that a germ f is finitely determined precisely when $\mu(f)$ is finite. We can be even more precise: if $\mu(f)$ is finite then f is $(\mu(f) + 1)$-determined.

5.4.2. Note that $J(f)$ "does not depend on the choice of coordinates". To be slightly more precise, let $(y_1, \ldots, y_n)$ be a system of local coordinates on $\mathbf{R}^n$ centred at 0, in which the germ f can be written

$$f(x_1, \ldots, x_n) = g(y_1, \ldots, y_n).$$

By differentiating the composition of functions we obtain

$$\frac{\partial g}{\partial y_i}(y_1, \ldots, y_n) = \sum \frac{\partial x_j}{\partial y_i} \frac{\partial f}{\partial x_j}(x_1, \ldots, x_n),$$

so that the $\partial g / \partial y_i$ belong to $J(f)$, and therefore $J(g)$ — "after replacing the y's by their values in terms of the x's"— is contained in $J(f)$. The reverse inclusion is obtained by considering the inverse change of coordinates. In particular, *two equivalent germs have the same Milnor number.*

We give an elementary example. The one-variable germ x^{r+1} has as its Jacobian ideal the set of multiples of x^r, that is the set of functions that vanish to order r at 0. We have already remarked that this subspace is complementary to the space of polynomials of degree $< r$, which is of dimension r. We therefore have $\mu(x^{r+1}) = r$. More generally, we have $\mu(f) = r$ for every one-variable germ which is of order exactly $r + 1$ at 0.

The Milnor number is the first measure of complexity of a singularity. The simplest cases ($\mu = 0$ and $\mu = 1$) are known to us already:

Proposition 5.4.3.

a) *The following conditions are equivalent:*
   0 is a critical point of f ; $J(f) \neq \mathcal{E}_n$; $J(f) \subset \mathfrak{m}_n$; $\mu(f) > 0$.
b) *The following conditions are equivalent:*
   0 is a nondegenerate critical point of f ; $J(f) = \mathfrak{m}_n$; $\mu(f) = 1$.

a) If the derivative $f'(0)$ is nonzero then one of the partial derivatives of f at 0 is nonzero, say the ith one; then $\partial f / \partial x_i$ is invertible in $\mathcal{E}_n$ and every element g of $\mathcal{E}_n$ can be written as $g = (g(\partial f / \partial x_i)^{-1})\partial f / \partial x_i$ and therefore belongs to $J(f)$; thus $J(f) = \mathcal{E}_n$ and $\mu(f) = 0$. On the other hand, if all the derivatives of f at 0 are zero then $J(f) \subset \mathfrak{m}_n$ by definition and so $\mu(f) > 0$.

b) If 0 is a nondegenerate critical point of f then f is equivalent to $g = \sum \pm x_i^2$ according to the Morse Lemma. But $J(g) = \mathfrak{m}_n$, so $J(f) = \mathfrak{m}_n$ and $\mu(f) = 1$. Conversely, if we have $\mu(f) = 1$ then 0 is a critical point of f and we have $J(f) = \mathfrak{m}_n$ by a). We can therefore write for each i

$$x_i = \sum_{j=1}^{n} a_{i,j} \frac{\partial f}{\partial x_j}.$$

Expanding the right hand side to first order we see that the matrix $(a_{i,j}(0))$ is inverse to the Hessian matrix of f at 0, and so 0 is nondegenerate.   □

For the next cases we shall use two lower bounds for the Milnor number.

Proposition 5.4.4.

a) *If* $f \in \mathfrak{m}_n^{r+1}$ *then* $n + \mu(f) \geq (n+1) \cdots (n+r)/r!$.
b) *If* f *has a critical point of corank* m *at* 0 *then* $\mu(f) > m(m+1)/2$.

a) Consider the vector space P_r, of dimension $d = (n+1) \cdots (n+r)/r!$, and the linear map $j^r : \mathcal{E}_n \to P_r$. It is surjective. Let V denote the vector subspace of P_r generated by the n elements $j^r(\frac{\partial f}{\partial x_i})$. We have $j^r(J(f)) \subset V$; indeed, if $g \in J(f)$ we can write $g = \sum g_i \frac{\partial f}{\partial x_i}$. However, since the $j^r(\frac{\partial f}{\partial x_i})$ are homogeneous of degree r this implies that

$$j^r(g) = \sum g_i(0) j^r(\frac{\partial f}{\partial x_i}) \in V.$$

Consequently

$$\mu(f) = \operatorname{codim}_{\mathcal{E}_n}(J(f)) \geq \operatorname{codim}_{P_r}(V) \geq d - n.$$

b) By the Splitting Lemma (Corollary 4.5.5), the germ f is equivalent to a germ f' of the form

$$f' = a \pm x_1^2 \pm \cdots \pm x_{n-m}^2 + h(x_{n-m+1}, \ldots, x_n),$$

where h is of order ≥ 3. Since $\mu(f) = \mu(f')$, it is enough to consider the case of f'. We then argue as in a), considering the linear map

$$u \mapsto j^2 u(0, \ldots, 0, x_{n-m+1}, \ldots, x_n)$$

from $\mathcal{E}_n$ into the space of polynomials of degree ≤ 2 in m variables. We obtain

$$\mu(f) = \mu(f') \geq \frac{1}{2}(m+1)(m+2) - m = \frac{1}{2}m(m+1) + 1$$

which gives the result.   □

In certain cases it is possible to calculate Milnor numbers using the following observation. Suppose that f can be written as

$$f(x_1, \ldots, x_n) = g(x_1, \ldots, x_m) + h(x_{m+1}, \ldots, x_n);$$

let $\mu(g)$ denote the Milnor number of the germ g in m variables, and let $\mu(h)$ be the Milnor number of the germ h in $n - m$ variables. Then $\mu(f) = \mu(g)\mu(h)$ (exercise). In the case dealt with above, for example, we had $\mu(g) = 1$ and so $\mu(f') = \mu(h)$.

In particular this multiplicative property of the Milnor number implies

$$\mu(x_1^{r_1+1} + \cdots + x_n^{r_n+1}) = r_1 \cdots r_n$$

which can also be proved directly by showing that the monomials $x_1^{\alpha_1} \cdots x_n^{\alpha_n}$ for $\alpha_i < r_i$ form a basis for a subspace complementary to the Jacobian ideal.

Once again let f denote a representative of the germ f. Suppose $\mu(f) > 0$; then 0 belongs to the critical locus of f.

Proposition 5.4.5. *If $\mu(f)$ is finite then 0 is an isolated point of the critical locus of f.*

Let x be one of the coordinate functions. The monomials x^i, $i > 0$, which are infinite in number, cannot be linearly independent modulo $J(f)$. Therefore in a neighbourhood of 0 there exists a relation of the form

$$\sum_{j=1}^{r} \lambda_j x^j + \sum_{i=1}^{n} g_i \frac{\partial f}{\partial x_i} = P(x) + g(x) = 0,$$

where the λ_i are not all zero. Let a be a critical point sufficiently close to 0. Then, since the $\partial f/\partial x_i(a)$ are zero we have $g(x(a)) = 0$ and so $P(x(a)) = 0$. However, as the polynomial P vanishes at 0 and is not identically zero it follows that $x(a)$ must be zero if it is sufficiently small.   □

The converse of this result is false. It would be true if taken in the complex setting rather than in the real setting that we have been considering up to now.

5.5 The Theorem on Sufficiency of Jets

5.5.1. To be able to state the next theorem in a convenient way we need a little algebraic terminology. Recall that an *ideal* of the ring $\mathcal{E}_n$ is by definition a subset I of $\mathcal{E}_n$ satisfying the following two conditions:

a) $(f \in I$ and $g \in I)$ implies $(f + g \in I)$,
b) $(f \in \mathcal{E}_n$ and $g \in I)$ implies $(fg \in I)$.

An ideal of $\mathcal{E}_n$ is in particular a vector subspace. The smallest ideal containing the elements $g_1, \ldots, g_n$, which we call the ideal *generated* by these elements, consists of those $f \in \mathcal{E}_n$ that can be written in the form $\sum f_i g_i$. Thus m_n is the ideal generated by the x_i; likewise $J(f)$ is the ideal generated by the partial derivatives of f.

If I and J are two ideals, the vector subspace $I + J$ consisting of sums $u+v$ with $u \in I$ and $v \in J$ is the smallest ideal containing I and J. Similarly, we let IJ denote the ideal consisting of finite sums of elements of the form uv for $u \in I$ and $v \in J$; we have $IJ \subset I \cap J$; if I is generated by $\{g_i\}$ and J by $\{h_j\}$ then IJ is generated by the products $g_i h_j$. This allows us to define the powers $I^2, I^3, \ldots$ of an ideal I. For example, m_n^{r+1} is just the ideal m_n to the power $r + 1$. By convention, I^0 is taken as $\mathcal{E}_n$, so that we have $I^r I^s \subset I^{r+s}$ for every pair of non-negative integers r and s.

5.5.2. We shall meet below several examples of conditions of the form

$$m_n^s \subset I + m_n^{s+1}, \tag{5.5.1}$$

where I is an ideal of $\mathcal{E}_n$ and $s \geq 0$ is an integer. This can be translated in several ways. The first way, which goes straight back to the definitions, is as follows: every monomial of total degree s in the x_i can be written as the sum of an element in I and a germ which vanishes to order $s + 1$ at 0. The second translation has the advantage of showing that in fact everything is taking place in the finite-dimensional space P_s of polynomials of degree $< s + 1$: we write each element f of I in the form $j^s + R^s f$ with $j^s f \in P_s$ and $R^s f \in m_n^{s+1}$. We then look at the subspace of P_s generated by the $j^s f$, and the condition means that it contains all the polynomials of degree s. As far as calculations are concerned, this means that by neglecting Taylor remainders at order s and setting the elements of I equal to zero we annihilate all the monomials of total degree s. One way or the other, it is clear that we may multiply both sides of (5.1) by a power of m_n: for every integer $t > 0$ the condition (5.1) implies

$$m_n^{s+t} \subset m_n^t I + m_n^{s+t+1}. \tag{5.5.2}$$

In fact the condition (5.1) is equivalent to the apparently stronger condition: $m_n^s \subset I$, as follows from Nakayama's Lemma (see [BL] for example). The interest of the form (5.1) is that it is easier to verify since, as we observed, it 'lives' in the finite-dimensional space P_s.

There exist several algebraic criteria for sufficiency, all of rather similar form. The following, due to Mather, is fairly convenient:

Theorem 5.5.3. *Let $f \in \mathcal{E}_n$ be a germ and let $r > 0$ be an integer. If*

$$m_n^{r+1} \subset m_n^2 J(f) + m_n^{r+2}, \tag{5.5.3}$$

then f is r-determined. □

Before giving some examples of applications, we make a few observations about this statement.

1) Note first of all two conditions that are clearly stronger than (5.3) (and which therefore imply r-determinacy of f), namely

$$m_n^r \subset m_n J(f) + m_n^{r+1}, \tag{5.5.4}$$

$$m_n^{r-1} \subset J(f) + m_n^r. \tag{5.5.5}$$

2) It is also worth observing that (5.3),(5.4) and (5.5) do not alter if we replace f by $j^r f$, or more generally by another germ g with $j^r g = j^r f$; indeed, we then have $f = g + h$ with $h \in m_n^{r+1}$ and hence $J(f) + m_n^r = J(g) + m_n^r$.

3) We point out incidentally that the sufficient conditions (5.3) and (5.4) are necessary, apart from a small gap. More precisely, let (D_r) denote the condition "f is r-determined" and (SD_r) the condition "f is strongly r-determined". It can then be shown that (5.3) is equivalent to (SD_r) and that (D_{r-1}) implies (5.4). We therefore have the following chain of implications, where in order to simplify the notation we write $m_n = m$ and $J(f) = J$ (and where the last ideal in each inclusion can be suppressed if wished):

$$\cdots \Rightarrow (D_{r-1}) \Rightarrow (m^r \subset mJ + m^{r+1}) \Rightarrow (m^{r+1} \subset m^2 J + m^{r+2}) \Leftrightarrow$$
$$\Leftrightarrow (SD_r) \Rightarrow (D_r) \Rightarrow \cdots$$

5.5.4. Here are some examples.

a) If 0 is not a critical point of f then $J(f) = \mathcal{E}_n$ and f is 1-determined.

b) If 0 is a nondegenerate critical point of f then $J(f) = m_n$, so (5.5) is satisfied for $r = 2$ and f is 2-determined.

c) For $n = 1$ and f vanishing exactly to order $r > 0$ we have $J(f) = m_n^{r-1}$ and f is r-determined.

In the following examples, we take $n = 2$ and write $m_n = m$, $x_1 = x$ and $x_2 = y$ to simplify the notation.

d) First take $f = j^3 f = x^3 + y^3$. Then $J(f)$ is generated by x^2 and y^2; hence $mJ(f)$ is generated by xx^2, yx^2, xy^2 and yy^2, and is therefore equal to m^3. Thus (5.4) is satisfied for $r = 3$ and f is 3-sufficient. Note that xy does not belong to $J(f) + m^3$, and so (5.5) is not satisfied for $r = 3$.

e) Take $f = x^2 y + ay^r$, with $r > 2$ and nonzero constant $a \in \mathbf{R}$. The ideal $J(f)$ is generated by xy and by $x^2 + ary^{r-1}$. Then $mJ(f)$ contains all the monomials of degree > 2 that are divisible by xy; it also contains $x^3 + arxy^{r-1}$ and hence x^3, and also $x^2 y + ary^r$ and hence y^r. Thus $mJ(f)$ contains m^r and f is r-sufficient.

f) Take $f = x^4 + y^4$. Then $J(f)$ is generated by x^3 and y^3; we deduce from this that $m^2 J(f)$ is equal to m^5 and (5.3) is satisfied for $r = 4$. Thus f is 4-sufficient. Note that $mJ(f) + m^5$ does not contain $x^2 y^2$, and therefore that (5.4) is not satisfied for $r = 4$ but only for $r = 5$.

g) More generally, take $f = x^r + y^r$, with $r > 4$. By the same calculation we have

$$m^{2r-3} = m^{r-2} J(f) \subset m^2 J(f),$$

and f is $(2r-4)$-sufficient. However, the monomial $x^{r-2}y^{r-2}$ does not belong to $mJ(f)$. Hence $mJ(f)$ does not contain m^{2r-4} and, by the converse given above, f is not $(2r-5)$-sufficient. Thus for example $x^5 + y^5$ is 6-sufficient but not 5-sufficient.

5.5.5. The above theorem therefore algebraicises the problem of classification of finitely determined germs of $\mathcal{E}_n$ up to equivalence. There exist germs which are not finitely determined (exercise), but these in a certain sense form an infinite-codimension submanifold of $\mathcal{E}_n$. From this it can be deduced using the Transversality Theorem that in a generic family of functions with a finite number of parameters all the germs are finitely determined.

5.6 Deformations of a Singularity

5.6.1. In our use of terminology from now on we shall be quite casual about mixing up germs and functions that represent them. An element of $\mathcal{E}_n$ will be denoted equally well by f or $f(x)$ or $f(x_1,\ldots,x_n)$. Let $f = f(x) = f(x_1,\ldots,x_n)$ be a germ in n variables. We shall call a p-parameter *deformation* of f (the term *unfolding* of f is also used) a germ $F = F(x,u) = F(x_1,\ldots,x_n,u_1,\ldots,u_p)$ in $n + p$ variables such that $F(x,0) = f(x)$. For example, $F = F(x,(s,t)) = x^3 + sx + t$ is a 2-parameter unfolding of $f = f(x) = x^3$.

This can be expressed another way, using suitable representatives for the germs. Consider open sets E of $\mathbf{R}^n$ and U of $\mathbf{R}^p$, each containing the origin, such that f is defined on E and F is defined on $E \times U$. For $x \in E$ and $u \in U$ we put $F_u(x) = F(x,u)$; this defines a family of functions parametrized by u. We take E small enough so that $F_0 = f$.

The situation of main interest to us is the case when f has a *critical point with finite Milnor number* at the origin and which is therefore isolated (4.5). We could clearly have chosen E small enough so that 0 is the unique critical point of f in E. Each of the functions F_u has (possibly) critical points in E and we wish to describe how they vary with u. The first remark is that, for u small enough, *these critical points are finite in number, there being fewer than* $\mu(f)$ *of them*. More precisely, for each value $u \in U$ of the parameters and each point $x \in E$ we consider the germ of F_u at x, which we shall denote by (F_u,x), and its Milnor number $\mu(F_u,x)$. To say that this number is nonzero means that x is a critical point of F_u. The finiteness property just stated is a consequence of the following more precise result, which we assert without proof:

Proposition 5.6.2. *If E and U are small enough then for every $u \in U$ we have*

$$\sum_{x \in E} \mu(F_u,x) \le \mu(f). \qquad \square$$

If we happen to be working over the field of complex numbers (and not, as we have been up to now, over $\mathbf{R}$) then there are two additions to this result:

a) there is equality in the formula above;
b) for a generic deformation and generic u all the $\mu(F_u, x)$ are equal to 0 or 1.

Thus we cause the critical point 0 of f to 'explode' into nondegenerate critical points. This gives a nice interpretation of the Milnor number, which directly generalizes the interpretation of the multiplicity of a root of a polynomial as the number of simple roots to which it gives birth under a small deformation.

5.6.3. Let us return to the definitions, and let $F = F(x, u)$ be a p-parameter deformation of $f = f(x)$. Suppose we are given

a) an integer $q \geq 0$,
b) a family $y(x, v)$ of n germs $y_i = y_i(x, v)$ in the $n + q$ variables $x_1, \ldots, x_n$ and $v_1, \ldots, v_q$ with $y(x, 0) = x$, that is to say $y_i(x, 0) = x_i$,
c) a family $w(v)$ of p germs $z_i(v)$ in q variables $v_1, \ldots, v_q$ with $w(0) = 0$.

We can then form the $(n + q)$-variable germ

$$G(x, v) = F(y(x, v), w(v)). \tag{5.6.1}$$

We have $G(x, 0) = F(y(x, 0), w(0)) = F(x, 0) = f(x)$, and so G is a q-parameter deformation of f which we call the *inverse image* of F under the transformation $(x, v) \mapsto (y, w) = (y(x, v), w(v))$. For example, the 3-parameter deformation $x^3 + ax^2 + bx + c$ of x^3 is an inverse image of the deformation $x^3 + sx + t$ since

$$x^3 + ax^2 + bx + c = (x + \frac{a}{3})^3 + (b - \frac{a^2}{3})(x + \frac{a}{3}) + (c - \frac{ab}{3} + \frac{2a^3}{27}).$$

Take representatives for f, F and G as above. Then for v and x small enough we have

$$G_v(x) = F_{w(v)}(y(x, v)).$$

The family of functions (G_v) can therefore be deduced (in a neighbourhood of $(0, 0)$) from the family of functions (F_u) by the reassignment $v \mapsto u = w(v)$ and, for each v, by the diffeomorphism $x \mapsto y = y(x, v)$. The study of the family G thus reduces to that of the family F. This remark is what makes the following definition important:

Definition 5.6.4. We say that F is a *versal* deformation of f if every deformation of f is the inverse image of F under a suitable transformation.

The adjective 'versal', which is not very beautiful, was forged by mathematicians from the word universal. The definition above is of a very general type: for every object G of a certain class (here, every deformation of f) we assert the existence of a transformation (here $(x, v) \mapsto (y, w)$) enabling it to be deduced from the fixed object F, which corresponds well to the idea of F being universal. However, usage requires that we reserve this adjective for the case when the transformation is unique, G being given. Hence we quite naturally suppress the prefix 'uni'.

Several variants of this definition can be found in the literature.

With the terminology above, Theorem 4.6.2 can simply be stated: the r-parameter deformation

$$x^{r+1} + a_{r-1}x^{r-1} + \cdots + a_0$$

of x^{r+1} is versal.

5.6.5. Let $F = F(x, u)$ be a p-parameter deformation of $f = f(x) \in \mathcal{E}_n$. For $i = 1, \ldots, p$, consider the p germs in $\mathcal{E}_n$ obtained by taking the values at $u = 0$ of the partial derivatives of F with respect to the u_j:

$$\partial_j F = \frac{\partial F}{\partial u_j}(x, 0) \in \mathcal{E}_n.$$

Let T_F denote the vector subspace of $\mathcal{E}_n$ generated by the $\partial_j F$; then of course $\dim(T_F) \leq p$. Now consider an inverse image G of F:

$$G(x, v) = F(y(x, v), w(v)).$$

Differentiating with respect to the v_i and putting $v = 0$ we obtain

$$\frac{\partial G}{\partial v_i}(x, 0) = \sum_{j=1}^{p} \frac{\partial w_j}{\partial v_i}(0, 0)\frac{\partial F}{\partial u_j}(x, 0) + \sum_{k=1}^{n} \frac{\partial f}{\partial x_k}(x)\frac{\partial y_k}{\partial v_i}(x, 0),$$

or

$$\partial_i G = \sum_{j=1}^{p} \frac{\partial w_j}{\partial v_i}(0, 0)\partial_j F + h, \quad h \in J(f). \tag{5.6.2}$$

It follows that the $\partial_i G$ belong to $T_F + J(f)$, and hence that T_G is contained in $T_F + J(f)$. Considering the family $G(x, v) = f(x) + vg(x)$, with $g \in \mathcal{E}_n$, we conclude that if F is versal then every element g of $\mathcal{E}_n$ belongs to $T_F + J(f)$. This necessary condition for versality is also sufficient:

Theorem 5.6.6. (Mather's Universal Deformation Theorem.) *Suppose $\mu(f)$ is finite. In order for the p-parameter deformation $F(x, u)$ of f to be versal it is necessary and sufficient that $T_F + J(f) = \mathcal{E}_n$.* $\square$

If F is versal we therefore have

$$p \geq \dim(T_F) \geq \mathrm{codim}(J(f)) = \mu(f).$$

We say that the deformation F is *universal* if in addition $p = \mu(f)$, that is to say the $\partial_i F$ form a basis of $\mathcal{E}_n$ modulo $J(f)$.

This usage of the word 'universal' goes against the philosophy indicated in the remark above. Some authors have coined the adjective *miniversal* for this precise usage, since we then have uniqueness to first order (exercise).

From the theorem we can deduce in particular:

Corollary 5.6.7. *Let* $g_1, \ldots, g_p$, *with* $p = \mu(f)$, *be germs which form a basis of* $\mathcal{E}_n$ *modulo* $J(f)$. *Then*

$$F(x, u) = f(x) + u_1 g_1(x) + \cdots + u_p g_p(x)$$

is a universal deformation of f.   □

Applying this for $n = 1$ to $f = x^{r+1}$ and noting that $1, x, \ldots, x^{r-1}$ form a basis modulo the Jacobian ideal of x^{r+1} we recover Theorem 4.6.2. Likewise, for a Morse point we recover Theorem 4.5.3.

5.6.8. It follows from the above corollary that every germ with finite Milnor number (which is to say, as we have already observed, every finitely-determined germ) does indeed possess universal deformations. Moreover, *two such deformations are equivalent.* For let $F(x, u)$ and $G(x, v)$ be two universal deformations of $f(x)$ both therefore having $p = \mu(f)$ parameters. By definition there exist families $y(x, v)$ and $w(v)$ with

$$G(x, v) = F(y(x, v), w(v)).$$

Take the relation (6.2). Since the $\partial_i G$ on the one hand and the $\partial_j F$ on the other hand each form bases of $\mathcal{E}_n$ modulo $J(f)$, the matrix of the $\partial w_j / \partial v_i(0, 0)$ is invertible. Hence the map $(x, v) \mapsto (y(x, v), w(v))$ is a local diffeomorphism of $\mathbf{R}^{n+p}$ at 0 transforming F into G.

As indicated above, we usually introduce a versal deformation $F(x, u)$ of a given function $f(x)$ in order to study the behaviour of a nearby function F_b, for $b \in U$ close to 0, in the neighbourhood of a point $a \in E$ sufficiently close to 0. Obviously the question is interesting only when a is a critical point of F_b. It is worth mentioning in this regard the following general and not very difficult result:

Theorem 5.6.9. (Openness of versality.) *Let* $F(x, u)$, $(x, u) \in E \times U$, *be a versal deformation of* $f(x)$, $x \in U$. *For* $a \in E$ *sufficiently close to 0 and* $b \in U$ *sufficiently close to 0 the family* $(x, u) \mapsto F(a + x, b + u)$ *is a versal deformation of the function* $x \mapsto F(a + x, b)$.   □

5.7 The Principles of Catastrophe Theory

We shall now give a rather loose description of the philosophy of catastrophe theory. Matters will be made progressively more precise as we study examples; the reader is advised to consult these if any uneasiness is experienced in this section.

5.7.1. Consider a system (mechanical, chemical, ...) which we call (Σ), having n degrees of freedom and with potential energy Φ. Thus Φ is a function on the *space E of configurations* that are possible for (Σ). In the spirit of everything that we have been doing up to now (and bearing in mind in particular the Transversality Theorem) it is natural to suppose that E is a manifold of dimension n and that Φ is a real C^∞ function on E. To study what happens in a neighbourhood of an equilibrium position of (Σ) we can take a system of local coordinates on E. We thus arrive at the situation where the state of (Σ) is described by n *state variables*, that is to say by a point $x = (x_1, \ldots, x_n)$ in an open subset E of $\mathbf{R}^n$, and where Φ is a C^∞ function on E, the *potential*. The stable equilibrium points of (Σ) are the relative minima of Φ and in particular they are critical points.

5.7.2. We aim to study the behaviour of (Σ) in the neighbourhood of one of its equilibrium points. We suppose in fact that the potential depends further on certain *control parameters* and we seek to describe the behaviour of (Σ) as these parameters are varied. More precisely, we assume that (Σ) is *strongly dissipative* or, which amounts to the same thing, that the control parameters vary very slowly[33] and we are interested only in the evolution of the equilibrium positions. The discussion in the Introduction (see also 9.3 below) concerning *hidden parameters* shows that in order for this problem to have physical meaning we have to assume that we are dealing with a family $u = (u_1, \ldots, u_p)$ of control parameters that is rich enough for the corresponding family of potentials Φ_u to be *stable*. Equivalently, we observe that it suffices to study the case, which will contain all the others, when the family Φ_u is a *universal deformation* of the initial potential Φ. Moreover, since two potentials that differ by an additive constant give the same results, we can even suppress one of the parameters of this deformation.

5.7.3. The scene is therefore set. We consider a germ $\Phi = \Phi(x) \in \mathcal{E}_n$ in n variables and a deformation $\Phi(x, u)$ of this germ with $p = \mu(f) - 1$ parameters $(u_1, \ldots, u_p)$ such that $\Phi(x, u) + u_0$ is a universal deformation of $\Phi(x)$. As usual, we represent the germ $\Phi(x, u)$ by a function of the same name, defined on an open set in $\mathbf{R}^{n+p}$ of the form $E \times U$ where E is an open set in $\mathbf{R}^n$ containing 0 and U is an open set in $\mathbf{R}^p$ containing 0. For every $u \in U$ we define a function

---

[33]We often refer to this as a *quasi-static* theory.

on E by $\Phi_u(x) = \Phi(x, u)$; we thus have $\Phi_0 = \Phi$. Consider the p-dimensional submanifold V of $E \times U$ consisting of the critical points of Φ_u:

$$V = \{(x, u) \in E \times U \mid \frac{\partial \Phi}{\partial x_i}(x, u) = 0, \; i = 1, \ldots, n\},$$

and the projection $\chi : (x, u) \mapsto u$ of V into U. We say that V is the *equilibrium manifold* and χ is the *catastrophe map*. By the Transversality Theorem, the set V is actually a submanifold of $E \times U$, but the map χ is not everywhere a submersion. It has a critical locus C, also called the *catastrophe set*; it consists of the points $(x, u) \in V$ such that the Hessian form of Φ_u at the point x (which is critical, by definition) is degenerate. The image B of C under χ is called the *bifurcation* set or locus.[34] See Fig. 5.2.

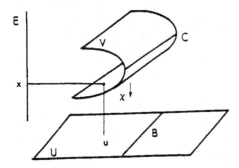

Fig. 5.2. Equilibrium manifold V and catastrophe map χ

5.7.4. The map χ is thus a local diffeomorphism at every point of $V \setminus C$. Suppose that at a given moment the system is in the equilibrium position x, for the value u of the parameters (so we have $(x, u) \in V$), and that the point (x, u) does not belong to the catastrophe locus C and therefore u does not belong to the bifurcation locus B. A priori, if we slowly move the point u in U the position of equilibrium x will tend to move gradually, following the local diffeomorphism $u \mapsto x$ which is locally inverse to χ. However, in certain cases we observe sudden changes in the equilibrium position x that Thom calls *catastrophes*. To determine the precise behaviour of the system some further information is required: we need to know what Thom calls a bifurcation *convention*. There are two conventions that are most common: the delay convention and the Maxwell convention.

The *delay convention* states that x will 'jump' only when it has to do so, which means that it must reach the critical locus C (so u then reaches the bifurcation locus B).

The absolute minimum convention or *Maxwell convention* (we shall see later why it has this name) requires that x remain always at the absolute

---

[34]Some authors call B (rather than C) the catastrophe set. – *Trans.*

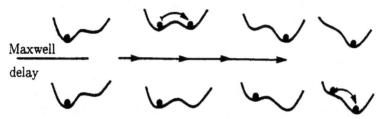

Maxwell

delay

Fig. 5.3. Maxwell convention versus delay convention

minimum of the potential Φ_u and therefore that x jumps as soon as a minimum lower than it appears[35]; this means that u reaches the *Maxwell set* consisting of points u for which Φ_u possesses two absolute minima. See Fig. 5.3. Observe an essential difference between these two conventions: the latter uses data of the potential Φ while the former uses only that of the equilibrium manifold V. Below we shall see a case (liquid-vapour phase transition) where information about V does not suffice to determine the behaviour of the system.

5.7.5. The initial number of degrees of freedom turns out in fact to have no importance: all that counts is the *corank* of the germ being considered. Recall the *Decomposition Theorem* (Corollary 4.5.5): if n denotes the corank of a germ h in N variables then after a suitable change of coordinates we can write

$$h = f(x_1, \ldots, x_n) + \sum_{i=n+1}^{N} \pm x_i^2.$$

We shall now see that f and h define exactly the same catastrophe. To do this we first prove the following lemma :

Lemma 5.7.6. *Let $f = f(x_1, \ldots, x_n)$ be a germ in n variables, and let z be a new variable. Consider the germ $f' = f \pm z^2$ in $n+1$ variables.*

a) *Let g' be a germ in $n+1$ variables. Put $g = g'(x_1, \ldots, x_n, 0)$. The conditions $g \in J(f)$ and $g' \in J(f')$ are equivalent.*
b) *The Milnor numbers of f and f' are the same. Suppose they are finite, and write $p = \mu(f) = \mu(f')$.*
c) *Let $g_1, \ldots, g_p$ be germs in n variables forming a basis modulo $J(f)$. Then, regarded as germs in $n+1$ variables, they form a basis modulo $J(f')$.*
d) *Let $g'_1, \ldots, g'_p$ be germs in $n+1$ variables forming a basis modulo $J(f')$. Put $g_i = g'_i(x_1, \ldots, x_n, 0)$. Then the germs $g_1, \ldots, g_p$ in n variables form a basis modulo $J(f)$.*

---

[35]Should we not then call this the *Minwell* convention? (Remark by Christopher Zeeman.)

The partial derivatives of f' are $\pm 2z$ and the $\partial f / \partial x_i$. Let g' and g be as in a). If g' belongs to $J(f')$ we may write

$$g = a_1(x, z)\frac{\partial f}{\partial x_1} + \cdots + a_n(x, z)\frac{\partial f}{\partial x_n} + b(x, z)z.$$

Taking $z = 0$ we see that g belongs to $J(f)$. Conversely, we can write $g' = g + zc(x, z)$; if g belongs to $J(f)$ it also belongs to $J(f')$ and g' belongs to $J(f')$. This proves a). From this we deduce c), d) and then b). $\square$

5.7.7. Returning to our original notation, consider the germ $\Phi(x) \in \mathcal{E}_n$ and the 'universal up to a constant' deformation $\Phi(x, u)$. Let $z_1, \ldots, z_m$ denote new variables, fix signs $\varepsilon_k = \pm 1$, $k = 1, \ldots, m$, and for $z \in \mathbf{R}^m$, $x \in E$ and $u \in U$ let

$$Q(z) = \sum_{k=1}^{m} \varepsilon_k z_k^2, \quad \Psi(z, x) = Q(z) + \Phi(x), \quad \Psi(z, x, u) = Q(z) + \Phi(x, u).$$

We have

$$\frac{\partial \Psi}{\partial z_k}(z, x, u) = 2\varepsilon_k z_k, \quad \frac{\partial \Psi}{\partial x_i}(z, x, u) = \frac{\partial \Phi}{\partial x_i}(z, x, u).$$

Consequently the equilibrium manifold for Ψ is simply $\{0\} \times V \subset \mathbf{R}^m \times E \times U$; we identify this with V by the map $(x, u) \mapsto (0, x, u)$. Then the catastrophe map is still $(x, u) \mapsto u$ and, moreover, the values of Φ and Ψ at the corresponding points are the same. Finally, if all the ε_k are equal to 1, these points are or are not minima at the same time, while if at least one of the ε_k is equal to -1 then Ψ has no minimum. All this shows, as promised, that the study of the catastrophe associated to Ψ reduces directly to that of the universal deformation Φ.

5.8 Catastrophes of Cusp Type

We shall give below (see Theorem 11.2) the classification of catastrophes according to increasing Milnor number μ. For $\mu = 0$ there is nothing to say (there is no equilibrium at all), nor is there for $\mu = 1$ (the initial equilibrium is a Morse point and there is no bifurcation, cf. Proposition 4.6.1). The cases $\mu = 2$ and $\mu = 3$ are the *fold* and *cusp* catastrophes that we have already met. As the fold is a little too simple to illustrate the general principles expounded above, we shall consider the case of cusps in some detail.

5.8.1. Suppose therefore that $\mu(\Phi) = 3$; the classification then says that, assuming useless variables have been suppressed (see 7.7), we may suppose that $n = 1$ and the potential Φ is equivalent to ax^4 with $a \neq 0$. First we take $a > 0$ and choose the state coordinate x so that we have $\Phi = x^4/4$. Then, up to an additive constant, we can take as universal deformation

$$\Phi_u(x) = \Phi(x, a, b) = \frac{x^4}{4} - a\frac{x^2}{2} + bx. \tag{5.8.1}$$

The state x runs through the state space $E = \mathbf{R}$, while the parameter $u = (a, b)$ belongs to the parameter space U. The equilibrium manifold is the surface

$$V = \{(x, a, b) \mid x^3 - ax + b = 0\}, \tag{5.8.2}$$

with $\chi(x, a, b) = (a, b)$ and it is immediate that the catastrophe locus C and the bifurcation locus B are the curves

$$C = \{(x, a, b) \mid a = 3x^2, \ b = 2x^3\}, \quad B = \{(a, b) \mid 4a^3 = 27b^2\}, \tag{5.8.3}$$

while the Maxwell set W is the straight line $\{(a, b) \mid b = 0, \ a > 0\}$ with corresponding values of x given by $x^2 = a$. See Fig. 5.4. The open set $U \setminus B$ has two connected components

$$U_+ = \{(a, b) \mid 4a^3 < 27b^2\}, \quad U_- = \{(a, b) \mid 4a^3 > 27b^2\}. \tag{5.8.4}$$

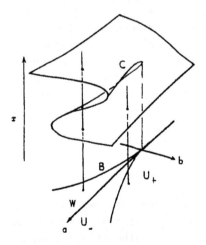

Fig. 5.4. The cusp catastrophe

5.8.2. When u belongs to the 'exterior' U_+ of B there exists a unique point (x, u) of V above u; the potential Φ_u possesses a unique critical point, namely x, and it is an absolute minimum. When u belongs to the 'interior' U_- of B there are three points of V above u; the potential Φ_u then has two relative minima separated by a relative maximum, so there are two possible choices for x. When u follows a curve in the parameter space U the destiny of x

depends on the convention that has been chosen. Suppose for example that u traces the straight line $a = 1$ which crosses B at the two points $b = \pm\beta$ with $\beta = 2/3\sqrt{3}$, and crosses W at the point $b = 0$. Then we have $x^3 - x + b = 0$ and the path of x is illustrated in Fig. 5.5a for the delay convention and in Fig. 5.5b for the Maxwell convention. In the former case x jumps only when the leaf of the surface which it is following disappears; notice the phenomenon of *hysteresis* that this generates. In the latter case x jumps as soon as the new minimum appearing on the other leaf corresponds to a lower value of Φ; this happens at $b = 0$.

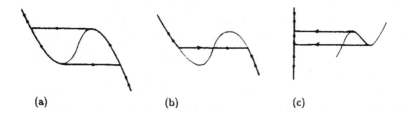

(a)   (b)   (c)

Fig. 5.5. Different dynamical paths under different conventions: (a) delay convention, (b) Maxwell convention, (c) dual cusp with delay convention.

5.8.3 Observe that there is an essential qualitative difference that allows us to identify experimentally the convention used by the system: the path followed by x, which is continuous in the two cases, is of class C^1 for the delay convention but has an angular point for the Maxwell convention.

Note finally that if we had originally chosen the other sign possible for Φ (then we say we have a *dual cusp*) we would have had to invert maxima and minima in the above description. In a dual cusp the bifurcation mechanism is quite different: Φ_u has a unique relative minimum for $u \in U_-$ and no relative minimum for $u \in U_+$. Along the route described previously, the delay convention gives the behaviour illustrated in Fig. 5.5c, while the Maxwell convention gives x constantly equal to $-\infty$.

5.9 A Cusp Example

We now give a very simple example that models a buckling phenomenon in the most rudimentary way possible. We see a cusp arising naturally.

5.9.1. Consider a heavy rod pinned without friction at its base, constrained to move in a vertical plane and supported in a vertical position above its hinge point by two opposing springs (see Fig. 5.6). If θ denotes the angle of the rod from the vertical and we suppose that the two springs are exactly symmetrical then the movement of the rod is governed by a differential equation of the form

$$\ddot{\theta} = -a\theta + g\sin\theta = -\frac{df}{d\theta}, \tag{5.9.1}$$

where a and g are two constants > 0 and where

$$f(\theta) = a\frac{\theta^2}{2} + g(\cos\theta - 1) = \frac{(a-g)}{2}\theta^2 + \frac{g}{24}\theta^4 + \cdots . \tag{5.9.2}$$

The equilibrium positions are the critical points of f, given by the condition $a\theta = g\sin\theta$. We shall consider g as fixed (it is a geometric parameter of the rod) and act on the control variable a which we suppose to be very large at first and then gradually reducing (symmetrically!) the tension in the supporting springs. As long as $a > g$ the only critical point is $\theta = 0$, and it is a Morse point (in fact the absolute minimum of f) and therefore a stable equilibrium point. When a becomes $< g$ this point becomes a local maximum and therefore corresponds to an unstable equilibrium; simultaneoulsy two symmetric local minima appear (see Fig. 5.7(a)). For $a = g$ we thus observe a *bifurcation of critical points*. Naturally, since it is impossible to realise the symmetry condition exactly, the rod does not remain vertical and we do not observe the phenomenon described above, but there is a so-called *symmetry-breaking*: the variation of the equilibrium position θ as a function of the control parameter a has the form indicated in Fig. 5.7(b).

5.9.2. Clearly we could introduce an asymmetry into the model right from the start and replace the differential equation (9.1) by

$$\ddot{\theta} = -b - a\theta + g\sin\theta = -\frac{d\Phi}{d\theta}, \tag{5.9.3}$$

with this time

$$\Phi(\theta) = b\theta + a\frac{\theta^2}{2} + g(\cos\theta - 1) = b\theta + \frac{(a-g)}{2}\theta^2 + \frac{g}{24}\theta^4 + \cdots \tag{5.9.4}$$

Now we have two control parameters a and b and in (a, b, θ)-space the surface $d\Phi/d\theta = 0$ is a pleated surface (Fig. 5.8) whose plane sections for $b = 0$ and $b \neq 0$ respectively are the curves described above.

5.9.3. All this is very fine, but how can we now be sure that we have the complete answer? The key is in Theorem 6.6, or in this particular case in Theorem 4.6.2. To see this, we have to return to the bifurcation point where the potential is of the form

$$f_0 = g(\cos\theta - 1 + \frac{\theta^2}{2}) = \frac{g}{24}\theta^4 + \cdots = x^4,$$

for a suitably chosen local coordinte $x = c\theta + \cdots$ (the fourth root of f_0 !). The family (f_λ) given by (9.2) with $\lambda = a - g$, that is

$$f_\lambda(x) = g(\cos\theta - 1 + \frac{\theta^2}{2}) + \lambda\frac{\theta^2}{2} = x^4 + (\lambda + \cdots)x^2, \tag{5.9.5}$$

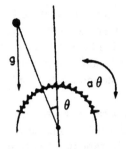

Fig. 5.6 Heavy rod with springs

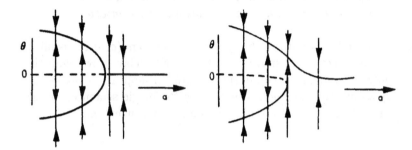

Fig. 5.7 Broken symmetry

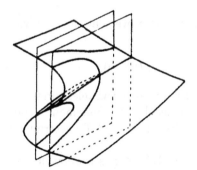

Fig. 5.8. Sections of the equilibrium surface yielding Fig. 5.7

is *universal among the symmetric deformations* of f_0 (always neglecting an additive constant). On the other hand, the family $(\Phi_{\lambda,b})$ given by (9.4), namely

$$\Phi_{\lambda,b}(x) = g(\cos\theta - 1 + \frac{\theta^2}{2}) + \lambda\frac{\theta^2}{2} + b\theta$$
$$= x^4 + (\lambda + \cdots)x^2 + (b + \cdots)x, \qquad (5.9.6)$$

is now *universal among all the deformations* of $f_0 = \Phi_{0,0}$, again up to an additive constant (Theorem 4.6.2). A small modification of hidden parameters[36] in the physical system under consideration will always be able to be 'absorbed' by the model (9.6), where it will be translated into small modifications of λ and b (that is a and b), while it will be compatible with the model (9.5) only if it maintains the symmetry.

In other words, it hardly matters that the additional parameter was introduced in order to take account of the asymmetry of the springs, or inaccurate alignment with the vertical, or the wind speed; the important point is that after it is introduced the potential is then a universal deformation of the function given by any particular value of the parameter. Indeed, *the parameter b is physically significant because it is mathematically indispensable.*

5.9.4. If we now follow a path in the space of parameters (a, b) the equilibrium point of the system will move according to the delay convention: the rod swings only when its initial equilibrium position disappears. Thus we can summarize qualitatively the whole situation by the magic sentence: we are simply dealing with a *cusp catastrophe* with the delay convention.

5.10 Liquid-Vapour Equilibrium

5.10.1. The van der Waals equation, which is an improvement on the Boyle-Mariotte law $PV = RT$ for a perfect gas, is

$$h(P, V, T) = (P + a/V^2)(V - b) - nkT = 0. \qquad (5.10.1)$$

In this equation P, V and T are the usual thermodynamic variables (pressure, volume and temperature), n is the number of molecules, k is Boltzmann's constant $(1 \cdot 380 \times 10^{-16} \, erg/deg)$ and a and b are two experimentally determined numbers > 0.

The equation $h = 0$ defines a surface S in $\mathbf{R}^3_+$ that is the graph of a map $(P, V) \mapsto T$. The projection $(P, V, T) \mapsto T$ from S to $\mathbf{R}_+$ has no critical point and S is represented traditionally by the family of curves obtained by giving T a fixed value (*isotherms*). The projection $(P, V, T) \mapsto (P, T)$ on the other hand does have critical points. These are obtained by writing $h = 0$ and $\partial h/\partial V = 0$; they are fold points except for a unique cusp point which can be found by writing $h = 0, \partial h/\partial V = 0$ and $\partial^2 h/\partial V^2 = 0$. This point is traditionally called (alas!) the *critical point*. It is the point with coordinates

$$P_c = \frac{a}{27b^2}, \quad V_c = 3b, \quad T_c = \frac{8a}{27bnk} . \qquad (5.10.2)$$

---

[36]See the discussion in the Introduction, of which this section is an explanation.

5.10.2. To eliminate unnecessary parameters we shall use as coordinates the *normalized pressure* p, the *normalized density* x and the *normalized temperature* t defined by

$$p = \frac{P - P_c}{P_c}, \quad x = \frac{V_c - V}{V}, \quad t = \frac{T - T_c}{T_c}. \tag{5.10.3}$$

Each of these normalized coordinates now runs through the interval $(-1, \infty)$ and the critical point is the point $(0, 0, 0)$. The original variables can be calculated to be

$$P = \frac{a}{27b^2}(1 + p), \quad V = \frac{3b}{x + 1}, \quad T = \frac{8a}{27bnk}(1 + t), \tag{5.10.4}$$

van der Waals' law now becomes

$$g(x, p, t) = x^3 + \frac{8t + p}{3}x + \frac{8t - 2p}{3} = 0, \tag{5.10.5}$$

and we recover the cusp situation, up to a linear transformation of parameters.

5.10.3. In fact it can be shown more precisely[37] that the actual state of a gas minimises the free energy which expressed in the original coordinates is

$$\begin{aligned}
\Phi &= \frac{6b}{a} \int_{3b}^{V} \left(P + a/V^2 - \frac{nkT}{V - b}\right) dV \\
&= \frac{6b}{a}\left(PV - 3bP - \frac{a}{V} + \frac{a}{3b} - nkT \log\frac{V - b}{2b}\right)
\end{aligned} \tag{5.10.6}$$

and in normalized coordinates

$$\Phi(x, p, t) = \frac{16(t + 1)}{9} \log\left(\frac{2 + 2x}{2 - x}\right) - \frac{2x(1 + p)}{3(1 + x)} - 2x. \tag{5.10.7}$$

In the original coordinates we have

$$\frac{\partial \Phi}{\partial V} = \frac{6b}{a}\left(P + \frac{a}{V^2} - \frac{nkT}{V - b}\right) = \frac{6b}{a(V - b)}h(P, V, T); \tag{5.10.8}$$

and in the normalized coordinates a tedious calculation gives

$$\frac{\partial \Phi}{\partial x} = \frac{6x^3 + (16t + 2p)x + (16t - 4p)}{3(1 + x)^2(2 - x)} = \frac{2}{(1 + x)^2(2 - x)}g(x, p, t). \tag{5.10.9}$$

This shows not only that the gas state is indeed a solution of van der Waals' equation, but also that when (10.5) determines several values of x for a given p and t (or when (10.1) gives several values of V for a given P and T) we have to apply the Maxwell convention and choose that value which corresponds to an *absolute minimum* of the potential Φ.

---

[37]See G.M.Bell and D.A.Lavis : *Thermodynamic phase changes and catastrophe theory* cited in [PS].

5.10.4. Note that

$$j^4\Phi = \frac{8t-2p}{3}x + \frac{2p-2t}{3}x^2 + \frac{2t-2p}{3}x^3 + (\frac{1}{4} + \frac{8p-5t}{12})x^4.$$

Thus $f(x) = \Phi(x,0,0)$ can be written $x^4/4 + \cdots$ and $J(f)$ is generated by x^3. Since the two partial derivatives

$$\frac{\partial\Phi}{\partial p}(x,0,0) = -\frac{2x}{3} + \frac{2x^2}{3} + \cdots , \qquad \frac{\partial\Phi}{\partial t}(x,0,0) = \frac{8x}{3} - \frac{2x^2}{3} + \cdots$$

together with 1 form a basis modulo $J(f)$, it follows from Theorem 6.6 that $\Phi(x,p,t) + c$ is indeed a three-parameter universal deformation of $f(x)$. We might note that, starting from van der Waal's equation in the form (10.5), we could equally well have considered the potential

$$\frac{x^4}{4} + \frac{8t+p}{6}x^2 + \frac{8t-2p}{3}x, \tag{5.10.10}$$

but this would not have given the correct Maxwell set (we would have found $p = 4t$), and therefore not the correct bifurcation.

Observe in passing that the Maxwell convention for Φ can be translated as the equality of the two areas bounded by the isotherm $(P + a/V^2)(V - b) = nkT$ and the straight line $P = P_0$ followed by the point representing the state (exercise). This is what is called *Maxwell's rule* in thermodynamics, and it is the equivalence of this rule with the absolute minimum rule for the thermodynamic potential Φ which led Thom to call the absolute minimum rule the 'Maxwell convention'.

5.10.5. By regarding the density x as the state variable and the variables p and t as control variables (or the volume V as state variable and the variables P and T as control variables) we can, as in the previous example, summarize the situation from the qualitative point of view by the one magic sentence: we are simply dealing with *a cusp catastrophe with the absolute minimum convention*.

5.11 The Elementary Catastrophes

In Thom's terminology these are the germs with Milnor number ≤ 5. According to Thom, the number of control parameters of a physical system is as a general rule ≤ 4 (often we are dealing with space-time; this number is also related to the 'Gibbs phase rule'). Therefore if we adopt the philosophy of genericity, the *elementary catastrophes* are those that can arise in generic families of functions that depend on at most four control parameters. Now, this condition means precisely that their Milnor number is ≤ 5 by virtue of the following theorem:

Theorem 5.11.1. *Let V be an open set in $\mathbf{R}^n \times \mathbf{R}^p$ with $p \leq 4$, and let $F(x, u)$ be a generic C^∞ function on V. Let (a, b) be a point of V; consider the germ $\phi(x) = F(a + x, b)$ and the $(p + 1)$-parameter family $\Phi(x, (u_0, u)) = F(a + x, b + u) + u_0$. Then we have $\mu(\phi) \leq p + 1 \leq 5$ and Φ is a versal deformation of ϕ.* □

These germs are given by:

Theorem 5.11.2. *Let f be a C^∞ function on a neighbourhood of a point a in $\mathbf{R}^n$, having a critical point at a with Milnor number $r \leq 5$ (hence $r \in \{1, 2, 3, 4, 5\}$). Then there exists a system of local coordinates $(x_1, \ldots, x_n)$ centred at a in which f can be written in one of the following forms:*

$$f = f(a) \pm x_1^{r+1} + \sum_{i=2}^{n} \pm x_i^2, \quad r \in \{1, 2, 3, 4, 5\},$$

$$f = f(a) + x_1^2 x_2 \pm x_2^{r-1} + \sum_{i=3}^{n} \pm x_i^2, \quad r \in \{4, 5\}.$$

Of course, the constant $f(a)$ plays no role; also we saw in 7.7 that adding a function of the form $\sum \pm z_j^2$, where the z_j are new variables, makes no essential change to the corresponding catastrophe. The above theorem thus gives us seven simple models, the *seven elementary catastrophes* of Thom, which (following Thom and Bernard Morin) are customarily given names which evoke the shape of the equilibrium manifold and the geometry of the catastrophe map (see 7.3):

A_2 : *fold* x^3
$A_3^\pm$: *cusp* x^4 and *dual cusp* $-x^4$
A_4 : *swallowtail* x^5
$A_5^\pm$: *butterfly* x^6 and its *dual* $-x^6$
D_4^+ : *hyperbolic umbilic* $x^2 y + y^3$
D_4^- : *elliptic umbilic* $x^2 y - y^3$
D_5 : *parabolic umbilic* $x^2 y + y^4$ and its *dual* $x^2 y - y^4$.

This leads us into a somewhat detailed study of two particular families of germs which contain all the germs in the above list.

5.11.3. For every integer $r \geq 1$ let $A_r^\pm$ denote the 1-variable germ $\pm x^{r+1}$. Its Milnor number is r and the germs $1, x, \ldots, x^{r-1}$ form a basis modulo the Jacobian ideal. A universal deformation of $A_r^\pm$ is

$$\pm x^{r+1} + u_1 x^{r-1} + \cdots + u_i x^{r-i} + \cdots + u_r.$$

Every 1-variable germ of order $r + 1$ at the origin is equivalent to $A_r^\pm$. When r is even, the germs A_r^+ and A_r^- are equivalent (by changing x to $-x$).

5.11.4. For every integer $r \geq 4$ let $D_r^{\pm}$ denote the 2-variable germ $x^2 y \pm y^{r-1}$. Its Milnor number is r and the germs $x, 1, y, \ldots, y^{r-2}$ form a basis modulo the Jacobian ideal. A universal deformation of $D_r^{\pm}$ is

$$x^2 y \pm y^{r-1} + u_1 x + u_2 y^{r-2} + \cdots + u_i y^{r-i} + \cdots + u_r.$$

When r is even the germ A_r^- is equivalent to the negative of A_r^+ (by changing x to $-x$).

We end this section by proving Theorem 11.2.

According to the Splitting Lemma 4.5.5, after a suitable coordinate change every germ h in N variables can be written

$$h(0) + f(x_1, \ldots, x_n) \pm x_{n+1}^2 \pm \cdots \pm x_N^2,$$

where $j^2 f$ is zero. By Lemma 7.6 we are reduced to studying those germs $f \in \mathcal{E}_n$ whose 2-jet is zero. Let $s+1$ denote the order of f; we have $s > 1$. Lemma 4.4 gives the bound $\mu(f) \geq m(n, s)$, with $m(n, s) = (n+1) \cdots (n+s)/s! - n$. The function $m(n, s)$ is strictly increasing in n and s. Since $m(3, 2) = 7$ and $m(2, 3) = 8$ we shall already obtain all the germs whose Milnor number is ≤ 6 by studying the cases $\{n = 1\}$ and $\{n = 2, s = 2\}$.

For $n = 1$ we already know the classification: taking a suitable coordinate we have $f = \pm x^{r+1}$, with $\mu(f) = r$.

For $n = 2$, we first consider the 3-jet $P = j^3 f$ of f. This is a polynomial of degree 3 which is homogeneous (since we have supposed $j^2 f = 0$) and nonzero (since we have supposed $s = 2$).

Lemma 5.11.5. *Let P be a real nonzero homogeneous polynomial of degree 3 in two variables. After a suitable linear change of coordinates P can be written in one of the following forms:*

$$P = x^2 y + y^3, \tag{5.11.1}$$
$$P = x^2 y - y^3, \tag{5.11.2}$$
$$P = x^2 y, \tag{5.11.3}$$
$$P = x^3. \tag{5.11.4}$$

The polynomial P is the product of three (complex) linear forms. There are four possible cases: $P = L^3$ where L is real; $P = L^2 L'$ where L and L' are real and non-proportional; $P = LL'L''$ where L, L' and L'' are real and non-proportional; $P = L\bar{L}L'$ where L is non-real and L' is real. The first two cases give (11.4) and (11.3) immediately. Let us deal with the fourth case, for example. In suitable coordinates the quadratic form $L\bar{L}$ can be written as $u^2 + v^2$. We have then $P = (au + bv)(u^2 + v^2)$ and we obtain the form (11.1) by rotation and magnification. □

In the first two cases we are dealing with sufficient jets (see 5.4 e)) and that is the end of the matter. The two others are handled using the two lemmas below which complete the proof.

Lemma 5.11.6. *If $j^3 f = x^2 y$ and $\mu(f)$ is finite then f is equivalent to $x^2 y \pm y^{r-1}$, with $r = \mu(f) \geq 4$.*

Lemma 5.11.7. *If $j^3 f = x^3$ then $\mu(f) > 5$.*

First we prove Lemma 11.6. For this we start by proving that if for $s \geq 3$ we have $j^s f = x^2 y$ then f is equivalent to a germ f' such that $j^{s+1} f' = x^2 y + a y^{s+1}$ with $a \in \mathbf{R}$. Indeed, we may write

$$j^{s+1} f = x^2 y + u(x,y)\frac{x^2}{2} + v(x,y)xy + a y^{s+1}$$

where u and v are two homogeneous polynomials of degree $s - 1$ and where $a \in \mathbf{R}$. Now let $f' = f(x - v(x,y), y - u(x,y))$; an immediate calculation gives $j^{s+1} f' = x^2 y + a y^{s+1}$.

We distinguish two cases. If $a \neq 0$ then f' is $(s+1)$-determined (5.4 e)), so f is equivalent to $x^2 y + a y^{s+1}$ and the result follows after suitable rescaling of x and y. If $a = 0$ then we begin again, replacing s by $s + 1$. But this cannot go on for ever, because if $j^s f = x^2 y$ then $j^{s-1}(\partial f/\partial x) = 2xy$ and $j^{s-1}(\partial f/\partial y) = x^2$ and it follows that $1, y, \ldots, y^{s-2}$ are linearly independent modulo $J(f)$ and therefore $s - 1 \leq \mu(f)$. □

Finally we prove Lemma 11.7. We have $j^4 f = x^3 + Q$ where Q is a homogeneous polynomial of degree 4. Every element of $J(f)$ can be written

$$(a + bx + cy + \cdots)(3x^2 + \frac{\partial Q}{\partial x}) + (d + \cdots)\frac{\partial Q}{\partial y} =$$

$$= a(3x^2 + \frac{\partial Q}{\partial x}) + 3bx^3 + 3cx^2 y + d\frac{\partial Q}{\partial y} + \cdots.$$

Therefore the image of $J(f)$ under j^3 in the 10-dimensional space of polynomials of degree ≤ 3 is a subspace spanned by 4 elements. Hence we have $\mu(f) = \operatorname{codim}(J(f)) \geq \operatorname{codim}(j^3(J(f))) \geq 10 - 4 = 6$, which completes the proof of Lemma 11.7 and also of Theorem 11.2. □

5.12 Catastrophes and Controversies

The major article by Whitney in 1955 (the main result of which we have already given in Theorem 4.7.4) can be regarded as the birth of catastrophe theory. The construction of the theory in its definitive form rests upon the Transversality Theorem (see Sect. 3.10) and on the algebraicisation theorems of Mather (see Sect. 4.11). Its change of status and partial transformation into a 'cultural' tool dates essentially from the essay [T1] of Thom (1973).

Since that date various aspects of catastrophe theory have been subjects of lively criticism and remain controversial. First of all there is the question of the *limitations of its original context.* It applies only to dynamical systems described by potentials and therefore associated to gradient fields. Even if we assume that the 'model' system does effectively arise from a potential, the very philosophy of the theory can apply only if the 'real' system remains in the subset (which is certainly not dense) consisting of systems of this type. This therefore requires a philosophical standpoint concerning 'nature', which cannot but raise certain objections.

A response to this criticism might be that even if catastrophe theory *stricto sensu* does indeed suffer from this limitation, it is only one particular technical illustration of ideas of much more general importance, built around the central notion of stability.

Independently of the technical illustration that we have been considering, the body of Thom's philosophy rests essentially on the notion of stability and more precisely on the *hypothesis of structural stability*: it is only stable phenomena that can be observed, *therefore* every observable phenomenon has to be modelled by a stable system, *therefore* every system has to be close to a stable system, *therefore* the stable systems must be dense. We saw in the previous chapter how, with a suitable defintion of stability and at the cost of considerable technical difficulties, Mather was eventually able to transform this hope into a theorem in the context of systems of potentials.

Now, and here is the rub, the analogous statement is certainly *false* in the context of general dynamical systems in dimension 3 or above (see Chapt. 9). From this point on, the controversy rebounds. Should we, as some do, conclude from this that Thom's philosophy does not have the universal character that it claims? Or, as Thom does, should we invert the argument and deduce that since nature (or at least observable nature) has to be structurally stable, we have simply shown that there exist theoretical systems which cannot model any physical system? Here is a revealing quotation ([T1], p.31; you never have the last word with a mathematician!):

"forms that are subjectively identifiable, forms that are provided with a name and are represented in language by a substantive, *are necessarily structurally stable...* "

which can be compared for example with [GH], p.259 (in a slightly different context):

"The logic which supports the stability dogma is faulty."

Other criticisms concern the attempts to apply catastrophe theory to the *social sciences*, led notably by Zeeman (see the examples in [PS]). We quote an extract from the polemical booklet [A6] by Arnol'd:

"I remark only that articles on catastrophe theory are distinguished by a sharp and catastrophic lowering of the level of demands of rigour and also of novelty of published results."

However, in order to avoid a common error of interpretation, we point out that catastrophe theory (in any case, in the form desired by Thom) does not claim any predictive purpose. It claims only to describe an unavoidable set of forms and, more precisely, to describe how these forms are created; it is not for nothing that the only word that is common to the titles of the two all-embracing works written by Thom on the subject is precisely *morphogenesis*.

To gain a deeper understanding of the few points indicated in this chapter we suggest reading [BL] or [CH] for mathematical details, and [PS], [ZE] or [GE] for a wide range of applications: the introduction and conclusion of [PS] (Sects. 1 and 18) are especially recommended. The surveys [A1] and [A2] help to put the subject in a wider context. Also [A6] or [A7] should certainly not be missed.

The ideas of Thom on stability and unfolding have now been developed widely in the context of local bifurcation theory, and have proved especially fruitful in understanding problems with symmetry and mechanisms of spontaneous symmetry-breaking: see [GS].

6 Vector Fields

6.1 Introduction

The notion of a *vector field* is a rather convenient device enabling us to model the evolution of systems with a finite number of degrees of freedom and governed by differential equations. We start by associating to each state of the system a 'representative' point, and the set of these points forms what in general we call the *phase space* of the system. This representation of the state of a system by a point in phase space must be rich enough so that knowing the point corresponding to the initial state will suffice to determine (in theory!) all the subsequent evolution. In mechanics, for example, where we deal with second order differential equations, the representative point will have to contain not just the position parameters for the different elements of the system (these parameters define a point in what is usually called the *configuration space*), but also the velocity parameters for these elements. Likewise, if the coefficients of the differential system depend on time (in this case we say that the system is *nonautonomous*) then the representative point will have to contain the time (we then often speak of an *extended* phase space). If some of the coefficients are regarded as adjustable then they also have to be included as coordinates for the representative point. The phase space P introduced in this way often has additional structure. For example, the notion of 'constraint' is translated by the fact that P is given as a subset (often a submanifold) of an ambient $\mathbf{R}^N$.

Once the phase space is fixed, the temporal evolution of the system is translated as a parametrized curve, namely the trajectory in P of the representative point. This curve is an *integral curve* of the vector field X on P which corresponds to the differential system: for each point x in P the system provides a vector $X(x)$ which will be the velocity vector of the representative point and which is 'tangent' to P (in mechanics, a vector tangent to the phase space is often called a "virtual dispacement respecting the constraints"). So that we can work in a simple and pleasant setting we shall assume that the phase space is a manifold and that the vector field under consideration is as regular as may be necessary (say C^∞ to fix the ideas).

The three basic problems will be solved by three fundamental theorems: *uniqueness* of the integral curve passing through a given point, *existence* of this integral curve over a maximal interval of time, and *regularity* of this curve.

Since initial data, parameters, etc. may be included in the representative point this last result will also imply the regularity of the system as a function of these auxiliary data.

The set of all the possible time evolutions, that is to say all the integral curves of the field X, is conveniently represented as the *integral flow* associated to X. It is the family (Φ_t) of maps (partially defined) from P to P such that $\Phi_t(x)$ is the point reached after time t starting from initial position x. In the nicest case when all the integral curves are defined on the whole of $\mathbf{R}$ (the field X is then said to be *complete*) the map Φ_t is everywhere defined for all t and depends *multiplicatively* on t, in the sense that $\Phi_t \circ \Phi_{t'} = \Phi_{t+t'}$. We thus obtain a so-called *one-parameter group of diffeomorphisms* of P. Conversely, starting from a one-parameter group of diffeomorphisms of P we construct a vector field that it comes from and which is often called its *infinitesimal generator*.

The global qualitative study of a vector field consists of determining its *phase portrait*, that is to say the topological structure of the set of integral curves. In this regard there are two special types of integral curves which immediately stand out: those which reduce to a point (*singular points*) and those which are periodic. In mechanics, the former correspond to equilibria and the latter to oscillations. We shall study them in more detail in Chapts. 8 and 9. Let us simply note that the singular points are where the vector field vanishes, and that at a non-singular point the 'local phase portrait' is very simple. It is given by the *Straightening-out Theorem*: in a suitable local coordinate system the integral curves are segments of parallel straight lines.

In detail the contents of this chapter are as follows. Sects. 2,3 and 4 give definitions and terminology, first in the special case of open sets of vector spaces and then in the general case of submanifolds. Sect. 5 is devoted to the Uniqueness Theorem which is very elementary and therefore it is convenient to dispose of it as soon as possible. After that, in Sects. 6 and 7, we are able to complete the general theory from the opening sections.

These first sections are in fact no more than warming-up exercises. The serious matters begin next. The Existence Theorem occupies Sects. 8,9 and 10 in its various forms: local existence and straightening-out, global existence and the principle of *a priori* bounds, integral flows. In Sect. 11 we present some general features of phase portraits, and end with some comments relating the 'continuous flows' previously introduced and the 'discrete flows' obtained by iteration of a diffeomorphism.

6.2 Examples of Vector Fields (R^n Case)

6.2.1 A *vector field* on an open subset U of a finite-dimensional vector space E is a map X from U into E. An *integral curve* of X, or equivalently a solution of the differential equation

$$\dot{x} = X(x), \quad x \in U , \tag{6.2.1}$$

is a differentiable parametrized arc $\gamma : I \to U$ such that for every $t \in I$ we have

$$\dot{\gamma}(t) = X(\gamma(t)). \tag{6.2.2}$$

We often say that U is the *phase space* of the problem considered. To solve (or to integrate) the differential equation (2.1) is to find all its integral curves. Note that if $\gamma : t \mapsto \gamma(t)$ is an integral curve of X then so is $t \mapsto \gamma(t + t_0)$ for every $t_0 \in \mathbf{R}$; therefore in order to find the integral curves γ passing through a given point a of U it is enough to solve the 'Cauchy problem' $\gamma(0) = a$.

6.2.2. For the constant arc $\gamma(t) = a$ to be an integral curve of X it is necessary and sufficient that $X(a) = 0$. The zeros of X are often called *singular points* or *singularities* of X, even though the map $X : U \to E$ need not have any singularity at such a point[38].

6.2.3. These general definitions encompass numerous special cases, some of which are worth making explicit. Suppose first of all that the phase space U is an open set in $\mathbf{R}^n$ and let $X_i(x_1, \ldots, x_n)$, $i = 1, \ldots, n$ denote the components of X at the point with coordinates $x_1, \ldots, x_n$. We can then write (2.1) in the form of a system of differential equations

$$\dot{x}_i = X_i(x_1, \ldots, x_n), \quad i = 1, \ldots, n , \tag{6.2.3}$$

and the integral curves $\gamma : I \to U$ of (2.1) are sequences $(\gamma_1, \ldots, \gamma_n)$ of functions differentiable on I such that for every $t \in I$ we have $(\gamma_i(t)) \in U$ and

$$\dot{\gamma}_i(t) = X_i(\gamma_1(t), \ldots, \gamma_n(t)), \quad i = 1, \ldots, n, \tag{6.2.4}$$

and the Cauchy problem consists of finding γ_i to satisfy (2.4) and $\gamma_i(0) = a_i$.

6.2.4. Now consider a *scalar n-th order differential equation* of the form

$$\frac{d^n x}{dt^n} = g(x, \frac{dx}{dt}, \ldots, \frac{d^{n-1} x}{dt^{n-1}}), \tag{6.2.5}$$

where the function g is defined on an open set U of $\mathbf{R}^n$, and associate to it the following first order differential system on U:

---

[38]See 6.1 for an explanation of the origin of this terminology.

$$\dot{x}_1 = x_2, \ \dot{x}_2 = x_3, \ \ldots, \ \dot{x}_n = g(x_1, \ldots, x_n). \tag{6.2.6}$$

The solutions γ of (2.5) and the solution systems (γ_i) of (2.6) correspond bijectively: to a solution γ of (2.5) we associate the sequence $\gamma_1 = \gamma$, $\gamma_2 = d\gamma/dt$, $\ldots$, $\gamma_n = d^{n-1}\gamma/dt^{n-1}$. The given n-th order equation is said to have been *reduced to first order*. It is essential to note that there is no point-wise transformation between the *configuration space* $\mathbf{R}$ described by the variable x and the *phase space* $\mathbf{R}^n$ described by $(x_1, \ldots, x_n)$: we associate to a parametrized arc of the first a parametrized arc of the second in such a way that the integral curves of the two vector fields correspond. This also explains why the Cauchy problem for (2.5) is written

$$\gamma(0) = a_1, \ \ldots, \ \frac{d^{n-1}\gamma}{dt^{n-1}}(0) = a_n.$$

In fact the phase space considered in this example is nothing other than the jet space $P^{n-1}(\mathbf{R}; \mathbf{R})$ introduced in 3.8.3.

This can all be generalized immediately to higher order differential systems. The study of a system of r equations of orders $n_1, \ldots, n_r$ reduces to the study of a vector field on a phase space of dimension $n_1 + \cdots + n_r$. Note in passing that there is some abuse in speaking of "the phase space" associated to an equation (or system) of higher order. For example, the second order equation $\ddot{x} + \omega^2 x = 0$ can be reduced to a first order system $\dot{x} = y$, $\dot{y} = -\omega^2 x$, but also more symmetrically to the system $\dot{x} = -\omega y$, $\dot{y} = \omega x$, or even to the single equation in the complex plane $\dot{z} = i\omega z$ (compare the example in 3.3 below). Note also that such a reduction to first order assumes that the given system can be 'solved' with respect to the derivatives of highest order. Thus for example the relation $\dot{x}^2 = 1 - x^3$ gives rise to two differential equations $\dot{x} = \pm\sqrt{1 - x^3}$.

6.2.5. There is another interesting case worth pointing out. Suppose the vector field X depends on additional parameters and we wish to study the way in which the solutions depend on these parameters. To fix the ideas, let Λ be an auxiliary vector space, with U an open set in $E \times \Lambda$ and $X : U \to E$ a map. Then the study of the 'parametrized' differential system

$$\dot{x} = X(x, \lambda), \quad (x, \lambda) \in U$$

can be reduced to that of the usual system

$$\dot{x} = X(x, \lambda), \ \dot{\lambda} = 0, \quad (x, \lambda) \in U,$$

and hence to the study of the vector field $(x, \lambda) \mapsto (X(x, \lambda), 0)$ on U.

6.2.6. Suppose the vector space E has a euclidean scalar product $(\xi \mid \eta)$. To each C^1 function f on U we associate its *gradient*, that is the vector field $\mathrm{grad}(f)$ such that

$$(\mathrm{grad}(f)(x) \mid \eta) = df(x) \cdot \eta$$

for all $x \in U$ and all $\eta \in E$. At every noncritical point x of f the gradient of f is orthogonal to the level hypersurface of f passing through x. The integral curves of this field are the *gradient lines* of f. They are orthogonal to the level hypersurfaces of f. When f is a quadratic form on E the vector field $\mathrm{grad}(f)$ is linear; it is the symmetric endomorphism associated to f (see 7.7.7).

6.2.7. Let us return for a moment to the actual definition of an integral curve $\gamma : I \to V$ of the vector field X.

a) First consider the defining relation (2.2) and suppose that X is C^r with $r \in [0, \infty]$. Then since γ is continuous by assumption, $X \circ \gamma$ is continuous and therefore $\dot{\gamma}$ is continuous and γ is C^1; continuing by induction we see that γ is automatically of class C^{r+1}.

b) The interpretation of the relation (2.2) itself poses no difficulty if t is in the interior of I. If t is an end-point of I we have to interpret $\dot{\gamma}(t)$ as a semi-derivative (to the right or the left). Finally, if I reduces to the point t then $\dot{\gamma}(t)$ is indeterminate and (2.2) has to be read as $\gamma(t) \in U$.

c) For every $a \in U$ there thus exists at least one integral curve passing through a : the map from $\{0\}$ into V taking the value a ! Of course we are clearly looking for more serious solutions and we shall see later (Theorem 8.1) that when X is C^1 every point has an integral curve passing through it and defined on an *open* interval.

d) Note that from the definition we can always concatenate integral curves. Let a, b and c be three real numbers with $a \le b \le c$ and let $\gamma : [a, b] \to U$ and $\delta : [b, c] \to U$ be two integral curves with $\gamma(b) = \delta(b)$. Define $\theta : [a, c] \to U$ by $\theta(t) = \gamma(t)$ for $t \in [a, b]$ and $\theta(t) = \delta(t)$ for $t \in [b, c]$. Then γ and δ have the same velocity vector for $t = b$, that is $X(\gamma(b)) = X(\delta(b))$, so θ is differentiable at b and is an integral curve of X.

6.3 First Integrals

We have already introduced the *Lie derivative* L_X associated to the vector field X on U. It is the first order differential operator which associates to a function ϕ (of class C^1, say) the function $L_X \phi$ such that

$$(L_X \phi)(a) = \langle X(a), d\phi(a) \rangle = \frac{d}{dt}\phi(a + tX(a))|_{t=0}$$

$$= \lim_{t=0} \frac{1}{t}(\phi(a + X(t)) - \phi(a)).$$

For $E = \mathbf{R}^n$ we have

$$L_X = \sum_{i=1}^{n} X_i \frac{\partial}{\partial x_i},$$

and in particular $L_X x_i = X_i$. If E has a scalar product $(\xi \mid \eta)$ then $L_X \phi = (X \mid \mathrm{grad}(\phi))$; in particular, if X is itself the gradient of a function f then we obtain $L_X \phi = (\mathrm{grad}(f) \mid \mathrm{grad}(\phi))$ and $L_X f = \|\mathrm{grad}(f)\|^2$.

The Lie derivative allows us, without solving the equation, to calculate the derivative at a point along an integral curve that passes through that point:

Proposition 6.3.1. *If γ is an integral curve of X we have $(\phi \circ \gamma)'(t) = (L_X \phi)(\gamma(t))$.*

This is immediate: we have

$$(\phi \circ \gamma)'(t) = \langle \dot{\gamma}(t), d\phi(\gamma(t)) \rangle = \langle X(\gamma(t)), d\phi(\gamma(t)) \rangle = (L_X \phi)(\gamma(t)).$$

$\square$

6.3.2. We say that ϕ is a *first integral*[39] if $L_X \phi = 0$; then ϕ is constant along each integral curve and the integral curves are therefore traced on the level sets $\{\phi = a\}$. If a is a noncritical point of the first integral ϕ then $X(a)$ is tangent at a to the level hypersurface $\phi^{-1}(\phi(a))$.

6.3.3. We now give a simple example: a *conservative mechanical system*. Let U be the configuration space; it is an open subset of a finite-dimensional vector space V equipped with a euclidean scalar product $(\xi|\eta)$. Let f be the potential; it is a C^1 function on U. The system we shall study is the second order vector equation

$$\ddot{x} + \mathrm{grad}(f)(x) = 0, \quad x \in U. \tag{6.3.1}$$

If (x_i) is a linear coordinate system associated to an orthogonal basis of V for which

$$((\xi_i) \mid (\eta_i)) = \sum_{i=1}^{n} m_i \xi_i \eta_i,$$

the equation (3.1) can be written in the usual form

$$m_i \ddot{x}_i + \frac{\partial f}{\partial x_i} = 0, \quad i = 1, \ldots, n. \tag{6.3.2}$$

Then (3.1) is reduced to first order by

---

[39]This old-fashioned and somewhat strange terminology evokes the fact that being given a first integral in general allows us to lower the order of the equation by 1; see 3.7.

$$\dot{x} = y, \quad \dot{y} = -\operatorname{grad}(f)(x), \tag{6.3.3}$$

where in orthogonal coordinates

$$\dot{x}_i = y_i, \ \dot{y}_i = -\frac{1}{m_i}\frac{\partial f}{\partial x_i}, \quad i = 1, \dots, n, \tag{6.3.4}$$

which creates the vector field X in the phase space $U \times V$ such that $X(x, y) = (y, \operatorname{grad}(f)(x))$. We see immediately that the energy $E = (y \mid y)/2 + f(x)$ is a first integral (*conservation of energy*) since

$$L_X E(x, y) = df(x) \cdot y - (y \mid \operatorname{grad}(f)(x)) = 0.$$

The critical points of the energy E are pairs (x, y) with $y = 0$ and $df(x) = 0$, namely the singularities of the vector field X. They correspond to equilibrium positions of the system. Away from these points we have the level surfaces $E^{-1}(\alpha)$ on which the integral curves are traced.

Note that there is another reduction to first order also used, where it is not *velocities* but *momenta* that are introduced. In orthogonal coordinates this comes down to considering $p_i = m_i \dot{x}_i$ and writing

$$\dot{x}_i = \frac{1}{m_i} p_i, \quad \dot{p}_i = -\frac{\partial f}{\partial x_i}, \quad i = 1, \dots, n.$$

A little reflection shows that in coordinate-free terms this amounts to placing ourselves in the phase space $U \times V^*$. Let $u : V^* \to V$ be the (bijective) linear map associated to the given scalar product such that by definition $\alpha \cdot \xi = \langle \xi | \alpha \rangle = (\xi | u(\alpha))$ for $\xi \in V$ and $\alpha \in V^*$. The associated first order system is then

$$\dot{x} = u(p), \quad \dot{p} = -df(x), \tag{6.3.5}$$

which creates the vector field Y such that $Y(x, p) = (u(p), -df(x))$. Let Q be the quadratic form on V^* defined by $Q(\alpha) = (u(\alpha)|u(\alpha)) = \langle u(\alpha), \alpha \rangle$, and put $H(x, p) = Q(p)/2 + f(x)$. Then $dH(x, p) \cdot (\xi, \alpha) = df(x) \cdot \xi + \alpha \cdot u(p)$ and (3.5) can also be written

$$\dot{x} = d_p H(x, p), \quad \dot{p} = -d_x H(x, p). \tag{6.3.6}$$

Thus we see the two usual presentations of mechanics with a finite number of degrees of freedom: the *Lagrangian* presentation where the phase space is the tangent bundle $U \times V$ of the configuration space U, and the *Hamiltonian* presentation where the phase space is the 'cotangent' bundle $U \times V^*$ of the configuration space. We shall not pursue this theme any further, but refer to [AM] or [A5].

6.3.4. The procedure for lowering the order generalizes to the case when we know several first integrals. To explain this procedure it is convenient to know how to *express a vector field in curvilinear coordinates* and more generally to *transform a vector field by a diffeomorphism*. Let E and F be two finite-dimensional vector spaces, with U, V open sets in E, F respectively, and with $f : U \to V$ a diffeomorphism. Suppose we are given a vector field X on U. We define the transformed vector field Y on V as follows. For each point $b \in V$ corresponding to a point $a = f^{-1}(b) \in U$, the vector $Y(b)$ is the image of $X(a)$ under the derived map $f'(a) : E \to F$. In formulae we have

$$Y(f(a)) = f'(a) \cdot X(a),$$

which can be written also as $X(f^{-1}(b)) = (f^{-1})'(b) \cdot Y(b)$. Thus X is the transform of Y by the inverse diffeomorphism $f^{-1} : V \to U$.

We also say that Y is the direct image (or *push-forward*) of X – and that X is the inverse image (or *pull-back*) of Y – by f, and write $Y = f_* X$, $X = f^* Y$.

The formulae for differentiating composed functions immediately give:

Proposition 6.3.5.

a) *For $\gamma : I \to U$ to be an integral curve of X it is necessary and sufficient that $f \circ \gamma$ be an integral curve of Y.*

b) *Let ϕ be a C^1 function on V . Then $L_X(\phi \circ f) = (L_Y \phi) \circ f$.*   □

6.3.6. As an example let us take for f the diffeomorphism from U to an open set V in $\mathbf{R}^n$ associated to a system of curvilinear coordinates $(x_1, \ldots, x_n)$ on U. Then by b) above we see that the components of the transformed field at the point $(x_1, \ldots, x_n)$ are none other than the $L_X x_i$, which we denote by X_i as in the case of linear coordinates. Dually, if we represent a parametrized arc $\gamma : I \to U$ by its ('curvilinear') components $\gamma_i = x_i \circ \gamma$, then it follows from a) that the integral curves are characterised by the relations

$$\dot{\gamma}_i(t) = X_i(\gamma_1(t), \ldots, \gamma_n(t)), \quad i = 1, \ldots, n.$$

Therefore, in conformity with the general principle stated in 1.7.2, the expressions given above in linear coordinates remain valid in curvilinear coordinates; for further details see 4.4 and 4.5 below.

6.3.7. After these remarks, we again take up the method of *reduction of order* encountered in 2.4. Suppose in the neighbourhood of a point a of U we have r first integrals $\phi_1, \ldots, \phi_r$ whose differentials at a are independent. We can then find a system of local coordinates $(x_1, \ldots, x_n)$ at a with $x_1 = \phi_1, \ldots, x_r = \phi_r$. In this system we have $X_1 = L_X x_1 = 0, \cdots, X_r = L_X x_r = 0$ and the differential equation can be written

$$\dot{x}_1 = 0, \ldots, \dot{x}_r = 0, \dot{x}_{r+1} = X_{r+1}(x_1, \ldots, x_n), \ldots, \dot{x}_n = X_n(x_1, \ldots, x_n).$$

Solving this system reduces to solving the system of $n - r$ equations

$$\dot{x}_i = X_i(\alpha_1, \ldots, \alpha_r, x_{r+1}, \ldots, x_n), \quad i = r + 1, \ldots, n,$$

where $\alpha_1, \ldots, \alpha_r$ are parameters. In geometric terms, the integral curves are traced on the level surfaces

$$V_{\alpha_1, \ldots, \alpha_r} = \bigcap_{i=1}^{r} x_i^{-1}(\alpha_i)$$

which have codimension r, are always tangent to the field X, and on which the functions $x_{r+1}, \ldots, x_n$ provide curvilinear coordinates. We have, as they say, *lowered the order of the equation by r*.

6.4 Vector Fields on Submanifolds

6.4.1. Not all the differential systems that we might meet can necessarily be expressed as vector fields on (open subsets of) vector spaces. In particular, this is the case when the natural phase space is not linear (for example, it may be a level manifold for certain first integrals, as we have just seen). We therefore need to generalize the definitions in the previous section to the case of phase spaces that are *submanifolds.* If V is a submanifold of the finite-dimensional vector space E, a (tangent) *vector field* on V is by definition a map X from V to E such that for every $x \in V$ we have $X(x) \in T_x V$. An *integral curve* of X, or equivalently a solution of the differential equation

$$\dot{x} = X(x), \quad x \in V, \tag{6.4.1}$$

is a differentiable parametrized arc $\gamma : I \to E$ such that $\gamma(t) \in V$ and

$$\dot{\gamma}(t) = X(\gamma(t)) \tag{6.4.2}$$

for every $t \in I$. The remarks made in 2.7 carry over to this more general setting just as they stand.

6.4.2. The definition of Lie derivative can likewise be transported to this new setting. If g is a C^1 function on V, the *Lie derivative* of g with respect to the field X is the function $L_X g$ on V defined as follows. For every $a \in V$ we have that $X(a)$ is an element of $T_a V$ and $T_a g$ is a linear form on $T_a V$; we then let

$$(L_X g)(a) = T_a g \cdot X(a) = \langle X(a), T_a g \rangle. \tag{6.4.3}$$

If g is the restriction to V of a function f on an open set U in E containing V and if X is the restriction to V of a vector field Y on U, then $L_X g$ is the restriction of the Lie derivative $L_Y f$. It follows in particular that if g is of class C^{r+1} and X is C^r, with $r \in [0, \infty]$, then $L_X g$ is C^r.

The following formulae concerning elementary properties of the Lie derivative are important to note:

$$L_X(g + g') = L_X g + L_X g', \tag{6.4.4}$$

$$L_X(g g') = g L_X g' + g' L_X g, \tag{6.4.5}$$

$$L_{X+X'} g = L_X g + L_{X'} g, \tag{6.4.6}$$

$$L_{gX} g' = g L_X g', \tag{6.4.7}$$

where X and X' denote two vector fields on V and g and g' are two functions on V.

6.4.3. Now we can extend what was said in 3.4 about diffeomorphisms. Let E and F be two finite-dimensional vector spaces, with V a submanifold of E and W a submanifold of F, and let $u : W \to V$ be a diffeomorphism (2.8.3). Let X denote a vector field on V. The vector field u^*X on W is defined as follows. For two points $a \in W$ and $b = u(a) \in V$ which correspond, the linear map $T_a u$ is a bijection from $T_a W$ to $T_{u(a)} V$, and we define $(u^*X)(a)$ by the equivalent equations $X(b) = T_a u \cdot (u^*X)(a)$ and $(u^*X)(a) = T_b u^{-1} \cdot X(b)$. As in Proposition 3.5, we see that the integral curves of X are the $u \circ \gamma$ where γ runs through all integral curves of u^*X. Likewise, for every function g on V we have

$$(L_X g) \circ u = L_{u^*X}(g \circ u). \tag{6.4.8}$$

6.4.4. Let us look at this construction in detail in the particular case of parametrizations and systems of local coordinates. Let $u : \Omega \to V$ be a local parametrization of V. By definition (2.9.4), Ω is therefore an open set in $\mathbf{R}^n$ and u is an embedding whose image is an open subset V' of V ; the map inverse to u is a system of local coordinates $(x_1, \ldots, x_n)$ with domain V'. Let $a \in V'$, corresponding to a point $m = (x_1(a), \ldots, x_n(a))$ in the open set Ω. The linear map $u'(m)$ tangent to u at the point m is a bijection from $\mathbf{R}^n$ to the tangent space $T_a V$. Consequently, the partial derivative vectors $(\partial u/\partial x_i)(m)$ form a basis for $T_a V$ and so we obtain n vector fields on V' that are traditionally denoted $\partial/\partial x_i$. Therefore by construction we have

$$\frac{\partial}{\partial x_i}(a) = \frac{\partial u}{\partial x_i}(m), \ a = u(m) .$$

The components of v in the basis formed by the $\frac{\partial}{\partial x_i}(a)$ are called the *components* of the tangent vector $v \in T_a V$ in the system of local coordinates $(x_1, \ldots, x_n)$ or in the parametrization u . Every vector field X on V' thus has components X_i considered either as functions on V' or on Ω :

$$X = \sum_{i=1}^{n} X_i \frac{\partial}{\partial x_i}. \tag{6.4.9}$$

If $\gamma : I \to V'$ is a parametrized arc traced on V' the components of γ are the $\gamma_i = x_i \circ \gamma$ and (4.2) can be written in local coordinates in the usual way as

$$\dot{\gamma}_i(t) = X_i(\gamma_1(t), \ldots, \gamma_n(t)), \ i = 1, \ldots, n. \tag{6.4.10}$$

Now let g be a function on W and put $h = g \circ u$, which can also be written as $g(a) = h(x_1(a), \ldots, x_n(a))$ for $a \in V'$, or $g = h(x_1, \ldots, x_n)$ for short. Then by the formula (4.8) we have

$$(L_X g)(a) = \sum_{i=0}^{n} X_i(a) \frac{\partial g}{\partial x_i}(a) \tag{6.4.11}$$

where, following the usual abuse of notation, we write $\partial g / \partial x_i$ instead of $\partial h / \partial x_i$. In particular we have $X_i = L_X x_i$; it follows that the map $X \mapsto L_X$ is injective, so this justifies writing (4.9) which is tantamount to putting $X = L_X$.

6.4.5. As an example, take V to be the open subset of $\mathbf{R}^2$ that is the complement of the origin, and for u take a local parametrization in polar coordinates. Thus Ω is an open set in $\mathbf{R}^2$, and we let r and θ denote the local coordinates on V so obtained. We have the usual formulae for local change of coordinates:

$$x = \cos \theta, \quad y = \sin \theta.$$

The vector field X can then be expressed in either of the two forms

$$X = (L_X x)\frac{\partial}{\partial x} + (L_X y)\frac{\partial}{\partial y} = (L_X r)\frac{\partial}{\partial r} + (L_X \theta)\frac{\partial}{\partial \theta}.$$

To pass from one description to the other we just have to apply the formulae above. For example, we obtain

$$L_X r = \frac{1}{r}(x L_X x + y L_X y), \quad L_X \theta = \frac{1}{r^2}(x L_X y - y L_X x).$$

6.5 The Uniqueness Theorem and Maximal Integral Curves

Theorem 6.5.1. (Uniqueness Theorem.) *Let V be a submanifold of the finite-dimensional vector space E, with X a C^1 vector field on V, and let $\gamma : I \to V$ and $\gamma' : I \to V$ be two integral curves of X. If there exists t_0 in $I \cap I'$ with $\gamma(t_0) = \gamma(t_0')$ then $\gamma(t) = \gamma'(t)$ for all $t \in I \cap I'$.*

Consider the set J (nonempty by assumption) consisting of those $t \in I \cap I'$ such that $\gamma(t) = \gamma(t')$. Since γ and γ' are continuous, J is closed in $I \cap I'$; since $I \cap I'$ is connected (it is an interval) it is enough to prove that J is open in $I \cap I'$. Therefore, let t_0 be a point of J ; put $a = \gamma(t_0) = \gamma'(t_0) \in V$. Let E be given a norm. Since X is C^1 there exist constants A and $\varepsilon > 0$ with

$$\|X(x) - X(y)\| \leq A\|x - y\| \quad \text{for} \quad \|x - y\| \leq \varepsilon. \tag{6.5.1}$$

Since γ and γ' are continuous and have the same value at t_0 there exists $\eta > 0$ with $\eta A < 1$ and with $\|\gamma(t) - \gamma'(t)\| \leq \varepsilon$ for $t \in I \cap I'$ and $|t - t_0| < \eta$. Hence for $t \in I \cap I'$ and $|t - t_0| < \eta$ we have

$$\|\dot{\gamma}(t) - \dot{\gamma}'(t)\| = \|X(\gamma(t)) - X(\gamma'(t))\| \leq A\|\gamma(t) - \gamma'(t)\|, \tag{6.5.2}$$

whence by integration

$$\|\gamma(t) - \gamma'(t)\| \leq A \Big| \int_{t_0}^{t} \|\gamma(s) - \gamma'(s)\| ds \Big|. \tag{6.5.3}$$

If m is an upper bound for $\|\gamma(t) - \gamma(t_0)\|$ for $t \in I \cap I'$ and $|t - t_0| < \eta$ we deduce from this that $m \leq A\eta m$ and so $m = 0$ since we supposed $A\eta < 1$. Thus $\gamma(t)$ and $\gamma'(t)$ coincide for $t \in I \cap I'$ and $|t - t_0| < \eta$ and J is indeed open in $I \cap I'$, which completes the proof.   □

Remark: for this proof to work it is enough to assume that the vector field X satisfies bounds of type (5.1), which we describe by saying that it is *locally Lipschitz*.

Definition 6.5.2. The integral curve $\gamma : I \to V$ of the C^1 vector field X is called *maximal* if it has the following property: for every integral curve $\delta : J \to V$ of X such that there is a point $t \in I \cap J$ with $\gamma(t) = \delta(t)$ we have $J \subset I$ (and therefore $\delta(t) = \gamma(t)$ for $t \in J$ by the Uniqueness Theorem).

Note that every integral curve defined on the whole of $\mathbf{R}$ is maximal and that if two maximal integral curves coincide at one point then they are equal. There is a special case worth noting: if a is a singular point of X then the constant map with value a is an integral curve that is clearly maximal; hence every integral curve that passes through a singularity is constant. Equivalently, every non-constant integral curve $\gamma : I \to V$ avoids all singularities, or in other words it has non-vanishing derivative everywhere and is therefore an immersion.

6.5.3. The Uniqueness Theorem serves essentially to "glue together" integral curves: suppose X is C^1 and consider two integral curves $\gamma : I \to V$ and $\delta : I \to V$ of X such that $\gamma(I)$ and $\delta(I)$ meet. Then there exist $t_0 \in I$ and $t_1 \in J$ with $\gamma(t_0) = \delta(t_1)$. Let $I' = J + (t_0 - t_1)$ and let $\gamma' : I' \to V$ be the integral curve $t \mapsto \delta(t + t_1 - t_0)$ so that $\gamma'(I') = \delta(J)$. Since $\gamma(t)$ and $\gamma'(t)$ are equal for $t = t_0$ they are equal for all of $I \cap I'$ (Theorem 5.1). Now letting $I'' = I \cup I'$ we define an integral curve $\gamma'' : I'' \to V$ extending γ and γ' by putting $\gamma''(t) = \gamma(t)$ for $t \in I$ and $\gamma''(t) = \gamma'(t)$ for $t \in I'$ and we have

$$\gamma''(I'') = \gamma(I) \cup \gamma'(I') = \gamma(I) \cup \delta(J).$$

Note that if γ is maximal we have $I' \subset I$ and $\delta(J) \subset \gamma(I)$. If γ and δ are maximal we have $\gamma = \gamma'$ and $\delta(J) = \gamma(I)$.

Corollary 6.5.4. *Suppose X is a C^1 vector field. The relation "x and y belong to the image of the same integral curve of X" is an equivalence relation on V.*

The only point that could cause a problem is the transitivity. However, with the notation above, if x and y belong to $\gamma(I)$ and if y and z belong to $\delta(J)$ then x and z belong to $\gamma''(I'')$.   □

The equivalence classes are called the *orbits* of X. The partition of V into equivalence classes is generally called the *phase portrait* of X. The phase portrait usually includes also the direction in which the orbits are traversed (see examples in the Figures). By definition, the image of an integral curve is contained in an orbit, and we saw above that the image of a maximal integral curve is an entire orbit. Conversely:

Corollary 6.5.5. *Suppose X is C^1 and let a be a point of V. There exists a maximal integral curve $\gamma_a : I_a \to V$ of X such that $0 \in I_a$ and $\gamma_a(0) = a$. It is uniquely determined and its image is the orbit of a.*

The last two assertions have already been proved. It is therefore enough to prove the existence of the sought-for curve. Consider the family $\mathcal{C}$ of all the integral curves $\gamma : I \to V$ such that $0 \in I$ and $\gamma(0) = a$. Let I_a be the union of all the intervals so obtained; it is an interval containing 0. For every $t \in I_a$ the curves of the family $\mathcal{C}$ that are defined at t (that is, whose corresponding interval contains t) all take the same value at t (Theorem 5.1); let $\gamma_a(t)$ be this value. Then $\gamma_a : I_a \to V$ is an integral curve of X and we have to verify that it is maximal. Thus let $\delta : J \to V$ be an integral curve of X such that there exists $t \in I_a \cap J$ with $\gamma_a(t) = \delta(t)$. Then as above we construct an integral curve defined on $I_a \cup J$ which extends γ_a and δ. But this curve is also a member of the family $\mathcal{C}$, which implies that $I_a \cup J$ is contained in I_a and so J is contained in I_a. Thus γ_a is indeed maximal, as we wished to prove. □

The interval I_a can be called the *lifetime interval* of a. If $I_a = \mathbf{R}$ for all $a \in V$, we often say that the vector field X is *complete*. We shall see later that every vector field on a compact manifold is complete (9.3).

For all that we know so far, we might have $I_a = \{0\}$ for every a ; we shall see below (Theorem 8.1) that fortunately this is not the case (provided X is not everywhere zero), but it is important to note that what we have just said about maximal integral curves in no way constitutes an existence result but is only a convenient way of presenting the Uniqueness Theorem.

6.6 Vector Fields and Line Fields. Elimination of the Time

6.6.1. Let V be a submanifold of a finite-dimensional vector space. A *line field D* on V is an assignment to each point x in V of a line $D(x)$ passing through the origin in the vector space $T_x V$. Likewise, a *half-line field* on V is an assignment to each point x in V of a half-line with end-point at the origin in $T_x V$. To every vector field that vanishes nowhere there is naturally associated a half-line field and therefore also a line field.

The points where a vector field vanishes thus appear as points where the associated half-line field and line field are not defined; this is the origin of the term 'singular points'.

The fields X and $-X$ define opposite half-line fields and the same line-field. For two vector fields X and Y to define the same half-line field it is necessary and sufficient that $Y = aX$ where a is a function that is everywhere positive. Likewise, vector fields X and Y define the same line field if $Y = bX$ where b is a function that is everywhere nonzero; if X and Y are assumed continuous then b will be continuous and in particular the sign of b will be constant if V is connected. A curve (1-dimensional submanifold) in V is said to be tangent to the line field D if for every $x \in C$ we have $T_x C = D(x)$. Let $\gamma : I \to V$ be an integral curve of the vector field X; if γ is an immersion then its image $\gamma(I)$ is tangent to the line field associated to X.

Let X be a vector field on V and let $\gamma : I \to V$ be an integral curve of X. If J is an interval of $\mathbf{R}$ and $u : J \to I$ is a C^1 diffeomorphism (that is a bijection with continuous non-vanishing derivative) then the arc $\delta = \gamma \circ u : J \to V$ satisfies $\dot{\delta}(t) = \frac{du}{dt}(t)X(\delta(t))$. Conversely:

Lemma 6.6.2. *Let X be a vector field on V and let $\delta : J \to V$ be a C^1 parametrized arc such that there exists a continuous nonvanishing function v on J with $\dot{\delta}(t) = v(t)X(\delta(t))$. Then there exists an integral curve $\gamma : I \to V$ of X and a diffeomorphism $u : J \to I$ such that $\delta = \gamma \circ u$.*

To see this, let u be a primitive of v. Then u is a C^1 function with nowhere vanishing derivative. By a classical result on functions of one variable, u is a diffeomorphism of J onto an interval I of $\mathbf{R}$ and it suffices to take as γ the composition $\delta \circ u^{-1}$. $\square$

It follows from this lemma that *two C^1 vector fields that define the same line field have the same orbits* (with possibly different parametrizations) and that *two C^1 vector fields that define the same half-line field have the same phase portrait.*

6.6.3. We frequently meet line fields in the following context. A *time-dependent vector field* on V is by definition a map $(x,t) \mapsto X(x,t)$ from an open set Ω of $V \times \mathbf{R}$ to E such that $X(x,t)$ belongs to $T_x V$ for all $(x,t) \in \Omega$. The solutions to the *nonautonomous differential equation*

$$\dot{x} = X(x,t), \quad (x,t) \in \Omega \tag{6.6.1}$$

are by definition those differentiable parametrized arcs $\gamma : I \to V$ such that for every $t \in I$ we have $(x,t) \in \Omega$ and

$$\dot{\gamma}(t) = X(\gamma(t),t). \tag{6.6.2}$$

The procedure called *elimination of the time* consists of associating to (6.1) the *autonomous* differential equation

$$\dot{x} = X(x,u), \quad \dot{u} = 1, \tag{6.6.3}$$

whose solution is equivalent to that of (6.1). This amounts to introducing on Ω (which we call the *extended phase space*) the vector field Y such that $Y(x,t) = (X(x,t),1)$ and its corresponding line field D. Observe that D enables Y and X to be reconstructed: $X(x,t)$ is the unique vector $v \in T_x V$ such that $(v,1)$ belongs to the line $D(x,t)$ in $(T_x V) \times \mathbf{R} = T_{(x,t)}(V \times \mathbf{R}) \subset E \times \mathbf{R}$. Every integral curve γ of Y is an immersion (because the projection of $\gamma(t)$ on $\mathbf{R}$ is of the form $t+$constant, which implies that γ is bicontinuous) and the graphs of the (maximal) solutions of (6.1) are exactly those (maximal) connected curves in V that are everywhere tangent to the field D.

For example, integrating the non-autonomous differential equation

$$\frac{dy}{dx} = \frac{f(x,y)}{g(x,y)},$$

where the functions f and g do not vanish simultaneously, reduces to integrating the autonomous differential equation

$$\frac{dx}{dt} = f(x,y), \quad \frac{dy}{dt} = g(x,y),$$

that is finding the (maximal) integral curves of the vector field $X(x,y) = (f(x,y), g(x,y))$.

6.6.4. Notice that the procedure for eliminating the time can already be applied to autonomous differential equations. The study of a usual (time-independent) vector field X on V can be viewed as the study of the non-singular vector field $Y = (X,1)$ on $V \times \mathbf{R}$ or even of the line field D generated on $V \times \mathbf{R}$. Let C be a connected curve that is everywhere tangent to D; let $pr_1 : C \to V$ and $pr_2 : C \to \mathbf{R}$ be the two projections. Then pr_2 is an embedding, with image an interval I, and $pr_1 \circ pr_2^{-1} : I \to V$ is an integral curve of X. The maximal curves C that correspond to singular points a of X are the straight lines $\{(a,t), t \in \mathbf{R}\}$ and pr_1 is then constant. For all the others (assuming X is of class C^1) the projection pr_1 is an immersion whose image is an orbit of X.

6.7 One-parameter Groups of Diffeomorphisms

6.7.1 Let V be a submanifold of a finite-dimensional vector space E and let Φ be a C^∞ map from $V \times \mathbf{R}$ into V. For $t \in \mathbf{R}$ we define a map $\Phi_t : V \to V$ by setting $\Phi_t(x) = \Phi(x,t)$. We say that Φ (or the family $(\Phi_t)_{t \in \mathbf{R}}$) is a *one-parameter group of diffeomorphisms*[40] of V if:

----
[40]of class C^∞; the extension to class C^r is immediate.

$$\Phi_0 = Id_V, \qquad (6.7.1)$$

that is $\Phi(x, 0) = x$, $x \in V$, and for all t and t' in $\mathbf{R}$

$$\Phi_t \circ \Phi_{t'} = \Phi_{t+t'}, \qquad (6.7.2)$$

that is $\Phi(\Phi(x, t'), t) = \Phi(x, t + t')$, $x \in V$. We see immediately that each Φ_t is a C^∞ bijection from V to V with inverse bijection Φ_{-t}. Fix $x \in V$. Then the map $t \mapsto \gamma_x(t) = \Phi(x, t) = \Phi_t(x)$ is a C^∞ parametrized arc $\gamma_x : \mathbf{R} \to V$ whose velocity vector at $t = 0$ we denote by $X(x)$:

$$X(x) = \frac{d}{dt}\Phi(x, t)|_{t=0} \in T_x V. \qquad (6.7.3)$$

In this way we obtain a C^∞ vector field on V, the *velocity field* of the one-parameter group Φ. From (7.2) we deduce $\gamma_x(t + s) = \Phi_s(\gamma_x(t))$; differentiating with respect to s at $s = 0$ we obtain the relation $\dot\gamma_x(t) = X(\gamma_x(t))$, which can be written as

$$\frac{d}{dt}\Phi_t(x) = X(\Phi_t(x)). \qquad (6.7.4)$$

Consequently the γ_x are the maximal integral curves of the vector field X (which is therefore complete) and knowledge of X suffices to reconstruct the γ_x and hence the one-parameter group Φ by integrating the corresponding differential equation. For this reason we say that the field X is the *infinitesimal generator* of Φ. Note in passing that for a function f (of class C^1) on V to be *invariant* under Φ, that is such that $f(\Phi_t(x)) = f(x)$ for every $t \in \mathbf{R}$ and every $x \in V$, it is necessary and sufficient that it satisfy the infinitesimal invariance condition $L_X f = 0$.

6.7.2. We give some elementary examples for $V = E$.

Let $e \in E$. The one-parameter group of translations $x \mapsto x + te$ has as its infinitesimal generator the constant vector field $X(x) = e$.

The one-parameter group of *homotheties* (scalar multiplications) $x \mapsto e^t x$ has as its infinitesimal generator the 'identity' vector field $X(x) = x$. Notice incidentally the following interpretation of Euler's identity: the condition $L_X f = mf$ is equivalent to $f \circ \Phi_t = e^{mt} f$, $t \in \mathbf{R}$.

6.7.3. In order for the one-parameter group (Φ_t) of diffeomorphisms of E to consist of linear transformations of E it is necessary and sufficient that its infinitesimal generator be a linear map from E to E.

To see this, let x and x' belong to E; let

$$a(t) = \Phi(x + x', t) - \Phi(x, t) - \Phi(x, t').$$

We have $a(0) = 0$,

$$\frac{da}{dt} = X(\Phi((x+x'),t)) - X(\Phi(x,t)) - X(\Phi(x',t))$$

and in particular

$$\frac{da}{dt}(0) = X(x+x') - X(x) - X(x').$$

If each Φ_t commutes with addition the map a is identically zero; then $da/dt(0)$ is zero and X commutes with addition. Conversely, suppose that $X(v+v') = X(v) + X(v')$ for all v and v' in E; then $X(0) = 0$ and $da/dt = X(a(t))$ and it follows by the Uniqueness Theorem that $a(t)$ is identically zero. We argue similarly for commutation with scalar multiplication.

In particular take $E = \mathbf{R}^2$. The one-parameter group of rotations

$$r_t = \begin{pmatrix} \cos t & -\sin t \\ \sin t & \cos t \end{pmatrix}$$

has as infinitesimal generator the matrix $\begin{pmatrix} 0 & -1 \\ 1 & 0 \end{pmatrix}$. Similarly the one-parameter group of hyperbolic rotations

$$h_t = \begin{pmatrix} \cosh t & \sinh t \\ \sinh t & \cosh t \end{pmatrix}$$

has as its infinitesimal generator the matrix $\begin{pmatrix} 0 & 1 \\ 1 & 0 \end{pmatrix}$.

6.7.4. We have seen how we associate to a one-parameter group of diffeomorphisms of V a complete vector field X on V. Conversely, suppose we are given a complete C^∞ vector field X on V. For every $x \in V$ there is the unique maximal integral curve $\gamma_x : \mathbf{R} \to V$ of X such that $\gamma_x(0) = x$. Put $\Phi(x,t) = \gamma_x(t)$. Then the property (7.1) is clear; the property (7.2) follows from the Uniqueness Theorem, since $t \mapsto \Phi(\Phi(x,t'),t)$ and $t \mapsto \Phi(x,t+t')$ are two integral curves of X that coincide for $t = 0$. Finally, we shall see (Theorem 10.1) that the map Φ is of class C^∞; we have therefore constructed a (unique) one-parameter group of diffeomorphisms of V with infinitesimal generator X.

Later (for example in Proposition 7.7.2) we shall observe that every linear vector field $X \in L(E;E)$ is complete; we denote by $\exp(tX)$, $t \in \mathbf{R}$, the associated one-parameter group of linear transformations. We shall return to this situation in detail in Sects. 7.6 and 7.7.

6.8 The Existence Theorem (Local Case)

The following theorem combines two local results: the so-called Cauchy-Lipschitz [41] Theorem that asserts the local existence of solutions of differential equations, and the fact that these solutions depend regularly on the initial conditions.

Theorem 6.8.1. *Let E be a finite-dimensional vector space with V a submanifold of E and X a vector field of class C^r, $r \in [1, \infty]$, on V. Let a be a point of V. There exists an open interval J of $\mathbf{R}$ containing 0, an open set U in V containing a and a C^r map f from $U \times J$ into V having the following two properties for every $x \in U$:*

$$f(x, 0) = x, \tag{6.8.1}$$

$$\frac{d}{dt} f(x, t) = X(f(x, t)), \quad t \in J. \tag{6.8.2}$$

In other words, for every $x \in U$ the map $t \mapsto f(x, t)$ from J into V is an (and hence the) integral curve of X starting at x. With the notation of Corollary 5.5, the result may be expressed as follows: we have $J \subset I_x$ for all $x \in U$ and moreover the function $\gamma_x(t) = f(x, t)$ is of class C^r in the pair (x, t).

By using a local parametrization for V in the neighbourhood of a we reduce the discussion to the case where V is an open subset of E and where $a = 0$. Let E be given a norm. Then V contains an open ball with centre 0; since such a ball is diffeomorphic to E it suffices to consider the case where $V = E$, that is where X is a C^r vector field on the whole space E. Let L denote the vector space of bounded maps α from $[0, 1]$ into E with the norm

$$\|\alpha\| = \sup_u \|\alpha(u)\|_E,$$

in which it is *complete*. Suppose first that the problem has been solved and let $\gamma : (-\varepsilon, \varepsilon) \to E$ be an integral curve of X with $\gamma(0) = x$. Then for $t \in (-\varepsilon, \varepsilon)$ we have

$$\gamma(t) = x + \int_0^t X(\gamma(s)) ds. \tag{6.8.3}$$

Let $\beta(x, t, u) = \gamma(tu) - x$ with $t \in [0, 1]$ and $u \in (-\varepsilon, \varepsilon)$. Then (8.3) can be written

---

[41] This local existence theorem is due to Cauchy (ca. 1820) for differential equations of class C^1 and to Lipschitz (1868) in the 'locally Lipschitz' case. Both use the method of approximation by piecewise linear functions, afterwards known as the Cauchy-Lipschitz method.

$$\beta(x,t,u) = \int_0^{tu} X\big(\gamma(s)\big)ds = \int_0^{tu} X\big(\gamma(ts)\big)d(ts)$$

$$= t\int_0^u X\big(x + \beta(x,t,s)\big)ds. \tag{6.8.4}$$

This suggests the following method: consider the expression

$$\alpha(u) = t\int_0^u X\big(x + \alpha(s)\big)ds, \tag{6.8.5}$$

relating three elements $x \in E, u \in \mathbf{R}$ and $\alpha \in L$, solve it by the Implicit Function Theorem[42] to find a map $(x,t) \mapsto \alpha(x,t) \in F$, define $\beta(x,t,u)$ to be $\alpha(x,t)(u)$ and then take $\gamma(t)$ to be $x + \beta(x,t,1)$. The trick therefore consists essentially of replacing the unknown interval $(-\varepsilon, \varepsilon)$ by the fixed interval $[0,1]$. We now have to verify that it works. Therefore, consider the map $F : E \times \mathbf{R} \times L \to L$ defined by

$$F(x,t,\alpha)(u) = \alpha(u) - t\int_0^u X\big(x + \alpha(s)\big)ds \tag{6.8.6}$$

(clearly the function $u \mapsto F(x,t,\alpha)(u)$ is bounded). It is not very difficult but not much fun to prove that it is of class C^r (in the appropriate sense, since the space E is not finite-dimensional); let us accept this. We have $F(x,0,\alpha) = \alpha$, and hence

$$F(0,0,0) = 0, \quad F'_\alpha(0,0,0) = Id_L. \tag{6.8.7}$$

The Implicit Function Theorem[43] implies that there exists an open set U in E containing 0, an open interval J of $\mathbf{R}$ and a unique C^r map β from $U \times J$ to L with $\beta(0,0) = 0$ and $F(x,t,\beta(x,t)) = 0$, that is

$$\beta(x,t)(u) = t\int_0^u X\big(x + \beta(x,t)(s)\big)ds. \tag{6.8.8}$$

For each $\lambda \in [0,1]$ consider the map h_λ from L to L which associates to $\alpha : u \mapsto \alpha(u)$ the map $h_\lambda\alpha : u \mapsto \alpha(\lambda u)$. An immediate calculation shows that $h_\lambda F(x,t,\alpha) = F(x,\lambda t, h_\lambda\alpha)$ and the uniqeness of the implicit function solution to (8.8) then implies that $h_\lambda\beta(x,t) = \beta(x,\lambda t)$, that is to say $\beta(x,t)(\lambda u) = \beta(x,\lambda t)(u)$, and so

$$\beta(x,t)(\lambda) = \beta(x,\lambda t)(1).$$

Putting this back into (8.8) we obtain

$$\beta(x,t)(1) = t\int_0^1 X\big(x + \beta(x,ts)(1)\big)ds = \int_0^t X\big(x + \beta(x,s)(1)\big)ds. \tag{6.8.9}$$

---

[42]In fact we shall use this theorem in a more general setting than in Chapt. 2, since the space E is not finite-dimensional.

[43]Here we are talking about the generalization of Corollary 2.7.3 to complete normed spaces. We could also go straight back to Theorem 1.4.1.

Setting $f(x,t) = x + \beta(x,t)(1)$, this gives

$$f(x,t) = x + \int_0^t X\big(f(x,s)\big)ds, \tag{6.8.10}$$

which is a form of (8.1) and (8.2). This completes the proof.   □

A particularly convenient form of the above theorem is the following:

Theorem 6.8.2. (Straightening-out Theorem.) *Let E be a finite-dimensional vector space, let V be a submanifold of E and let X be a C^r vector field, $r \in [1, +\infty]$, on V. Let a be a point of V with $X(a) \neq 0$. Then there exists a system of C^r local coordinates $(x_1, \ldots, x_n)$ on V at a such that $L_X x_1 = 1$ and $L_X x_i = 0$ for $i = 2, \ldots, n$.*

In these coordinates ('straightening-out' coordinates) the integral curves of X are the solutions of the system of differential equations

$$\dot{x}_1 = 1, \ \dot{x}_2 = 0, \ldots, \ \dot{x}_n = 0, \tag{6.8.11}$$

and are therefore the straight lines

$$x_1 = a_1 + t, \ x_2 = a_2, \ \ldots, \ x_n = a_n, \tag{6.8.12}$$

and the map f of Theorem 8.1 is given by

$$f(x_1, x_2, \ldots, x_n, t) = (x_1 + t, x_2, \ldots, x_n). \tag{6.8.13}$$

Note that the x_i, $i \geq 2$, are $n - 1$ first integrals of X, defined on a neighbourhood of a and functionally independent. See Fig. 6.1.

Fig. 6.1. Straightening-out in local coordinates

Turning to the proof, we reduce as usual to the case when V is an open set in E. Let E' be a vector subspace of E of codimension 1 and not containing $X(a)$. Consider the map $f : U \times I \to V$ given by Theorem 8.1; let U' be the open subset of E' given by $a + U' = (a + E') \cap U$ and let $g : U' \times I \to V$ be the map $(y,t) \mapsto f(a + y, t)$. We have on the one hand $g(y, 0) = a + y$ for $y \in U'$

and on the other hand $(dg/dt)(0,t) = X(g(0,t))$, so $(dg/dt)(0,0) = X(a)$: the tangent map to g at 0 sends $(\xi,\theta) \in E' \times \mathbf{R}$ to $\xi + \theta X(a) \in E$. Since the subspaces E' and $\mathbf{R}X(a)$ of E are complementary by assumption, $g'(0)$ is bijective. Applying the Local Inversion Theorem we deduce that g is a local diffeomorphism. Choose a basis $(e_2, \ldots, e_n)$ for E'; then the map

$$(x_1, x_2, \ldots, x_n) \mapsto g(x_2 e_2 + \cdots + x_n e_n, x_1) = f(a + x_2 e_2 + \cdots + x_n e_n, x_1)$$

is a local parametrization for E at a. By construction the integral curves of X in this local parametrization are given by the formulae (8.12), which implies the theorem. □

The same argument allows us to prove a slightly more precise form of the theorem as follows. Choose a submanifold V' of V of codimension 1, passing through a and transverse to $X(a)$ at a (that is to say $X(a)$ does not lie in the hyperplane $T_a V'$ of $T_a V$). Then we can construct the system of local coordinates in such a way that x_1 is a local equation for V' at a.

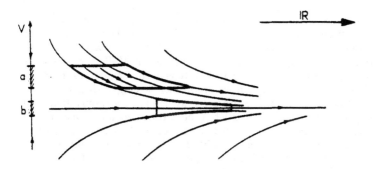

Fig. 6.2. Straightening-out in $V \times \mathbf{R}$

6.8.3. As we have seen, Theorem 8.2 imples the Local Existence Theorem 8.1 in a neighbourhood of a *non-singular* point a. To deduce this in the neighbourhood of a singular point it is enough, by 'elimination of the time', to replace V by $V \times \mathbf{R}$ and X by the field $(X,1)$ on $V \times \mathbf{R}$ that has no singular points. However, it is important to make the following essential remark, illustrated by Fig. 6.2 in which the integral curves of X and of $(X,1)$ are shown for a field on $\mathbf{R}$. In a small neighbourhood U of a non-singular point a the 'spatially' local situation is also local 'in time': every integral curve that enters U leaves after a finite time (which is even bounded independently of the choice of the curve). In contrast, the local situation in the neighbourhood of a singular point b is not 'local in time', as certain integral curves remain for an infinitely long time in U. In technical jargon, the integral curves in the extended phase space project *properly* into the phase space in the neighbourhood of non-singular points, but not in the neighbourhood of singular points (which provide asymptotes for the curves in the $\mathbf{R}$-direction).

6.9 The Existence Theorem (Global Case)

It is often possible to determine the lifetime interval of a maximal integral curve using the following result. Let V be a submanifold of the finite-dimensional vector space E and let X be a C^1 vector field on V.

Theorem 6.9.1. *Let $\gamma : I \to V$ be a maximal integral curve for X.*

a) *The interval I is open; write $I = (a, b)$.*
b) *If b is finite then $\gamma(t)$ leaves every compact set in V as t tends to b: for every compact subset K of V there exists ε with $a < b - \varepsilon < b$ and $\gamma(b - \varepsilon, b) \cap K = \emptyset$.*
c) *If a is finite then $\gamma(t)$ leaves every compact set in V as t tends to a: for every compact subset K of V there exists ε with $a < a + \varepsilon < b$ and $\gamma(a, a + \varepsilon) \cap K = \emptyset$.*

Remarks.

1) In the terminology of Sect. 2.10 (see 2.10.2, Proposition 2.10.4) the assertions b) and c) can be expressed as follows: if b (resp. a) is finite, the map γ is proper in a neighbourhood of b (resp. a): there is no sequence (t_i) of points of I which tends to a (resp. b) and such that the $\gamma(t_i)$ have a limit in V. This may also be stated as: the map $t \mapsto (\gamma(t), t)$ from I to $V \times \mathbf{R}$ is a proper map (exercise).
2) With this formulation we see that in we are in fact dealing with a statement about non-autonomous equations. More precisely, let V be a submanifold of a finite-dimensional vector space E and let X be a time-dependent vector field of class C^1 defined on an open set Ω in $V \times \mathbf{R}$. Let $\gamma : I \to \Omega$ be a maximal solution of the differential equation $\dot{x} = X(x, t)$. Then I is open and the map $t \mapsto (\gamma(t), t)$ from I into the extended phase space Ω is a proper map (exercise).

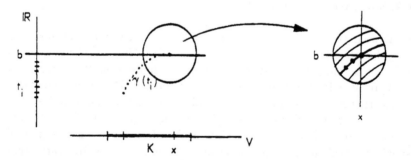

Fig. 6.3. Local existence of solutions prevents $\gamma(t)$ terminating

Now we turn to the proof of the theorem. It is a consequence of the Local Existence Theorem. The proof of a) is easy: let $t \in I$, with $x = \gamma(t)$. By Theorem 8.1 there exists an integral curve δ of X, defined on an interval $(-\varepsilon, \varepsilon)$ and such that $\delta(0) = x$. Then γ and δ may be glued together (see 5.3) and we see that I must contain $(t - \varepsilon, t + \varepsilon)$. Thus I is open, as we wished to show. Now let us prove b) by contradiction. Thus suppose b is finite and that there exists a compact subset K of V and a sequence of points t_i of I tending to b and such that $\gamma(t_i)$ belongs to K for all i. By the Bolzano-Weierstrass property we may suppose, by replacing the given sequence by a subsequence if necessary, that the $\gamma(t_i)$ tend to a point x of V. Now apply Theorem 8.1 and take i large enough so that $b - t_i \in J$ and $\gamma(t_i) \in U$. Glueing together the two integral curves γ and $t \mapsto f(\gamma(t_i), t - t_i)$ (which coincide for $t = t_i$) as above, we see that the maximal curve γ is defined for $t = b$. This contradiction proves b). See Fig. 6.3. The proof of c) is analogous, or can also be obtained by applying b) to the field $-X$. $\qquad\square$

6.9.2. There is a neat form of b) and c) as follows, often called the *principle of a priori bounds*. Fix a point x of V and consider an interval J of $\mathbf{R}$ containing 0. Suppose a compact subset K of V is given with the following property: for every compact interval J' with $0 \in J' \subset J$ and every integral curve $\gamma : J' \to V$ with $\gamma(0) = x$ we have $\gamma(J') \in K$. Then the maximal integral curve γ_x is defined on the whole of J. Somewhat fancifully we can say that the existence of an 'a priori bound' for the solution on its unknown interval of definition implies existence of the solution. To see this, suppose that the interval of definition (a, b) of the maximal integral curve γ_x does not contain J. We then have either $b < +\infty$ and $[0, b] \subset J$, or $a > -\infty$ and $[a, 0] \subset J$. We deal with the first case; taking $J' = [0, b']$ with $b' < b$ it follows from the assumptions that $\gamma_x(b') \in K$ and therefore $\gamma_x([0, b)) \subset K$, which contradicts b).

There is a particular case of the above which is fundamental:

Corollary 6.9.3. *If V is compact then X is complete: the lifetime interval for each point x of V is the whole of $\mathbf{R}$.*

This follows immediately from b), taking $K = V$. More generally:

Corollary 6.9.4. *If the vector field X vanishes outside a compact subset of V then it is complete.*

For this, let K be a compact subset outside which X vanishes. Let $\gamma : I \to V$ be a maximal integral curve of X. If γ takes any value outside K then this value is a singularity and γ is constant and so $I = \mathbf{R}$. If all the values of γ are in K then $I = \mathbf{R}$ by b) and c). $\qquad\square$

6.9.5. We now give some other examples of the use of the principle of *a priori* bounds.

1) If X has a first integral f with compact level sets (for example, if f is proper) then X is complete: we apply the principle with $K = f^{-1}(f(x))$.

2) In the conservative mechanical system 3.3, suppose the potential f is bounded below. Then every solution is defined on the whole of $\mathbf{R}$ (exercise).
3) Every solution of the pendulum equation $\ddot{x} + k\dot{x} + \sin x = 0$, with $k \geq 0$, is defined on the whole of $\mathbf{R}$ (exercise).

6.10 The Integral Flow of a Vector Field

Consider a submanifold V of a finite-dimensional vector space E and a vector field X on V of class C^r, $r \in [1, \infty]$. For each $x \in V$ let $\gamma_x : I_x \to V$ denote the maximal integral curve of X such that $\gamma_x(0) = x$ (see Corollary 5.5). Let Δ denote the set of pairs $(x, t) \in V \times \mathbf{R}$ such that γ_x is defined, that is

$$\Delta = \bigcup_{x \in V} \{x\} \times I_x.$$

For $(x, t) \in \Delta$ let $\Phi(x, t) = \gamma_x(t)$. By definition the maximal integral curve of X starting at 0 is therefore $t \mapsto \Phi(x, t)$.

Theorem 6.10.1.

a) *The set Δ is an open subset of $V \times \mathbf{R}$, and $\Phi : \Delta \to V$ is of class C^r.*
b) *For every $x \in V$ we have $(x, 0) \in \Delta$ and $\Phi(x, 0) = x$.*
c) *Let $(x, t) \in \Delta$ and let $t' \in \mathbf{R}$. In order that $(\Phi(x, t), t')$ belong to Δ it is necessary and sufficient that $(x, t + t')$ does so, and we then have $\Phi(\Phi(x, t), t') = \Phi(x, t + t')$.*

The property b) is clear. Let us prove c). Write $y = \Phi(x, t) = \gamma_x(t)$; then $t' \mapsto \Phi(y, t')$ and $t' \mapsto \Phi(x, t + t')$ are two maximal integral curves of X starting at y and therefore they coincide. Note finally that a) is true 'locally': by Theorem 8.1 for every $a \in V$ there exists an open set U of V containing a and an interval J of $\mathbf{R}$ containing 0 such that $U \times J$ is contained in Δ and the restriction f of Φ to $U \times J$ is of class C^r. Then a) follows without much difficulty apart from rewording (exercise).   □

Definition 6.10.2. The map Φ is called the *integral flow* of the field X.

To say that the field is complete means that Δ is equal to the whole of $V \times \mathbf{R}$, and then Φ is a one-parameter group of diffeomorphisms with infinitesimal generator X (see 7.1). In the general case, for $t \in \mathbf{R}$ let U_t denote the set of those $x \in V$ for which $(x, t) \in \Delta$ and write $\Phi_t(x) = \Phi(x, t)$ for $x \in U_t$. By construction we have the equivalences

$$((x, t) \in \Delta) \iff (x \in U_t) \iff (t \in I_x),$$

and under these conditions

$$\Phi(x, t) = \Phi_t(x) = \gamma_x(t).$$

By definition also

$$\frac{d}{dt}\Phi_t(x) = X(\Phi_t(x)). \qquad (6.10.1)$$

By a), each U_t is open in V and each Φ_t is of class C^r. By b) we have

$$U_0 = V \quad \text{and} \quad \Phi_0 = Id_V. \qquad (6.10.2)$$

Finally, c) translates as follows: suppose $x \in U_t$ so that $\Phi_t(x)$ is defined; then $\Phi_{t'}(\Phi_t(x))$ is defined if and only if the same is true for $\Phi_{t+t'}(x)$ and then they coincide. In particular, taking $t' = -t$ we deduce that Φ_t is a diffeomorphism (of class C^r) from U_0 onto U_t, and Φ_{-t} is its inverse.

Let a be a singular point of X. Then $\Phi_t(a)$ is defined for all t; hence a belongs to the open set U_t for every $t \in \mathbf{R}$.

6.10.3. Using the integral flow we are able to give an interesting interpretation of the Lie derivative. Since the maximal integral curves of X are $t \mapsto \Phi_t(x)$ and the Lie derivative can be calculated by differentiating along the integral curves (Proposition 3.1), for every function ϕ (of class C^1) on V we have:

$$(L_X\phi)(\Phi_t(x)) = \frac{d}{dt}\phi(\Phi_t(x)). \qquad (6.10.3)$$

If we write

$$(\Phi_t^* u)(x) = u(\Phi_t(x))$$

(defined for $x \in U_t$) for any function u on V, then (10.3) can also be written

$$(\Phi_t^* L_X\phi)(x) = \frac{d}{dt}(\Phi_t^*\phi)(x),$$

or equivalently

$$L_X = \Phi_{-t}^* \circ \frac{d}{dt} \circ \Phi_t^*. \qquad (6.10.4)$$

As Arnol'd says, L_X is the *angler's derivative*: "the angler, sitting on the river bank, differentiates objects that are carried along by the flow".

6.10.4. In the neighbourhood of a nonsingular point a the integral flow is particularly simple in 'straightening-out' coordinates: with the notation of Theorem 8.2 we have

$$\Phi_t(x_1, x_2, \ldots, x_n) = (x_1 + t, x_2, \ldots, x_n),$$

so that the flow coincides locally with the translation flow $x \mapsto x + te_1$.

6.10.5. Let us denote the integral flow associated with the vector field X more explicitly as $\Phi^X = (\Phi^X_t)_{t \in \mathbf{R}}$. Let E' be another finite-dimensional vector space, with V' a submanifold of E' and $u : V' \to V$ a diffeomorphism. Consider the vector field u^*X on V' that is the inverse image of X, and its associated integral flow Φ^{u^*X}. Then for all t we have

$$\Phi^{u^*X}_t = u^{-1} \circ \Phi^X_t \circ u,$$

as follows immediately from the fact that u transforms the integral curves of u^*X into those of X. In particular, for $V' = V$ we obtain the following fact: if $u^*X = X$ (that is to say, if X is invariant under the diffeomorphism u) then the integral flow of X commutes with u.

6.10.6. We say that a subset A of V is *invariant* under the flow (Φ_t) if, for every $x \in A$ and every $t \in \mathbf{R}$ for which $\Phi_t(x)$ is defined, we have $\Phi_t(x) \in A$. This means also that A is a union of orbits of X. If A is a submanifold of V it means that X is tangent to A in the following sense: the vector field X is *tangent* to a submanifold W of V (or W is everywhere tangent to X) if for every $x \in W$ we have $X(x) \in T_xW$; the restriction of X to W is then a vector field on W (while in general it is merely a 'vector field tangent to V along W').

6.11 The Main Features of a Phase Portrait

6.11.1. Let E be a finite-dimensional vector space, with V a submanifold of E and with X a vector field of class C^r, $r \in [1, \infty]$, on V. Recall that the *phase portrait* of X is the partition of V consisting of the orbits of X, each one labelled with its direction of flow. We therefore forget the law of movement along the orbits, and concentrate on properties that are topological and hence somewhat coarse. Essentially we consider those properties that are invariant under homeomorphism; we shall make this precise a little later (with the notion of orbital equivalence). As far as the situation in the neighbourhood of a given point is concerned, we already have the Straightening-out Theorem 8.2 which totally describes the local phase portrait when the point is nonsingular: it is (up to diffeomorphism — which is better than we wish) a system of parallel line segments. The case of a singular point is much more complicated, and we shall return to it in Chapt. 8.

We begin with the individual study of orbits. We can already classify the maximal integral curves γ into three broad classes, according to whether γ is constant, non-constant but not injective, or injective. The first case corresponds to the singular points. In the second case the orbit is a 'circle' (compact submanifold diffeomorphic to S^1):

Proposition 6.11.2. *Let* $\gamma : I \to V$ *be a maximal integral curve of* x *and let* $\Omega = \gamma(I)$ *be the corresponding orbit. Suppose* γ *is neither constant nor injective. Then* $I = \mathbf{R}$ *and* γ *is periodic. If* $T > 0$ *is the period of* γ, *the map* $p : (\cos\theta, \sin\theta) \mapsto \gamma(\theta T/2\pi)$ *is an embedding of the circle* S^1 *into* V *with image* Ω.

Let u and $u + v$ with $v > 0$ be two distinct points of I such that $\gamma(u) = \gamma(u + v)$. Then γ and $t \mapsto \gamma(t + v)$ are two maximal integral curves which coincide for $t = u$; they are therefore equal, which implies that $I = \mathbf{R}$ (since $t \in I$ is equivalent to $t + v \in I$) and $\gamma(t) = \gamma(t + v)$ for all t. Let T be the smallest $v > 0$ satisfying this property (it exists, because γ is assumed non-constant). The above shows that the restriction of γ to $[0, T)$ is injective. Hence the map p is injective. Since its tangent map is everywhere nonzero, it is an embedding (Proposition 2.10.5). $\qquad\square$

Such an orbit is said to be *periodic*, and the number T is the *period* of the orbit. We sometimes say 'closed orbit' instead of periodic orbit, which can cause confusion since an orbit can be a closed subset of V without, however, being periodic.

The singular points and the periodic orbits are the only compact orbits. In fact:

Proposition 6.11.3. *Let* $\gamma : I \to V$ *be a maximal integral curve of* X. *If the map* γ *is injective, then its image* $\gamma(I)$ *is not compact.*

Suppose $\gamma(I)$ is compact. First, it follows from Theorem 9.1 that $I = \mathbf{R}$. Moreover, from the sequence $(\gamma(0), \gamma(1), \dots)$ we can extract a sequence which converges to an element $\gamma(c)$, say, of $\gamma(\mathbf{R})$. For each $d \in \mathbf{R}$ the map Φ_{d-c} is continuous and the sequence of the $\gamma(i + d - c) = \Phi_{d-c}(\gamma(i))$ converges to the point $\Phi_{d-c}(\gamma(c)) = \gamma(d)$. Thus we have proved that each point of $\gamma(\mathbf{R})$ is the limit of a sequence of points $\gamma(t_i)$ where the sequence of the $t_i \to +\infty$.

Armed with this fact, we now consider the sequence of compact intervals $K_n = [-n, +n]$. We have that $\gamma(\mathbf{R})$ is the union of the compact sets $\gamma(K_n)$; equivalently, in the compact space $\gamma(\mathbf{R})$ the intersection of the countable family of open sets $\gamma(\mathbf{R} \setminus K_n)$ is empty. By Baire's Theorem (3.2.4) this implies that at least one of the open sets is not dense in $\gamma(\mathbf{R})$, which means that there exists an integer n and a point of $\gamma(\mathbf{R})$ which has $\gamma(K_n)$ as a neighbourhood. But the first part of the proof then implies that there exist arbitrarily large elements t in $\mathbf{R}$ with $\gamma(t) \in \gamma(K_n)$. This contradicts the assumed injectivity of γ and completes the proof. $\qquad\square$

6.11.4. Now we analyze this third type of orbit in a little more detail. Let γ be an injective maximal integral curve with interval of definition $I = (a, b)$ and image $\Omega = \gamma(I)$. We have just seen that Ω is not compact. Moreover, it is an 'immersed curve', the image of the injective immersion γ. Let us study its closure $\overline{\Omega}$. By definition, every point of $\overline{\Omega}$ is the limit of a sequence of points $\gamma(t_i)$ where the t_i belong to I. By replacing this sequence by a suitable sub-sequence we can suppose that the t_i tend either to a point c of I or to a or b. In the first case we would have $x = \gamma(c) \in \Omega$; in the second case, a is necessarily $-\infty$ in virtue of Theorem 9.1; in the third case we likewise have $b = +\infty$. In the last two cases x is classically called a *limit point of the orbit* Ω; traditionally we specify 'α-limit' or 'ω-limit' according to whether the parameter tends to $-\infty$ or $+\infty$. We prefer to call them *past limit point* and *future limit point*. The closure of Ω is therefore the union (not necessarily disjoint) of three subsets: Ω itself, the set of past limit points (*past limit set*) and the set of future limit points (*future limit set*). If the sign of X is changed, the direction of flow along the orbits is reversed and the two limit sets are interchanged. The limit sets are closed, by construction. They are invariant under the flow, and therefore unions of orbits. They are automatically connected when they are compact (exercise).

Let x be a point whose orbit Ω is not compact. The map $t \mapsto \Phi_t(x)$ is then injective, and Ω is the union of the two *semi-orbits* of x, the future semi-orbit and the past semi-orbit, consisting of points $\Phi_t(x)$ with $t \geq 0$ and $t \leq 0$ respectively, with intersection just the point x. The closure of the future (past) semi-orbit of x is the union of this semi-orbit and the future (past) limit set of Ω. By abuse of language, we extend this terminology to the case of compact orbits: if the orbit Ω of x is compact the two semi-orbits of x are equal to Ω.

Now we make explicit the equivalence relations that were alluded to above. Let X and X' be vector fields (of class at least C^1) on V and V' respectively.

Definition 6.11.5. The vector fields X and X' are said to be *orbitally equivalent* (we also say *topologically equivalent*) if there exists a homeomorphism $h : V \to V'$ which maps each orbit of X to an orbit of X' preserving the direction of flow.

In terms of flows, this last condition is expressed by a relation of the form

$$h(\Phi_t^X(x)) = \Phi_{u(t,x)}^{X'}(h(x)),$$

where for each x the function $t \mapsto u(t, x)$ is strictly increasing. If h is of class C^1 this means also that the two vector fields $h_*(X)$ and X' define the same half-line field (see 6.1) on V', that is there exists a function $\alpha > 0$ on V' with $h_*(X) = \alpha X'$.

If Ω is an orbit of X in V then $h(\Omega)$ is an orbit of X' in V' and h maps the future and past limit sets of Ω to the corresponding limit sets of Ω'. In particular, the singular points, compact orbits and closed orbits of X are sent to analogous objects for X'. Likewise, each semi-orbit of a point x of V is sent to the corresponding semi-orbit of the point $h(x)$ of V'.

6.11.6. If A is a subset of V and A' a subset of V', we say more generally that X and X' are *orbitally equivalent in a neighbourhood of A and A'* if there exists an open set U of V containing A, an open set U' of V' containing A', and a homeomorphism $h : U \to U'$ which takes every orbit of X in U to an orbit of X' in V', preserving the sense of flow.

This is perhaps the moment to dispel a false notion. If U is an open set in V and if Ω is an orbit of X in V, then $\Omega \cap U$ is invariant under the flow of the field induced by X on U, and is therefore a union of orbits of this field; but it is not necessarily a single orbit. In fact, the orbits of X in U are the connected components of the intersections $\Omega \cap U$ where Ω runs through the orbits of X in V. It is in this sense only that we may say that the phase portrait of X in U is "the trace on U of the phase portrait of X in V".

6.12 Discrete Flows and Continuous Flows

6.12.1. As we have seen, the study of vector fields amounts to the study of their integral flows. In particular, complete vector fields (see after Corollary 5.5) correspond to one-parameter groups of diffeomorphisms. In the theory, there is no inconvenience in restricting ourselves to this case; either the vector field is already complete (because the phase space is compact, for example) or it can be extended to a complete vector field on a larger phase space (such as a suitable 'compactification'), or we are interested in a local situation and are able to modify the field far away from the relevant region in order to make it complete, or finally we are interested in the phase portrait and can replace the field by a complete field having the same phase portrait (this always exists: exercise). Moreover, in the general case statements and proofs differ from those of the complete case only by purely technical modifications that are essentially banal and always irritating. In what follows we shall therefore restrict ourselves to the 'complete vector fields / one-parameter groups of diffeomorphisms' context.

6.12.2 Besides these one-parameter groups that are called 'continuous'[44], referring not to the notion of continuity of maps but to the fact that the parameter t varies in $\mathbf{R}$, it is natural to consider 'discrete' one-parameter groups; these are the families $n \mapsto \Phi_n$ where n runs through $\mathbf{Z}$, where for each n the map Φ_n is a diffeomorphism (or homeomorphism, or ...) of the phase space V under consideration, and where we have $\Phi_0 = Id_V$ and $\Phi_{n+m} = \Phi_n \circ \Phi_m$ for every pair of integers n and m. If we put $u = \Phi_1$ then u is a diffeomorphism of V and we have $\Phi_n = u^n$ for each n. Conversely, we associate to each diffeomorphism u of V the 'discrete flow' $n \mapsto u^n$, for which

---

[44]The objects that are nowadays called Lie groups were for a long time known as *continuous groups*.

u is the *generator* in the usual sense. Note the parallels: "continuous flow — discrete flow", "complete vector field — diffeomorphism", "infinitesimal generator — generator", "integral flow – group generated".

6.12.3. It is generally the case that notions developed for continuous flows have analogues for discrete flows, that is (to be precise) for diffeomorphisms. There is also an obvious way of deriving a discrete flow from a continuous flow $t \mapsto \Phi_t$: it is enough to choose a nonzero value t_0 of the parameter t (often we take $t_0 = 1$) and to consider the diffeomorphism $u = \Phi_{t_0}$, that is the discrete flow $n \mapsto u^n = \Phi_{nt_0}$. Each element of the continuous flow is the composition of an element of the discrete flow and one of the diffeomorphisms Φ_t for $0 \leq t < |t_0|$; the latter form a bounded family, and it is often possible to deduce certain properties of the continuous flow from the analogous properties of the discrete flow.

From this point of view, the study of discrete flows is more general than that of continuous flows. Note in fact that the diffeomorphisms that belong to a continuous flow are not arbitrary: in the set of all diffeomorphisms they belong to a parametrized arc which passes through the identity diffeomorphism, and therefore they share with the latter those of its properties that are preserved under continuous deformation.

In technical terms we can make this a little more precise: the group of diffeomorphisms of V has a topology, and the 'continuous' one-parameter groups are traced out in the path-connected component of the identity diffeomorphism.

6.12.4. There exists another much more important relation between these two types of flow. It comes from a basic construction for the study of closed orbits which is due to Poincaré and which we shall explain in Chapt. 9. Here we shall simply give a construction that proceeds from the same body of ideas and which goes in the opposite sense, from diffeomorphisms to vector fields (but altering the relevant phase space). This construction, which is in essence very simple, is not easily expressed in the context of submanifolds, so we make an exception to our principles and move into the setting of 'abstract manifolds'.

Starting with a diffeomorphism u of a manifold W, we are going to construct a manifold V and a vector field X on V. In the product $[0,1] \times V$ we identify each point $(1,x)$ of $\{1\} \times W$ with the point $(0, u(x))$ of $\{0\} \times W$. In this way we obtain a manifold V which naturally projects to the circle C obtained by identifying the points 0 and 1 of $[0,1]$. Let p denote this projection. The identity map of $(0,1) \times W$ in V is an embedding whose image is the open set complementary to $p^{-1}(0)$; we likewise define an embedding of $(-1/2, 1/2) \times W$ into V whose image is the open set complementary to $p^{-1}(1/2)$ by associating to the point (t,x) the point of V corresponding to $(1 + t, x)$ if $t \leq 0$ and to $(t, u(x))$ if $t \geq 0$. More generally, consider the map $q : \mathbf{R} \times W \to V$ which associates to a pair (t,x) the point of V that is the image of the point $(t - n, u^n(x))$, where n is the unique integer such that

$n \leq t < n + 1$. Then q is a local diffeomorphism whose restriction to every open set $(a, b) \times W$ with $b - a < 1$ is an embedding. The composed map $p \circ q : \mathbf{R} \times W \to C$ associates to (t, x) the class of t modulo 1. For every (t, x) in $\mathbf{R} \times W$ and every n in $\mathbf{Z}$ we have $q(t + n, x) = q(t, u^n(x))$. Conversely, the relation $q(a, y) = q(b, z)$ is equivalent to the existence of an integer n with $a = b + n$ and $z = u^n(y)$. This realizes V as the quotient of the manifold $\mathbf{R} \times W$ by the equivalence relation $(t + n, x) \sim (t, u^n(x))$, $n \in \mathbf{Z}$.

A vector field X is defined on V by taking the field whose components in each of the two charts considered are $\partial/\partial t$ and 0, or alternatively as the 'image' under q of the field $(\partial/\partial t, 0)$. The integral curves of X are by construction the maps $t \mapsto q(t, x)$ and X is complete. Identify W with a submanifold of V by the map $x \mapsto q(0, x)$. The orbit of a point x of W meets W again successively at the points $u(x), u^2(x), \ldots$ after x and the points $u^{-1}(x), u^{-2}(x), \ldots$ before x. The trace on W of the orbit Ω of x is thus the set of images of x under the discrete flow formed by the powers of u, or in other words the orbit of x under this discrete flow. Note that in order to reconstruct Ω we have to know not only $W \cap \Omega$ but also the order in which the elements succeed each other on Ω.

It is easy to set up a dictionary between the properties of Ω and those of $W \cap \Omega$. For example, to say that Ω is periodic is equivalent to saying that $W \cap \Omega$ is finite; the period of Ω is then the number of elements of $W \cap \Omega$. In order for Ω to be closed in V it is necessary and sufficient that $W \cap \Omega$ be closed in W; more generally, the set of limit points of Ω in V is the union of the orbits of the limit points of $W \cap \Omega$ in W.

7 Linear Vector Fields

7.1 Introduction

Our reasons for studying linear differential equations are twofold.

Firstly, whenever we wish to analyze the local phase portrait of a vector field in the neighbourhood of a singular point it is natural to linearize the problem, so that we are then investigating the phase portrait of a linear vector field (for which, incidentally, the local study at the origin and the global study are the same thing). In this chapter we shall see that such an approach does not work without considerable difficulties and that the original situation is not as close to the linear approximation as we might naïvely expect. Nevertheless, studying the behaviour of the integral flows of linear vector fields in some detail is justifiable, if only to extract from them those properties which have some chance of being preserved under perturbations. This leads us once more into the ideas of stability and genericity that we have already met in connection with other questions.

Secondly, although the study of linear vector fields is technically easier than the local study of vector fields in the neighbourhood of a singular point, it already exhibits in simple form the most important structures of the general case and provides a good introduction to them avoiding continual recourse to the general theorems of the previous chapter.

For this reason we give a direct description of the integral flow of a linear field using the *exponential map* constructed as the sum of the usual series, and in this particular case we give proofs of general theorems (Propositions 7.2 and 7.3) that are independent of the proofs in Chapt. 6. In order to analyze the behaviour of the exponential flow in sufficient detail we make use of auxiliary results on endomorphisms of finite-dimensional vector spaces that we recall at the outset, together with proofs in most cases. Rather than giving a very explicit description closely tied to the linear structure (notably the Jordan form) we prefer a less precise study that suffices for what follows and has the advantage of intoducing one of the techniques from the nonlinear case (Lyapunov functions : see Chapt. 8, Sect. 6).

Given our general perspective (linearization) we deliberately work in the real context, even though from the point of view of linear algebra the situation is simpler in a complex setting (particularly regarding the description of the image of the exponential map).

Moreover, as we shall see later, the study of singular points is very close to that of closed orbits, of which they are a limiting case. This latter study is carried out using the so-called *Poincaré map* (see Sect. 9.1) which is a diffeomorphism (on a certain auxiliary manifold transverse to the orbit under consideration) that has a fixed point, and whose linear approximation is an automorphism of a suitable vector space (the tangent space to the transversal). This obliges us to study, as well as exponentials of endomorphisms, those automorphisms that are not necessarily exponentials. Thus every definition has two versions, according to whether we apply it 'before' or 'after' the exponential. For example, take the key property of *hyperbolicity*. Before, it is expressed as "none of the eigenvalues of the endomorphism u is purely imaginary"; after, by "none of the eigenvalues of the automorphism v is on the unit circle". If u satisfies the first condition then $\exp(u)$ satisfies the second, but the latter still has meaning for arbitrary v. In the literature, "u is hyperbolic" and "v is hyperbolic" are said indiscriminately, which is convenient but not really acceptable. In general we shall opt for the 'global' setting, and say for example "v is hyperbolic" but "u has hyperbolic exponential flow".

The essential point of the chapter is the adaptation of the 'stability and genericity' philosophy to the context of linear maps. Here we see that a sufficiently general linear flow or automorphism (specifically, one belonging to the open dense set defined by the hyperbolicity condition) can take only a finite number of forms, each stable under small perturbations. In order to define what "to have the same form" means we need considerably to relax the original structures and, in particular, dispense with the linearity: instead of the usual similarity or conjugacy relation $u' = h \circ u \circ h^{-1}$ where h is linear, we have to substitute the analogous but much weaker relation of *topological conjugacy* in which h may be an arbitrary homeomorphism.

This topological classification is summarized as follows. An endomorphism with hyperbolic exponential flow can be decomposed into a direct sum of two 'pieces', one of which has exponential flow topologically conjugate to that of $y \mapsto -y$ (the case called *stable, contracting* or *attracting*) and the other of which has exponential flow topologically conjugate to that of $z \mapsto z$ (the *unstable, expanding* or *repelling* case). The endomorphism is therefore determined up to topological similarity by the dimensions of the *stable* and *unstable* subspaces associated with it. In an n-dimensional space this gives $n+1$ classes corresponding to pairs of integers whose sum is n, going from the purely attracting case $(n, 0)$ to the purely repelling case $(0, n)$. For hyperbolic automorphisms the situation is a little more complicated and there are $4n$ classes in dimension n. We mention in passing that the adjectives 'stable' and 'unstable' have multiple and (alas!) contradictory meanings.

We shall see in the next chapter that the operation of linearization does not work very well outside the topological context when the dimension is greater than 2. Thus it is not worthwhile making a finer classification of endomorphisms in dimension greater than 2. However, this can be carried out in dimension 2, where we find a classical bestiary (nodes, foci, centres, saddles, ...).

The detailed plan of the chapter is as follows. In Sects. 2 to 4 we recall some constructions that will be useful later; in particular in Sect. 3 we show how there is a decomposition of the space on which an endomorphism acts associated to a partition of its eigenvalues. The construction of the exponential flow and the verification of its elementary properties occupy Sects. 6 and 7, which may be read independently of the rest of the chapter (except for 6.7, 6.8 and 6.9 which use results from Sect. 4). In Sect. 8 we characterize the image of the exponential map. The structure of hyperbolic flows is given in Sect. 9, parallel to that of hyperbolic automorphisms which, not requiring the exponential, is handled in Sect. 5. The topological classification is the subject of Sects. 10 and 11. The special case of dimension 2 is studied in Sect. 12.

7.2 The Spectrum of an Endomorphism

7.2.1. Let E be a finite-dimensional vector space and let u be an endomorphism of E. Recall that the *characteristic polynomial* of u is the polynomial

$$P_u(T) = \det(u - T1) \in \mathbf{R}[T]. \tag{7.2.1}$$

Its degree is equal to the dimension of E; its dominant term is $(-T)^{\dim(E)}$ and its constant term is $\det(u)$. We call the roots of $P_u(T)$, *real or complex*, the *eigenvalues* of u. If λ is an eigenvalue of u then so also is its complex conjugate $\bar{\lambda}$. The set of eigenvalues of u is called the *spectrum* of u and denoted $\mathrm{Sp}(u)$. Note that $\mathrm{Sp}(u) \neq \emptyset$ as long as E is not just $\{0\}$, and that P_u can be written

$$P_u(T) = \prod_{\lambda \in \mathrm{Sp}(u)} (\lambda - T)^{m(\lambda)} = \prod_{\lambda \in \mathbf{C}} (\lambda - T)^{m(\lambda)}, \tag{7.2.2}$$

where for each $\lambda \in \mathbf{C}$ the integer $m(\lambda)$ is the *multiplicity* of the complex number λ as an eigenvalue of u (zero when λ is not an eigenvalue).

In fact when we talk about the spectrum of u we usually mean the family of eigenvalues with each one repeated a number of times equal to its multiplicity.

Recall that two endomorphisms $u \in \mathrm{End}(E)$ and $u' \in \mathrm{End}(E')$ are called (linearly) *similar* or *conjugate* if there exists an isomorphism $h : E \to E'$ of vector spaces such that $u' \circ h = h \circ u$. Two conjugate endomorphisms have the same characteristic polynomials, and therefore the same spectra.

For λ real, to say that λ is an eigenvalue of u means that the endomorphism $u - \lambda1$ of E is not injective, that is there exist nonzero eigenvectors for the eigenvalue λ. When λ is not real, we are unable to consider the endomorphism $u - \lambda1$ directly. In this case, to avoid discussing 'complexification' of the vector space E (which we leave as an exercise) we use the following *ad hoc* lemma which covers the two cases. Let

$$\lambda = \sigma + i\tau = \rho e^{i\theta},$$

and define $R(T)$ by

$$R(T) = (T - \lambda)(T - \bar{\lambda}) = T^2 - 2\sigma T + \rho^2.$$

Lemma 7.2.2. *The following conditions are equivalent:*

a) *λ is an eigenvalue of u;*
b) *there exists an $\mathbf{R}$-linear map $j : \mathbf{C} \to E$ with $j(1) \neq 0$ and $u \cdot j(z) = j(\lambda z)$ for all $z \in \mathbf{C}$;*
c) *the endomorphism $R(u) = u^2 - 2\sigma u + \rho^2 1$ is not injective.*

We first prove the equivalence of a) and b) by translating into the language of matrices. Condition a) means that $\det(A - \lambda I)$ is zero, that is to say there is a (possibly complex) nonzero column vector X with $AX = \lambda X$. Writing X in the form $U - iV$ where U and V are real we obtain the equivalent condition $(AU = \sigma U + \tau V,\ AV = -\tau U + \sigma V)$. We may assume that U is nonzero, since if U is zero then V is not and we have $\tau = 0$ and $AV = \sigma V = \lambda V$; we then interchange the roles of U and V.

On the other hand, to be given an $\mathbf{R}$-linear map j from $\mathbf{C}$ to E is the same as being given the two vectors $j(1) = x$ and $j(i) = y$. The condition $j(\lambda z) = \lambda j(z)$ translates into

$$u \cdot x = \sigma x + \tau y, \quad u \cdot y = -\tau x + \sigma y. \qquad (7.2.3)$$

Note by the way the equivalent polar form

$$u \cdot (x \cos \alpha + y \sin \alpha) = \rho(x \cos(\alpha + \theta) + y \sin(\alpha + \theta)) \qquad (7.2.4)$$

for every $\alpha \in \mathbf{R}$. We have thus shown the equivalence of a) and b).

If τ is zero then the endomorphism $R(u)$ in c) is the square of $u - \lambda 1$; it is therefore injective or not according to whether λ is an eigenvalue of u or not, which proves the equivalence of a) and c). If τ is nonzero an immediate calculation shows that (2.3) is equivalent to $(y = (u \cdot x - \sigma x)/\tau,\ R(u) \cdot x = 0)$, and again b) and c) are equivalent. $\qquad \square$

For example, suppose that E is of dimension 2 and that the eigenvalues of u are not real. Choose one of them and call it λ, so the other is $\bar{\lambda}$. Then Lemma 2.2 implies that there exists an isomorphism of $\mathbf{R}$-vector spaces between E and $\mathbf{C}$ which transforms u into the complex scalar multiplication $z \mapsto \lambda z$.

7.2.3. We say that a vector subspace F of E is *invariant*[45] under u if $u \cdot x$ belongs to F for every $x \in F$, that is if the restriction $u|_F$ of u to F is an endomorphism of F. By taking a basis of E that contains a basis of F we obtain for u a block triangular matrix one of whose diagonal blocks is the matrix for $u|_F$. Thus $P_{u|_F}(T)$ divides $P_u(T)$ and we have $\text{Sp}(u|_F) \subset \text{Sp}(u)$. If F and F' are two complementary invariant subspaces then the polynomial P_u is the product of $P_{u|_F}$ and $P_{u|_{F'}}$, and we have $\text{Sp}(u) = \text{Sp}(u|_F) \cup \text{Sp}(u|_{F'})$.

Let $P \in \mathbf{R}[T]$ be a polynomial in one variable T with real coefficients:

$$P(T) = \sum_{i=0}^{r} a_i T^i.$$

For all $u \in \text{End}(E)$ we define $P(u) \in \text{End}(E)$ by

$$P(u) = \sum_{i=0}^{r} a_i u^i,$$

that is to say

$$P(u) \cdot x = \sum_{i=0}^{r} a_i (u^i \cdot x), \quad x \in E.$$

It is immediate that $(PQ)(u) = P(u)Q(u)$.

Proposition 7.2.4.

a) *For every polynomial $P \in \mathbf{R}[T]$ the spectrum of $P(u)$ consists of the $P(\lambda)$ for $\lambda \in \text{Sp}(u)$: we have*

$$\text{Sp}(P(u)) = P(\text{Sp}(u)). \tag{7.2.5}$$

b) *Let $F = \text{Ker}(P(u))$. Then the subspace F is invariant under u and the spectrum of the restriction of u to F is the set of the roots of P in $\text{Sp}(u)$:*

$$\text{Sp}(u|_F) = \{\lambda \in \text{Sp}(u)|P(\lambda) = 0\}. \tag{7.2.6}$$

First we prove that $P(\text{Sp}(u))$ is contained in $\text{Sp}(P(u))$. Let λ be an eigenvalue of u. If $j : \mathbf{C} \to E$ is the map provided by Lemma 2.2, we have $u^n \cdot (j(z)) = j(\lambda^n z)$ for every n, and so $P(u) \cdot j(z) = j(P(\lambda)z)$. Applying Lemma 2.2 in the opposite direction, we deduce that $P(\lambda)$ is an eigenvalue of $P(u)$.

Let us now turn to b). If x belongs to F we have $P(u) \cdot x = 0$ and hence $P(u) \cdot (u \cdot x) = u \cdot (P(u) \cdot x) = 0$ so $u \cdot x \in F$. The vector subspace F is therefore

---

[45]Traditionally F is more often said to be *stable* under u, but it is better to avoid this terminology here in order to avoid confusion with a notion that will be introduced under the same name in 9.3.

invariant under u. Let v denote the restriction of u to F. Since $P(v) = 0$ by construction, every eigenvalue of v is a root of P by a) above; moreover, we have already seen that it is an eigenvalue of u. Conversely, let $\lambda \in \mathrm{Sp}(u)$ be such that $P(\lambda) = 0$ and let $j : \mathbf{C} \to E$ be as in Lemma 2.2. We have seen above that $j(P(\lambda)z) = P(u) \cdot j(z)$. Since $P(\lambda) = 0$ the image of j is contained in F and λ is indeed an eigenvalue of v, again by Lemma 2.2 applied this time to the endomorphism v of F.

It remains to finish the proof of a) by showing that every eigenvalue μ of $P(u)$ is of the form $P(\lambda)$ where $\lambda \in \mathrm{Sp}(u)$. Let $R(T) = (T - \mu)(T - \bar{\mu})$ and let $G = \mathrm{Ker}(R(P(u)))$. Then G is invariant under u and under $P(u)$; by Lemma 2.2 it is not simply $\{0\}$. Let λ be an eigenvalue of the restriction of v to G. It is an eigenvalue of u. Moreover, since $P(\lambda)$ is an eigenvalue of $P(u)$ in G it is a root of R. We therefore have either $\mu = P(\lambda)$ or $\mu = \overline{P(\lambda)} = P(\bar{\lambda})$, which completes the proof. $\square$

We now look at some special cases of b). To say that $\mathrm{Ker}(P(u))$ does not reduce to $\{0\}$ is to say that $\mathrm{Sp}(u|_F)$ is nonempty, therefore there exists a root of P in $\mathrm{Sp}(u)$. Taking P of the form $(T - \lambda)(T - \bar{\lambda}) = T^2 - 2\sigma T + \rho^2$ we recover Lemma 2.2. Conversely, to say that the endomorphism $P(u)$ is invertible means that no root of P belongs to $\mathrm{Sp}(u)$.

We can obtain the following more general result by a proof quite analogous to that of Proposition 2.4 a). Let P and Q be polynomials. Suppose that Q does not vanish on $\mathrm{Sp}(u)$, so that $Q(u)$ is invertible. Then the spectrum of $P(u)Q(u)^{-1}$ consists of the $P(\lambda)/Q(\lambda)$ as λ runs through $\mathrm{Sp}(u)$. The simplest case of this statement is the following:

Lemma 7.2.5. *In order for u to be invertible it is necessary and sufficient that $\mathrm{Sp}(u)$ does not contain 0. In this case, $\mathrm{Sp}(u^{-1})$ consists of the λ^{-1} for $\lambda \in \mathrm{Sp}(u)$.*

The first assertion is clear. Suppose u is invertible. We have

$$P_{u^{-1}}(T) = \det(u^{-1} - T1) = \det(-Tu^{-1})\det(u - T^{-1}1)$$
$$= (-T)^{\dim(E)} \det(u^{-1}) P_u(T^{-1}).$$

The lemma follows immediately. $\square$

Finally we note a topological result:

Proposition 7.2.6. *Let $m_1, \ldots, m_k$ be integers > 0 with sum equal to $\dim(E)$, and let $U_1, \ldots, U_k$ be open subsets of $\mathbf{C}$. The set A of those u in $\mathrm{End}(E)$ which have m_i eigenvalues belonging to U_i (counting multiplicities) for $i = 1, \ldots, k$ is an open subset of $\mathrm{End}(E)$.*

Let S be the subset of $\mathrm{End}(E) \times \mathbf{C}^n$, with $n = \dim(E) = m_1 + \cdots + m_k$, consisting of families $(u, \lambda_1, \ldots, \lambda_n)$ such that

$$\det(u - T1) = (\lambda_1 - T) \cdots (\lambda_n - T).$$

This is a closed set since the two sides of the defining condition depend continuously on the point considered. Let $p : S \rightarrow \text{End}(E)$ denote the projection on the first factor. This is a proper map (2.10.1). To see this, note that the inverse image of a compact set K in $\text{End}(E)$ under p is closed in $\text{End}(E) \times \mathbb{C}^n$ (being the intersection of S and $K \times \mathbb{C}^n$) and it is bounded since by Theorem 4.4 below, for example, we have $|\lambda_i| \leq \|u\|$ for $(u, \lambda_1, \ldots, \lambda_n) \in S$. Hence (by Lemma 2.10.3) the image under p of every closed subset of S is closed in $\text{End}(E)$.

Now let $\mathcal{F}$ denote the set of maps from $\{1, \ldots, n\}$ to $\{1, \ldots, k\}$ which take the value i a total of m_i times, for $i = 1, \ldots, k$. Let U be the union of the $U_{f(1)} \times \cdots \times U_{f(n)}$ as f runs through $\mathcal{F}$. To say that an element u of $\text{End}(E)$ belongs to A is to say that $p^{-1}(u)$ is contained in $\text{End}(E) \times U$. Hence the complement of A in $\text{End}(E)$ is the image under p of the intersection of S with the closed complement of $\text{End}(E) \times U$. It is therefore a closed subset of $\text{End}(E)$, as we wished to prove. $\qquad\square$

7.3 Space Decomposition Corresponding to Partition of the Spectrum

We retain the previous notation: E is a finite-dimensional vector space and u is an endomorphism of E. It is clear that there exists a polynomial P not identically equal to zero such that $P(u) = 0$; indeed, the powers u^n of u cannot be linearly independent in the space of endomorphisms of E because the latter is finite-dimensional. In fact the characteristic polynomial of u has this property (see [HS] for example).

Theorem 7.3.1. (Cayley-Hamilton Theorem.) *Let $P_u(T) \in \mathbf{R}[T]$ be the characteristic polynomial of the endomorphism u. Then $P_u(u) = 0$.* $\qquad\square$

Note that the polynomial $Q(T) = \prod_{\lambda \in \text{Sp}(u)} (T - \lambda)^{\dim(E)}$ is a multiple of $P_u(T)$ and therefore vanishes on u. We can replace P_u by Q in applications where we need a polynomial which vanishes on u and has all its roots in $\text{Sp}(u)$. Here is an example of such an application that will be useful later on:

Lemma 7.3.2. *Let u, v and w be three endomorphisms of E with $wv = vu$ and $\text{Sp}(u) \cap \text{Sp}(w) = \emptyset$; then $v = 0$.*

To see this, observe that $w^n v = v u^n$ for all $n \in \mathbf{N}$ and so $P(w)v = vP(u)$ for every polynomial P. Taking P to be the characteristic polynomial of u (or the polynomial Q introduced above) we deduce that $P_u(w)v = 0$. However, $P_u(w)$ is invertible since no eigenvalue of w is a root of P_u. $\qquad\square$

Lemma 7.3.3. *Let $P \in \mathbf{R}[T]$ be a polynomial with real coefficients such that $P(u) = 0$. Suppose we are given two polynomials Q and R in $\mathbf{R}[T]$ with $P = QR$. If Q and R have no common roots, the two subspaces $F = \mathrm{Ker}(Q(u))$ and $F' = \mathrm{Ker}(R(u))$ are invariant under u and are complementary.*

For this we note that there exist A and B in $\mathbf{R}[T]$ with $QA + RB = 1$ ('Bézout's identity') and therefore $Q(u)A(u) + R(u)B(u) = 1$. For all $x \in E$ we have $x = 1 \cdot x = Q(u)A(u) \cdot x + R(u)B(u) \cdot x$. But $R(u) \cdot (Q(u)A(u) \cdot x) = A(u)P(u) \cdot x = 0$ so $Q(u)A(u) \cdot x$ belongs to F'; likewise $R(u)B(u) \cdot x$ belongs to F and we have proved that $E = F + F'$. If $x \in F \cap F'$ then $x = A(u) \cdot (Q(u) \cdot x) + B(u) \cdot (R(u) \cdot x) = 0$; hence $F \cap F'$ reduces to $\{0\}$ and the subspaces F and F' are complementary. $\qquad\square$

For every subset S of $\mathbf{C}$ invariant under complex conjugation, let

$$Q_S = \prod_{\lambda \in \mathrm{Sp}(u) \cap S} (T - \lambda I)^{\dim(E)} , \quad E_S(u) = \mathrm{Ker}(Q_S(u)) ;$$

note that Q_S has real coefficients, making the second formula meaningful. We have $E_\emptyset(u) = \{0\}$ and $E_{\mathbf{C}}(u) = E$. If $S \subset S'$ then $E_S(u) \subset E_{S'}(u)$.

Proposition 7.3.4.

a) *$E_S(u)$ is the largest invariant subspace F of E with $\mathrm{Sp}(u|_F) \subset S$.*

b) *The dimension of $E_S(u)$ is the sum of the multiplicites of the eigenvalues of u which belong to S.*

c) *We have $E_S(u) + E_{S'}(u) = E_{S \cup S'}(u)$ and $E_S(u) \cap E_{S'}(u) = E_{S \cap S'}(u)$.*

Write $E_S, \ldots$ instead of $E_S(u), \ldots$. By Proposition 2.4 b) the restriction of u to E_S has its spectrum contained in S. Conversely, let F be a subspace which is invariant under u and is such that, putting $v = u|_F$, we have $\mathrm{Sp}(v) \subset S$. Then the characteristic polynomial of v has degree $\leq \dim(E)$ and has all its roots in S and therefore divides Q_S; applying the Cayley-Hamilton Theorem to v we deduce that $Q_S(v) = 0$ and so F is contained in E_S. This proves a). Now we prove c). First let $F = E_S \cap E_{S'}$. Clearly $E_{S \cap S'} \subset F$. The subspace F is invariant under u and $\mathrm{Sp}(u|_F)$ is contained in S and in S'; the inclusion $F \subset E_{S \cap S'}$ thus follows from a). Also, the inclusion $E_S + E_{S'} \subset E_{S \cup S'}$ is clear. To prove the reverse inclusion, let T denote the set of those $\lambda \in S'$ which do not belong to S and let $F = E_{S \cup S'} = E_{S \cup T}$ and $v = u|_F$. Then v annihilates the product of the polynomials Q_S and Q_T and these have no common root. Applying Lemma 3.3 to $v \in \mathrm{End}(F)$ we obtain $F = F_S(v) + F_T(v)$; hence F is contained in $E_S + E_{S'}$, which proves c). Finally we prove b). Let W denote the complement of S. By c), E is the direct sum of E_S and E_W. The polynomial P_u thus decomposes into the product of the characteristic polynomials P_S and P_W of the restrictions of u to E_S and E_W. These two polynomials are therefore obtained by regrouping the factors $(\lambda - T)^{m(\lambda)}$ of P_u according to whether λ belongs to S or to W. The assertion b) then follows from the fact that the dimension of E_S is the degree of P_S. $\qquad\qquad\qquad\qquad\square$

7.3.5. We deduce from c) in particular that if S and S' are two subsets of **C** (invariant under complex conjugation) that are disjoint and whose union contains $\mathrm{Sp}(u)$, then the subspaces $E_S(u)$ and $E_{S'}(u)$ are invariant under u and complementary in E. Clearly we can iterate this operation: *to every partition (S_i) of* $\mathrm{Sp}(u)$ *into subsets invariant under complex conjugation there is an associated decomposition of E into a direct sum of subspaces $E_{S_i}(u)$ which are invariant under u.*

7.3.6. We note a special case of such a decomposition, namely that in which we separate the eigenvalue 0 (if it exists) from the others. In this way we obtain a decomposition of E into the direct sum of two invariant subspaces K and L. The restriction of u to L has all its eigenvalues nonzero, and is therefore an automorphism of L. The restriction of u to K has all its eigenvalues zero and there exists $m \in \mathbf{N}$ with $(u|_K)^m = 0$; here we say $u|_K$ is *nilpotent*. In fact the dimension of K is the multiplicity $m(0)$ of the eigenvalue 0 (if u is invertible we put $m(0) = 0$) and we have $(u|_K)^{m(0)} = 0$. A special case worth singling out is when the restriction is not only nilpotent but is zero, which means that K is the kernel of u. In this case we have a direct construction:

Proposition 7.3.7. *The following conditions are equivalent and are automatically satisfied when 0 is an eigenvalue of multiplicity 1:*

(*i*)   *the subspaces $\mathrm{Ker}(u)$ and $\mathrm{Im}(u)$ are complementary;*
(*ii*) *for all $x \in E$ the condition $u^2 \cdot x = 0$ implies $u \cdot x = 0$;*
(*iii*) *E is the direct sum of two subspaces K and L invariant under u and such that $u(K) = 0$ and $u(L) = L$.*

Assuming (iii), let x be an element of E; write $x = y + z$ with $y \in K$ and $z \in L$ so that $u \cdot x = u \cdot z$. Since the restriction of u to L is surjective it is bijective and we obtain $\mathrm{Im}(u) = L$, $\mathrm{Ker}(u) = K$, which gives (i). Conversely, (i) implies that $u(\mathrm{Im}(u)) = u(E) = \mathrm{Im}(u)$ and this implies (iii) with $K = \mathrm{Ker}(u)$, $L = \mathrm{Im}(u)$. Since $\dim(\mathrm{Im}(u)) = \mathrm{codim}(\mathrm{Ker}(u))$, the condition (i) is equivalent to $\mathrm{Ker}(u) \cap \mathrm{Im}(u) = \{0\}$, which is a form of (ii). Finally, if $m(0) = 1$ the subspace K obtained by the general construction is of dimension 1. Since it contains $\mathrm{Ker}(u)$ which does not reduce to $\{0\}$ it coincides with $\mathrm{Ker}(u)$, which imples (iii). $\qquad\qquad\square$

7.4 Norm and Eigenvalues

As usual, let E be a finite-dimensional vector space and let $\mathrm{End}(E)$ be the vector space of its endomorphisms. When $E = \mathbf{R}^n$ the space $\mathrm{End}(E)$ is the space $M_n(\mathbf{R})$ of square matrices of order n.

7.4.1. Suppose now that we are given a norm $x \mapsto \|x\|$ on E. For $u \in \mathrm{End}(E)$ we define $\|u\| \in [0, \infty]$ by

$$\|u\| = \sup_{x \neq 0} \frac{\|u \cdot x\|}{\|x\|} = \sup_{\|x\|=1} \|u \cdot x\|.$$

For u in $\mathrm{End}(E)$ and x in E we have by definition

$$\|u \cdot x\| \leq \|u\|\|x\|. \tag{7.4.1}$$

For u and v in $\mathrm{End}(E)$ and λ in $\mathbf{R}$ we immediately have

$$\|\lambda u\| = |\lambda|\|u\|, \quad \|u + v\| \leq \|u\| + \|v\|, \quad \|uv\| \leq \|u\|\|v\|. \tag{7.4.2}$$

In particular, $u \mapsto \|u\|$ is a norm on the vector space $\mathrm{End}(E)$. If as before we let 1 denote the identity map Id_E, we have $\|1\| = 1$ when E is not just zero, and $\|1\| = 0$ when $E = \{0\}$. From this and (4.2) we obtain

$$\|u^p\| \leq \|u\|^p, \quad p \in \mathbf{N}. \tag{7.4.3}$$

Let u be an automorphism of E. Write $c_u = \|u^{-1}\|^{-1}$. For every $x \in E$ we have $\|x\| = \|u^{-1} \cdot (u \cdot x)\| \leq c_u^{-1}\|u \cdot x\|$, and so $\|u \cdot x\| \geq c_u\|x\|$. Also it is clear that c_u is the smallest constant satisfying this property, or in other words it is the greatest lower bound of $\|u \cdot x\|/\|x\|$ for $x \neq 0$. We often say that c_u is the *conorm* of u.

A first very elementary relation between norm and eigenvalues is given by the following lemma:

Lemma 7.4.2. *Let λ be an eigenvalue of u. There exists a nonzero element x of E and constants A and B with $0 < A \leq B$ and $A|\lambda|^n \leq \|u^n \cdot x\| \leq B|\lambda|^n$ for every integer $n \geq 0$.*

To see this, let $\lambda = |\lambda|e^{i\theta} \in \mathrm{Sp}(u)$. Apply Lemma 2.2. There exist x and y in E with $x \neq 0$ and (from the formula (2.4))

$$u^n \cdot x = |\lambda|^n (x \cos(n\theta) + y \sin(n\theta)).$$

The function $\alpha \mapsto \|x \cos(\alpha) + y \sin(\alpha)\|$ is continuous, periodic and everywhere positive on $\mathbf{R}$. It is bounded above (by B, say) and its greatest lower bound A is positive. $\qquad\square$

Definition 7.4.3. Let $u \in \mathrm{End}(E)$. The greatest lower bound of the real numbers $\|u^n\|^{1/n}$ for all integers $n > 0$ is called the *spectral radius* of u and denoted by $\rho(u)$:

$$\rho(u) = \inf_n \|u^n\|^{1/n}.$$

To paraphrase: we have

$$\rho(u)^n \leq \|u^n\|$$

for every integer $n > 0$, and $\rho(u)$ is the largest number that satisfies all these inequalities. The name 'spectral radius' has to do with part d) of the following theorem: $\rho(u)$ is the radius of the smallest disc centred at 0 which contains the spectrum of u.

The basic properties of the spectral radius are collected together in the following theorem:

Theorem 7.4.4.

a) *The sequence $\|u^n\|^{1/n}$ tends to $\rho(u)$ as n tends to infinity.*

b) *$\rho(u)$ does not depend on the norm chosen on E. More generally, for every norm $v \mapsto \|v\|_1$ on the vector space $\mathrm{End}(E)$ (arising from a norm on E) the sequence $\|u^n\|_1^{1/n}$ converges to $\rho(u)$.*

c) *$\rho(u)$ is the greatest lower bound of the norms of u with respect to all the norms on E; we may even restrict ourselves to considering euclidean norms on E.*

d) *$\rho(u)$ is the maximum of the $|\lambda|$ for λ in $\mathrm{Sp}(u)$.*

The proof of this theorem will be quite long. First of all we deal with a). Let $a_n = \|u^n\|$. By (4.2) we have $a_{n+p} \leq a_n a_p$. Part a) is then a consequence of the following lemma:

Lemma 7.4.5. *Let (a_n) be a sequence of numbers ≥ 0 such that $a_{n+p} \leq a_n a_p$ for all n and p. Then the sequence $(a_n^{1/n})$ converges, and its limit is also its greatest lower bound.*

Let A denote the greatest lower bound of the $a_n^{1/n}$. Our task is to prove that for every $B > A$ we have $a_n^{1/n} \leq B$ for all sufficiently large n. However, by definition of greatest lower bound, there exists an integer m with $a_m^{1/m} = C < B$. Let n be an integer ≥ 0. By euclidean division we can write $n = qm + r$ with $0 \leq r < m$ and we have

$$a_n \leq a_m^q a_r = C^{qm} a_r = C^n a_m^{-r/m} a_r \leq C^n D$$

where D is any upper bound for the (finite) set of the $a_m^{-r/m} a_r$ for $0 \leq r < m$. Hence $a_n^{1/n} \leq D^{1/n} C$. If n is sufficiently large so that $(B/C)^n$ is greater than D we have $a_n^{1/n} \leq B$. $\qquad\square$

7.4.6. Now we turn to part b) of Theorem 4.4. Let $v \mapsto \|v\|_1$ be a norm on the vector space $\mathrm{End}(E)$; it is equivalent to the norm above, and there exist constants A and B with $0 < A \leq B$ and $A\|v\| \leq \|v\|_1 \leq B\|v\|$ for every endomorphism v. We deduce

$$A^{1/n}\|u^n\|^{1/n} \leq \|u^n\|_1^{1/n} \leq B^{1/n}\|u^n\|^{1/n}.$$

When n tends to infinity, $A^{1/n}$ and $B^{1/n}$ tend to 1 and we deduce b).

7.4.7. Next we prove part c). If $x \mapsto \|x\|_1$ is an arbitrary norm on E, then according to b) we may calculate $\rho(u)$ using this norm and in particular we have $\rho(u) \leq \|u\|_1$. To prove c) we have to show that for every $a > \rho(u)$ there exists a euclidean norm on E for which the norm of u is $\leq a$. Fix a positive definite quadratic form Q on E. Since the spectral radius of u may be calculated with the norm $x \mapsto \sqrt{Q(x)}$ and is $< a$, there exists an integer $m > 0$ with

$$Q(u^m \cdot x) \leq a^{2m}Q(x), \quad x \in E.$$

Define a quadratic form Q_1 by

$$Q_1(x) = \sum_{i=0}^{m-1} a^{2(m-i-1)}Q(u^i \cdot x) = a^{2(m-1)}Q(x) + \cdots + Q(u^{m-1} \cdot x).$$

Clearly Q_1 is positive definite. Moreover

$$Q_1(u \cdot x) = a^2 Q_1(x) + Q(u^m \cdot x) - a^{2m}Q(x) \leq a^2 Q_1(x),$$

and the norm of u relative to the euclidean norm $x \mapsto \sqrt{Q_1(x)}$ is indeed $\leq a$. This completes the proof of c).

7.4.8. Finally we come to part d) of Theorem 4.4.

The most natural proof uses *holomorphic functions*. We start by placing ourselves in the complex setting and considering the function $z \mapsto (1 - zu)^{-1}$. This is a holomorphic function (with values in the complex vector space of endomorphisms) defined away from the points $1/\lambda$, $\lambda \in \mathrm{Sp}(u)$ which are its poles. The radius of convergence of its Taylor series at the origin is therefore the greatest lower bound of the $1/|\lambda|$. But the general term of this series is $z^n u^n$; since the $\|u^n\|^{1/n}$ tend to $\rho(u)$, Cauchy's test tells us that this radius of convergence is also $1/\rho(u)$ and this proves d). We give a more elementary proof below, but in any case we cannot avoid going via complex numbers.

First we prove that the modulus of every eigenvalue λ of u is bounded above by $\rho(u)$. Take x and A as in Lemma 4.2. We have

$$A|\lambda|^n \leq \|u^n \cdot x\| \leq \|u^n\| \, \|x\|,$$

and so $|\lambda| \leq (\|x\|/A)^{1/n}\|u^n\|^{1/n}$. Letting n tend to infinity we obtain $|\lambda| \leq \rho(u)$, as claimed.

To complete the proof of d) it suffices to show that if $|\lambda| < a$ for every eigenvalue λ of u then $\rho(u) \leq a$. Replacing u by u/a this comes down to showing that if $|\lambda| < 1$ for every eigenvalue λ of u then the u^n form a bounded family in $\mathrm{End}(E)$. Let $P(T)$ be a polynomial with real coefficients that annihilates u and all of whose roots belong to $\mathrm{Sp}(u)$, for example the characteristic polynomial of u. For each integer $n > 0$ let $R_n(T)$ denote the remainder when T^n is divided by $P(T)$. We may write $T^n = A(T)P(T) + R_n(T)$ and we therefore have $u^n = R_n(u)$. To prove the u^n form a bounded family it is enough to prove that the coefficients of the R_n are bounded. This leads to the following purely algebraic lemma:

Lemma 7.4.9. *Let $P(T)$ be a polynomial belonging to $\mathbf{C}[T]$ all of whose roots lie in the interior of the unit disc. Let V denote the vector space of polynomials of $\mathbf{C}[T]$ of degree strictly less than that of P, and let $r : \mathbf{C}[T] \to V$ be the map which associates to each polynomial its remainder after division by P. Then the $r(T^n)$, for $n \in \mathbf{N}$, form a bounded subset of V.*

Let $\lambda_1, \ldots, \lambda_k$ be the roots of P, with multiplicities $m_1, \ldots, m_k$. The degree of P is therefore the sum d of the m_i, and V is a complex vector space of dimension d. Consider the map $v : \mathbf{C}[T] \to \mathbf{C}^d$ which associates to each polynomial Q the family of the $Q^{(m)}(\lambda_i)$ with $m < m_i$. Clearly, the kernel of v consists of the multiples of P. Hence we first of all have that $v(r(Q)) = v(Q)$. Moreover, this also implies that the restriction of v to V is injective and therefore bijective (which is the smart way to view Lagrange interpolation). To prove the lemma it is enough to prove that the family of the $v(T^n)$ is bounded in $\mathbf{C}^n$. However, this is clear since the components of the $v(T^n)$ are λ_i^n and $n(n-1) \cdots (n-m+1)\lambda_i^{n-m}$, with $i = 1, \ldots, k$ and $m = 1, \ldots, m_i - 1$, and the λ_i have absolute value < 1. $\qquad\square$

7.4.10. We conclude with a few properties of the spectral radius. Clearly $\rho(\lambda u) = |\lambda|\rho(u)$. For every integer $n > 0$ we have $\rho(u^n) = \rho(u)^n$; this follows essentially from Theorem 4.4 a) (or 4.4 d) and Lemma 2.4 a)). To say that $\rho(u)$ is zero is to say, by 4.4 d), that all the eigenvalues of u are zero, that is that u is nilpotent: (see 3.6).

7.5 Contracting, Expanding and Hyperbolic Endomorphisms

As usual, let E be a finite-dimensional vector space and let u be an endomorphism of E.

Proposition 7.5.1. *The following conditions are equivalent:*

(i) $\rho(u) < 1$;

(ii) $|\lambda| < 1$ *for every eigenvalue* λ *of* u;

(iii) *there exists a norm on* E *with respect to which* $\|u\| < 1$;

(iv) *for every* $x \in E$ *the sequence* $u^n \cdot x \to 0$ *as* $n \to \infty$.

The equivalence of properties (i),(ii) and (iii) comes from Theorem 4.4. We have $\|u^n \cdot x\| \leq \|u\|^n \|x\|$ and so (iii) implies (iv). Finally, (iv) implies (ii) by Lemma 4.2. $\square$

Definition 7.5.2. The endomorphism u is called *contracting* if it satisfies the equivalent conditions of Proposition 5.1.

Let $u \in \text{End}(E)$. We use the construction in Proposition 3.4, taking S to be the unit disc $\{z \in \mathbf{C} \mid |z| < 1\}$. We obtain an invariant subspace of E, denoted by $E_c(u)$, whose dimension is equal to the sum of the multiplicities of the eigenvalues λ of u with $|\lambda| < 1$. This is by definition *the largest invariant subspace of* E *on which the restriction of* u *is contracting*. It is also *the set of those* $x \in E$ *such that* $u^n \cdot x \to 0$ *as* $n \to \infty$; the latter is an invariant subspace on which the restriction of u is contracting, and which contains the former. We say that $E_c(u)$ is the *contracting subspace* for u.

Proposition 7.5.3. *The following conditions are equivalent:*

(i) $|\lambda| > 1$ *for every eigenvalue* λ *of* u;

(ii) u *is invertible and* u^{-1} *is contracting;*

(iii) *there exists a norm on* E *and a constant* $A > 1$ *for which* $\|u \cdot x\| \geq A\|x\|$ *for all* $x \in E$;

(iv) *for every nonzero element* x *in* E *the sequence* $u^n \cdot x \to \infty$ *as* $n \to \infty$.

We have (i)$\Rightarrow$(ii) by Lemma 2.5. If (ii) holds, we apply Proposition 5.1 to u^{-1}, put $A = \|u^{-1}\|^{-1} > 1$ and let $x \in E$. Then

$$\|x\| = \|u^{-1} \cdot (u \cdot x)\| \leq A^{-1}\|u \cdot x\|,$$

which gives (iii). If (iii) holds we have $\|u^n \cdot x\| \geq A^n \|x\|$ which gives (iv). Finally, suppose (iv) is satisfied; if there existed $\lambda \in \text{Sp}(u)$ with $|\lambda| \leq 1$ then Lemma 4.2 would contradict (iv); this proves (i) and completes the proof. $\square$

Definition 7.5.4. The endomorphism is said to be *expanding* if it satisfies the equivalent conditions of Proposition 5.3.

Let $u \in \text{End}(E)$. As above, we define the *expanding subspace* for u to be the largest invariant subspace $E_e(u)$ of E on which the restriction of u is expanding. If u is invertible, it is the contracting subspace for u^{-1}. The space E decomposes into the direct sum of $E_c(u), E_e(u)$ and the invariant subspace corresponding to eigenvalues of u which lie on the unit circle. Hence $E_c(u)$

and $E_e(u)$ intersect in $\{0\}$ only; in order for them to be complementary it is necessary and sufficient that u have no eigenvalue on the unit circle.

Proposition 7.5.5. *The following conditions are equivalent:*

(i) $|\lambda| \neq 1$ *for every eigenvalue of* u;
(ii) *the subspaces* $E_c(u)$ *and* $E_e(u)$ *are complementary;*
(iii) *there exist complementary subspaces* E' *and* E'' *of* E, *invariant under* u, *such that the restriction of* u *to* E' *is a contracting endomorphism and the restriction of* u *to* E'' *is an expanding endomorphism.*

Moreover, under the assumptions of (iii) we necessarily have that $E' = E_c(u)$ *and* $E'' = E_e(u)$.

We have seen that (i) is equivalent to (ii) and it is clear that (ii) implies (iii). If (iii) holds we have $E' \subset E_c(u)$ and $E'' \subset E_e(u)$; as E' and E'' are complementary and $E_c(u) \cap E_e(u) = \{0\}$ this imples (ii) and the final statement.   □

Definition 7.5.6. The endomorphism u is said to be *hyperbolic* if it satisfies the equivalent conditions of Proposition 5.5.

Proposition 7.5.7. *In the space* $\mathrm{End}(E)$, *the set of hyperbolic endomorphisms* u *such that* $\dim(E_c(u))$ *and* $\dim(E_e(u))$ *have given values is an open set, the set of contracting (resp. expanding) endomorphisms is an open set, and the set of hyperbolic endomorphisms is an open and dense set.*

The fact that these different sets are open follows directly from Proposition 2.6. To prove the density of the set of hyperbolic endomorphisms it is sufficient to note that for every automorphism u and every real number α the endomorphism αu, which is arbitrarily close to u, is hyperbolic as long as u does not belong to the finite set consisting of the $1/|\lambda|$ for $\lambda \in \mathrm{Sp}(u)$.   □

7.6 The Exponential of an Endomorphism

Let E be a finite-dimensional vector space. Equip E with a norm $x \mapsto \|x\|$ and let the space $\mathrm{End}(E)$ of its endomorphisms be given the associated norm $u \mapsto \|u\|$ as in 4.1.

Lemma 7.6.1. *Let* u *be an endomorphism of* E. *The series whose general term is* $u^p/p!$ *converges in* $\mathrm{End}(E)$.

By (4.3), we have $\|u\|^p \leq \|u\|^p$ and consequently

$$\sum_{p=0}^{\infty} \|\frac{u^p}{p!}\| \leq \sum_{p=0}^{\infty} \frac{\|u\|^p}{p!} \leq e^{\|u\|} < +\infty.$$

The given series is absolutely convergent and therefore convergent. □

Definition 7.6.2. The element

$$e^u = \exp(u) = \sum_{p=0}^{\infty} \frac{u^p}{p!} \tag{7.6.1}$$

of $\text{End}(E)$ is called the *exponential* of the endomorphism u.

We have $\exp(0) = 1$. For every x in E we have

$$\exp(u) \cdot x = \sum_{p=0}^{\infty} \frac{u^p \cdot x}{p!}. \tag{7.6.2}$$

If we choose a basis for E in which u is represented by the matrix A then the matrix of $\exp(u)$ is

$$e^A = \exp(A) = \sum_{p=0}^{\infty} \frac{A^p}{p!}.$$

7.6.3. Here are a few elementary observations.

a) The calculation above gives the (very rough) upper bound

$$\|\exp(u)\| \leq e^{\|u\|}. \tag{7.6.3}$$

b) For every automorphism v of E we have $vu^pv^{-1} = (vuv^{-1})^p$ for all p. From this we deduce

$$\exp(vuv^{-1}) = v\exp(u)v^{-1}. \tag{7.6.4}$$

c) Let F be a subspace of E that is invariant under u; then F is invariant under $\exp(u)$ and we have $\exp(u|_F) = \exp(u)|_F$.

Proposition 7.6.4. *Let u and v be two elements of $\text{End}(E)$ which commute. Then $v\exp(u) = \exp(u)v$ and $\exp(u + v) = \exp(u)\exp(v) = \exp(v)\exp(u)$.*

Since u and v commute we have $vu^p = u^pv$. Summing the exponential series we deduce $v\exp(u) = \exp(u)v$. Moreover, still because u and v commute, we may apply the binomial formula

$$\frac{(u+v)^n}{n!} = \sum_{p+q=n} \frac{u^p}{p!} \frac{v^q}{q!}.$$

It remains only to sum both sides to obtain the result. □

Corollary 7.6.5.

a) *The endomorphism $\exp(u)$ is invertible, and we have $\exp(u)^{-1} = \exp(-u)$.*
b) *For all $n \in \mathbf{Z}$ we have $\exp(nu) = \exp(u)^n$.*
c) *For all t and t' in $\mathbf{R}$ we have*

$$\exp\big((t + t')u\big) = \exp(tu)\exp(t'u). \tag{7.6.5}$$

Since tu and $t'u$ commute, part c) follows directly from the Proposition. In particular, we have $\exp(u)\exp(-u) = \exp(0) = 1$ and likewise $\exp(-u)\exp(u) = 1$; this implies a). Finally, b) follows from a) and c). $\square$

Let us give some examples of direct calculation of exponentials. First, the exponential of a scalar multiplication is a scalar multiplication: for $\lambda \in \mathbf{R}$ we have

$$\exp(\lambda 1) = e^\lambda 1. \tag{7.6.6}$$

More generally, if we have $u \cdot x = \lambda x$ then $u^p \cdot x = \lambda^p x$ and hence $\exp(u) \cdot x = e^\lambda x$.

Now suppose that $u^2 = \lambda^2 1$. Then we have $u^{2p} = \lambda^{2p} 1$ and $u^{2p+1} = \lambda^{2p} u$. This immediately gives $\exp(u) = \cosh(\lambda)1 + \lambda^{-1}\sinh(\lambda)u$. Likewise, if $u^2 = -\lambda^2 1$ we obtain $\exp(u) = \cos(\lambda)1 + \lambda^{-1}\sin(\lambda)u$. Thus we have

$$\exp\begin{pmatrix} 0 & \lambda \\ \lambda & 0 \end{pmatrix} = \begin{pmatrix} \cosh\lambda & \sinh\lambda \\ \sinh\lambda & \cosh\lambda \end{pmatrix}, \quad \exp\begin{pmatrix} 0 & -\lambda \\ \lambda & 0 \end{pmatrix} = \begin{pmatrix} \cos\lambda & -\sin\lambda \\ \sin\lambda & \cos\lambda \end{pmatrix}. \tag{7.6.7}$$

Since the matrices $\begin{pmatrix} \sigma & 0 \\ 0 & \sigma \end{pmatrix}$ and $\begin{pmatrix} 0 & \tau \\ \tau & 0 \end{pmatrix}$ commute, we deduce:

$$\exp\begin{pmatrix} \sigma & -\tau \\ \tau & \sigma \end{pmatrix} = \begin{pmatrix} e^\sigma \cos\tau & -e^\sigma \sin\tau \\ e^\sigma \sin\tau & e^\sigma \cos\tau \end{pmatrix}. \tag{7.6.8}$$

Proposition 7.6.6. *The eigenvalues of $\exp(u)$ are e^λ for $\lambda \in \mathrm{Sp}(u)$.*

Let $\lambda \in \mathrm{Sp}(u)$. By Lemma 2.2 there exists an $\mathbf{R}$-linear map $j : \mathbf{C} \to E$ with $j(1) \neq 0$ and $u \cdot j(z) = j(\lambda z)$. We deduce that $u^n \cdot j(z) = j(\lambda^n z)$ and $\exp(u) \cdot j(z) = j(e^\lambda z)$. Therefore e^λ belongs to $\mathrm{Sp}(\exp(u))$ by Lemma 2.2. Conversely, let $\mu \in \mathrm{Sp}(\exp(u))$; put $R(T) = (T - \mu)(T - \bar\mu)$ and $F = \mathrm{Ker}(R(\exp(u)))$. Since $uR(\exp(u)) = R(\exp(u))u$, the subspace F is invariant under u. Applying Lemma 2.4 b) to $\exp(u)$, we see that F is not just $\{0\}$ and that $\mathrm{Sp}(\exp(u)|_F) = \{\mu, \bar\mu\}$. Let $\lambda \in \mathrm{Sp}(u|_F) \subset \mathrm{Sp}(u)$. By applying the above to F we see that e^λ belongs to the spectrum of $\exp(u)|_F$ and is therefore equal to μ or to $\bar\mu$. Thus μ is equal to e^λ or to $e^{\bar\lambda}$. $\square$

The spectral radius of $\exp(u)$ is the maximum of the absolute values of its eigenvalues (Theorem 4.4); we have just seen that these are the exponentials of the eigenvalues of u. If we let $\Re(z)$ denote the real part of the complex number z we thus have:

Corollary 7.6.7. *The spectral radius of* $\exp(u)$ *is* e^b *where* b *is the maximum of the* $\Re(\lambda)$ *as* λ *runs through* $\mathrm{Sp}(u)$. $\square$

Corollary 7.6.8. *Let* β *be a real number such that* $\Re(\lambda) < \beta$ *for all* $\lambda \in \mathrm{Sp}(u)$. *There exists* $\delta > 0$ *such that*

$$\| \exp(tu) \| \leq \delta e^{\beta t}, \quad t \geq 0. \tag{7.6.9}$$

Since the spectral radius of $\exp(u)$ is $< e^\beta$ there exists an integer $n > 0$ with $\| \exp(u)^n \| \leq e^{\beta n}$. Let $t \geq 0$. There exists an integer $m \geq 0$ with $mn \leq t \leq (m+1)n$. Let $a = t - mn \in [0, n]$. We have

$$\| \exp(tu) \| \leq \| \exp(u)^n \|^m \| \exp(au) \| \leq e^{\beta mn} \| \exp(au) \| \leq e^{\beta t} e^{-\beta a} \| \exp(au) \|.$$

By 6.3a, $\| \exp(au) \|$ is bounded above by $e^{n\|u\|}$; moreover, $e^{-\beta a}$ is bounded above by $\sup(1, e^{-\beta})$. $\square$

Corollary 7.6.9. *Let* α *and* β *be real numbers such that* $\alpha < \Re(\lambda) < \beta$ *for all* $\lambda \in \mathrm{Sp}(u)$. *There exist* $\gamma > 0$ *and* $\delta > 0$ *such that*

$$\gamma e^{\alpha t} \|x\| \leq \| \exp(tu) \cdot x \| \leq \delta e^{\beta t} \|x\|, \quad x \in E, \quad t \geq 0. \tag{7.6.10}$$

The second inequality is merely a reformulation of the previous corollary. Applying it to the endomorphism $-u$ and the number $-\alpha$ we obtain the existence of $\gamma > 0$ with

$$\| \exp(-tu) \cdot y \| \leq \gamma^{-1} e^{-\alpha t} \|y\|, \quad y \in E, \quad t \geq 0 \,;$$

taking $y = \exp(tu) \cdot x$ we obtain the left hand side of (6.10). $\square$

7.7. One-parameter Groups of Linear Transformations

Proposition 7.7.1. *The exponential map* $\exp : \mathrm{End}(E) \to \mathrm{End}(E)$ *is of class* C^∞. *Its derivative at* 0 *is the identity map of* $\mathrm{End}(E)$.

We shall accept the first statement, which follows from general theorems on series of maps. The second statement follows from the sequence of bounds

$$\| \exp(u) - 1 - u \| = \| \sum_{p=2}^\infty \frac{u^p}{p!} \| \leq \sum_{p=2}^\infty \frac{\|u\|^p}{p!}$$
$$= e^{\|u\|} - 1 - \|u\| \leq \|u\|^2 e^{\|u\|}. \qquad \square$$

It is interesting to calculate the derivative of the exponential map at every point of $\mathrm{End}(E)$ (exercise).

Let u be fixed in $\mathrm{End}(E)$. The map $t \mapsto \exp(tu)$ from $\mathbf{R}$ into $\mathrm{End}(E)$ is of class C^∞, we have $\exp(0) = 1$ and

$$\exp((t + t')u) = \exp(tu)\exp(t'u). \tag{7.7.1}$$

Moreover, by the previous proposition, the derivative at 0 of this map is the image under the identity map of $\mathrm{End}(E)$ of the element $d(tu)/dt$, namely u. Therefore differentiating (7.1) with respect to t' at $t' = 0$ we obtain

$$\frac{d}{dt}\exp(tu) = u\exp(tu) = \exp(tu)u. \tag{7.7.2}$$

Hence in the terminology of 6.7.1:

Proposition 7.7.2. *The map $t \mapsto \exp(tu)$ is the one-parameter group of linear transformations of E with infinitesimal generator u.*   □

We shall say that $(\exp(tu))_{t\in\mathbf{R}}$ is the *exponential flow with generator u*. It is the integral flow of the differential equation $dx/dt = u \cdot x$. More generally, by using the *method of variation of constants* it is possible to solve equations of the form $dx/dt = u \cdot x + b(t)$. We obtain the following proposition:

Proposition 7.7.3. *Let I be an open interval of $\mathbf{R}$ and let $b : I \to E$ be a continuous map. Let $u \in \mathrm{End}(E), x_0 \in E$ and $t_0 \in I$. The unique solution in I to the differential equation*

$$\frac{dx}{dt} = u \cdot x + b(t), \quad x \in E$$

taking the value x_0 at $t = t_0$ is

$$x(t) = e^{(t-t_0)u} \cdot x_0 + \int_{t_0}^{t} e^{(t-s)u} \cdot b(s)\, ds.$$

To show this, put $y(t) = e^{(t_0-t)u} \cdot x(t) \in E$. We have $y(t_0) = x_0$ and

$$\frac{dy}{dt} = e^{(t_0-t)u} \cdot \left(\frac{dx}{dt} - u \cdot x\right) = e^{(t_0-t)u} \cdot b(t).$$

Hence

$$y(t) = x_0 + \int_{t_0}^{t} e^{(t_0-s)u} \cdot b(s)\, ds,$$

and finally

$$x(t) = e^{(t-t_0)u} \cdot y(t) = e^{(t-t_0)u} \cdot x_0 + \int_{t_0}^{t} e^{(t-t_0)u} \cdot \left(e^{(t_0-s)u} \cdot b(s)\right) ds,$$

which gives the stated formula.   □

The description of the exponential as the solution of the differential equation (7.2) enables us to demonstrate many of its properties. Here is an example:

Proposition 7.7.4. *We have*

$$\det(\exp(u)) = e^{\operatorname{Tr}(u)}. \tag{7.7.3}$$

Let $\delta(t) = \det(\exp(tu))$. By (7.1) we have $\delta(s+t) = \delta(s)\delta(t)$. Moreover, for s close to 0 we have

$$\delta(s) = \det(1 + su + o(s)) = 1 + s\operatorname{Tr}(u) + o(s),$$

the last equality being obtained by expanding the determinant. From this we deduce $\delta(s+t) - \delta(t) = \delta(t)(s\operatorname{Tr}(u) + o(s))$, and therefore

$$\frac{d\delta}{dt}(t) = \delta(t)\operatorname{Tr}(u).$$

Since $\delta(0) = 1$ this implies $\delta(t) = e^{t\operatorname{Tr}(u)}$ and finally $\det(\exp(u)) = \delta(1) = e^{\operatorname{Tr}(u)}$.   □

As we have seen in 6.7.3 and 6.7.4, the exponential flows $(\exp(tu))_{t\in\mathbf{R}}$ provide all the one-parameter groups of linear transformations of E as u runs through $\operatorname{End}(E)$. We can even prove a stronger result, directly and without the assumption of differentiability:

Proposition 7.7.5. *Let I be an open interval in $\mathbf{R}$ containing 0 and let $\phi : I \to \operatorname{End}(E)$ be a continuous map such that $\phi(0) = 1$ and $\phi(t+t') = \phi(t)\phi(t')$ for t, t' and $t+t'$ in I. Then there exists a unique endomorphism $u \in \operatorname{End}(E)$ such that $\phi(t) = \exp(tu)$ for all $t \in I$.*

The uniqueness of u is clear, since it is necessarily the derivative of ϕ at the origin. Next note that it is sufficient to show that ϕ is differentiable: if this has been done, then putting $u = d\phi/dt(0)$ and differentiating the given relation we obtain $d\phi(t)/dt = u\phi(t)$, $t \in I$. Then putting $\psi(t) = \exp(-tu)\phi(t)$ we find

$$\frac{d\psi}{dt} = \exp(-tu)\frac{d\phi}{dt} - \exp(-tu)u\phi(t) = 0,$$

and so $\psi(t) = \psi(0) = 1$, that is to say $\phi(t) = \exp(tu)$.

Now let us prove the differentiability of ϕ. Fix $t \in I$ and let $a > 0$ in I, small enough so that $t + a$ belongs to I. Integrating the relation $\phi(t + s) = \phi(t)\phi(s)$ for $s \in [0, a]$ we obtain

$$\int_t^{t+a} \phi(s)ds = \phi(t) \int_0^a \phi(s)ds.$$

However, $\int_0^a \phi(s)ds$ is of the form $a(\phi(0) + o(a)) = a(1 + o(a))$. For a suf-ficiently small this is an invertible element of $\mathrm{End}(E)$. Letting $v \in \mathrm{Aut}(E)$ denote its inverse, we obtain

$$\phi(t) = v \int_t^{t+a} \phi(s)ds,$$

which shows that ϕ is differentiable and completes the proof.   □

The argument above remains valid when we assume only that ϕ is locally inte-grable. In fact, every group homomorphism $\phi : \mathbf{R} \to \mathrm{Aut}(E)$ that is measurable is already automatically of class C^∞.

7.7.6. Here now is a particularly instructive example. Take E to be the space of polynomials of degree $< n$ in one variable X. For $t \in \mathbf{R}$ and $P \in E$ define an endomorphism $\phi(t)$ of E by $(\phi(t) \cdot P)(X) = P(X + t)$. Immediately we see $\phi(t + t') = \phi(t)\phi(t')$ and therefore we are dealing with a one-parameter group of linear transformations of E. By the very definition of the derivative, the infinitesimal generator u of this one-parameter group is the differentiation operator d/dX. Hence

$$P(X + t) = \exp(t\frac{d}{dX}) \cdot P(X) = \sum_{p=0}^\infty \frac{t^p}{p!}(\frac{d}{dX})^p \cdot P(X),$$

and we recognize Taylor's formula derived by an amusing proof. If as a basis of E we take the monomials

$$e_p = \frac{X^{p-1}}{(p-1)!}, \quad p \in [1, n],$$

we have $u \cdot e_1 = 0$ and, for $p > 1$,

$$u \cdot e_p = \frac{d}{dX}\frac{X^{p-1}}{(p-1)!} = \frac{X^{p-2}}{(p-2)!} = e_{p-1}.$$

The matrix for u in the basis (e_i) is therefore a 'Jordan matrix'. The formulae above give us the matrix of its exponential: we have

$$\exp(tu) \cdot e_p = \frac{(X+t)^{p-1}}{(p-1)!} = \sum_{i=1}^p \frac{t^{p-i}}{(p-i)!}\frac{X^{i-1}}{(i-1)!} = \sum_{i=1}^p \frac{t^{p-i}}{(p-i)!}e_i.$$

7.7.7. We conclude with two examples of linear systems constructed using gradients of quadratic forms. Suppose we are given a euclidean scalar product $(x \mid y)$ and a quadratic form $f(x)$ on a vector space E. Let $a \in \mathrm{End}(E)$ denote the symmetric endomorphism such that $f(x) = (a \cdot x \mid x)/2$, and therefore $\mathrm{grad}(f)(x) = a \cdot x$. As is well known, there exist orthonormal bases for $(x \mid y)$ which are orthogonal for f, that is to say they consist of eigenvectors of a. Hence E can be decomposed into the orthogonal direct sum of 1-dimensional subspaces invariant under a.

First we are interested in the linear system $\dot{x} = \mathrm{grad}(f)(x)$, that is $\dot{x} = a \cdot x$. This therefore reduces to the differential equations $\dot{x} = \lambda x$ where λ runs through the eigenvalues of a. To say that the $\lambda \in \mathrm{Sp}(a)$ are < 0 (that is, in the terminology introduced later, that the corresponding flow is contracting) means that f is negative definite, or equivalently that 0 is the unique absolute maximum of f.

Now consider the Lagrangian linear system $\ddot{x} + \mathrm{grad}(f)(x) = 0$, which can also be wrtitten $\ddot{x} + a \cdot x = 0$. This reduces to the first order system $\dot{x} = y$, $\dot{y} = -a \cdot x$ which introduces the endomorphism$(x, y) \mapsto (y, -a \cdot x)$ of $E \times E$. The system decomposes as above into the direct sum of equations in one space dimension having one of the three following types: $\ddot{x} + \omega^2 x = 0$, $\ddot{x} - \alpha^2 x = 0$, $\ddot{x} = 0$. These equations can be integrated immediately: the first reduces to the first order system $\dot{x} = -\omega y$, $\dot{y} = \omega x$, the second to the system $\dot{x} = \alpha y$, $\dot{y} = \alpha x$ and the third to $\dot{x} = y$, $\dot{y} = 0$. We recover the square 2×2 matrices whose exponentials we calculated earlier. To say that f is positive definite now means that only the first type of equation is involved, which is to say that all solutions of the system are bounded.

7.8 The Image of the Exponential

As before, let E be a finite-dimensional vector space. Our aim is to determine the image of the exponential map $\exp : \mathrm{End}(E) \to \mathrm{End}(E)$. We have seen that it lies in the open set $\mathrm{Aut}(E)$ consisting of the automorphisms of E. By Proposition 7.1 and the Local Inversion Theorem, this image contains a neighbourhood of $1 \in \mathrm{End}(E)$.

In fact it can be proved directly that every $v \in \mathrm{End}(E)$ with $\|v - 1\| < 1$ is of the form $\exp(u)$: it suffices to take as u the sum of the absolutely convergent series

$$\mathrm{Log}(1 + (v - 1)) = \sum_{n=0}^{\infty} \frac{(-1)^n}{n+1}(v - 1)^{n+1}.$$

7.8.1. By Proposition 7.4, the image of the exponential is contained in the open set $\mathrm{Aut}^+(E)$ consisting of automorphisms with determinant > 0. We have $\exp(\mathbf{R}) = \mathbf{R}^+$ and hence $\exp(\mathrm{End}(E)) = \mathrm{Aut}^+(E)$ when the dimension of E is 1. *When* $\dim(E) \geq 2$ *this no longer holds.* For example, take a diagonal matrix $v \in M_n(\mathbf{R})$ with diagonal $(\lambda_1, \ldots, \lambda_n)$. If v is of the form $\exp(u)$ where $u \in M_n(\mathbf{R})$, then by Proposition 6.6 the λ_i are the exponentials of the eigenvalues (real or complex) of u. If μ is a non-real eigenavlue of u then so is $\bar{\mu}$ and the two eigenvalues e^μ and $e^{\bar{\mu}}$ of v are complex conjugates and therefore equal since the eigenvalues of v are real by assumption. Thus if the λ_i are all distinct then the eigenvalues of u are necessarily real and all $\lambda_i > 0$. Therefore it suffices to take the λ_i distinct and with product > 0 but not all $\lambda_i > 0$ (which is possible for $n > 1$). For example, the matrix $v = \begin{pmatrix} -1 & 0 \\ 0 & -e \end{pmatrix}$ is not an exponential. On the other hand, we have $\begin{pmatrix} -1 & 0 \\ 0 & -1 \end{pmatrix} = \exp\begin{pmatrix} 0 & -\pi \\ \pi & 0 \end{pmatrix}$.

Thus we see that the real matrix v is the product of the exponentials of the real matrices $\begin{pmatrix} 0 & 0 \\ 0 & 1 \end{pmatrix}$ and $\begin{pmatrix} 0 & -\pi \\ \pi & 0 \end{pmatrix}$ but is not itself the exponential of a real matrix. When $\dim(E) > 1$ the image of the exponential is not closed under multiplication and it is vain to hope for a general multiplication formula of the type $\exp(a)\exp(b) = \exp(H(a,b))$. Such a formula exists only for a and b small; it involves a "non-commutative series" in a and b, with restricted domain of convergence, called the *Campbell-Hausdorff* series.

7.8.2. The situation is simpler in the complex setting: every **C**-linear automorphism of a complex vector space is the exponential of a suitably chosen **C**-linear endomorphism. In the real context where we are working, the situation is more subtle: *an automorphism is an exponential if and only if it is a square.* This condition is clearly necessary since $\exp(u) = \exp(u/2)^2$. (Note incidentally that saying an automorphism of E belongs to the image of the exponential means also that it belongs to a one-parameter group). The sufficiency comes from the following more precise theorem which we shall assume:

Theorem 7.8.3. *Let* $v \in \mathrm{Aut}(E)$. *Then there exists* $u \in \mathrm{End}(E)$ *with* $uv = vu$ *and* $\exp(u) = v^2$. $\square$

Since v commutes with u it also commutes with $\exp(u/2)$ (Proposition 6.4). If we let $s = \exp(-u/2)v = v\exp(-u/2)$ we obtain:

Corollary 7.8.4. *Let* $v \in \mathrm{Aut}(E)$. *Then there exist* $u \in \mathrm{End}(E)$ *and* $s \in \mathrm{Aut}(E)$ *with* $us = su$, $s^2 = 1$ *and* $v = s\exp(u) = \exp(u)s$. $\square$

Since $T^2 - 1 = (T-1)(T+1)$ it follows from Lemma 3.3 that the subspaces $\mathrm{Ker}(s-1)$ and $\mathrm{Ker}(s+1)$ are complementary. They are invariant under both u and v. By construction, s takes the value 1 on one and -1 on the other. Hence:

Corollary 7.8.5. *Let* $v \in \mathrm{Aut}(E)$. *There exist two complementary subspaces* F^+ *and* F^- *of* E, *invariant under* v, *and endomorphisms* $u_+ \in \mathrm{End}(F^+)$ *and* $u_- \in \mathrm{End}(F^-)$ *with* $v|_{F^+} = \exp(u_+)$ *and* $v|_{F^-} = -\exp(u_-)$. $\square$

The difficulty arises from the fact that the element -1 of $\mathrm{Aut}(F^-)$ (even when it is an exponential – which is the case when F^- has even dimension) cannot be written as $\exp(c)$ where $cw = wc$ for every $w \in \mathrm{End}(F^-)$; otherwise we could write $-\exp(u_-) = \exp(c + u_-)$. In the complex case we are saved by the existence of the element $\pi i 1$.

To illustrate this situation, we shall prove the complete result for dimension 2 by a direct analysis independent of the above results.

Proposition 7.8.6. *Suppose E has dimension 2 and let $v \in \mathrm{Aut}(E)$.*

a) *If the eigenvalues of v are positive and distinct, or if v has a repeated positive eigenvalue but is not of the form $\mu 1$, then there exists a unique $u \in \mathrm{End}(E)$ such that $\exp(u) = v$.*

b) *If the eigenvalues of v are not real, or if v is of the form $\mu 1$ with $\mu > 0$, there exist infinitely many elements $u \in \mathrm{End}(E)$ with $\exp(u) = v$. More precisely, there exist w and r in $\mathrm{End}(E)$ with $rw = wr$, $r \neq 0$, $\exp(w) = v$, $\exp(r) = 1$ and such that the relation $\exp(u) = v$ is equivalent to $u = w + nr$, $n \in \mathbf{Z}$.*

c) *In the other cases, v is not an exponential.*

Let $u \in \mathrm{End}(E)$ and $v \in \mathrm{Aut}(E)$ with $v = \exp(u)$. We have $\det(v) = e^{\mathrm{Tr}(u)}$ by Proposition 7.4. Let $\alpha = \mathrm{Tr}(u)/2 = \mathrm{Log}(\det(v))/2$. We may write $u = \alpha 1 + u'$ and $v = e^\alpha v'$. Then $\exp(u') = v'$ and we are reduced to solving the equation $\exp(u) = v$ in the case when $\mathrm{Tr}(u) = 0$ and $\det(v) = 1$, which we shall assume to be the case from now on. The eigenvalues of u are equal and opposite in sign, and the eigenvalues of v are their exponentials (Proposition 6.6). There are then three mutually exclusive possible cases:

1) the eigenvalues of u are real and distinct; then the eigenvalues of v are positive and distinct;

2) the eigenvalues of u are zero but u is not zero; then (since $u^2 = 0$ and therefore $v = 1 + u$) the eigenvalues of v are equal to 1 but v is not the identity;

3) the eigenvalues of u are of the form $\pm i\tau$ (with also $u = 0$ if $\tau = 0$). By Lemma 2.2 there then exists an **R**-linear bijection $\mathbf{C} \to E$ which transforms u into complex multiplication by $i\tau$, so transforms v into scalar multiplication by $e^{i\tau}$. Hence either the eigenvalues of v are of modulus 1 and different from 1, or $v = 1$.

In case 1) the eigenspaces of u and v are the same, the eigenvalues of u are the logarithms of those of v, and u is uniquely determined by v. In case 2) we have $u = v - 1$ and u is also uniquely determined by v. Finally, we consider case 3). Let $e^{i\tau}$ denote one of the eigenvalues of v. We can identify E with $\mathbf{C}$ in such a way that v is complex multiplication by $e^{i\tau}$. Then, letting w and r denote complex multiplication by $i\tau$ and $2i\pi$ respectively, the relation $\exp(u) = v$ is equivalent to $u = w + nr$ with $n \in \mathbf{Z}$.   □

Note that this example shows that the image of the exponential map is not open in $\text{End}(E)$ when $\dim(E) > 1$.

7.9 Contracting, Expanding and Hyperbolic Exponential Flows

Let E be a finite-dimensional vector space.

Theorem 7.9.1. *Let $u \in \text{End}(E)$. The following conditions are equivalent:*

(i)   *every eigenvalue of u has real part < 0;*
(ii)   *there exist constants r and $s > 0$ with $\| \exp(tu) \| \leq re^{-st}$ for all $t \geq 0$ (an arbitrary norm having been chosen on E);*
(iii) *for every $x \in E$, the vector $\exp(tu) \cdot x \to 0$ as $t \to +\infty$;*
(iv)   *the endomorphism $\exp(u)$ is contracting (Definition 5.2).*

It is clear that (ii) implies (iii) and that (iii) implies (iv). We have (iv)$\Rightarrow$(i) by Proposition 6.6. Finally, suppose that (i) is satisfied. There exists a constant $s > 0$ such that $\lambda < -s$ for every eigenvalue λ of u, and (ii) follows from Corollary 6.8. $\square$

Definition 7.9.2. The endomorphism u is said to have *contracting exponential flow* if it satisfies the equivalent conditions of Theorem 9.1.

In anticipation of 8.2.4 we say also that 0 is an *attracting* singular point for the linear vector field u.

Likewise we say that the exponential flow of u is *expanding*, or that 0 is a *repelling* singular point for the linear vector field u if the exponential of $-u$ is contracting, that is if $\exp(u)$ is expanding (Definition 5.4) or if the eigenvalues of u all have real part > 0. We say the exponential flow is *hyperbolic* if u has no purely imaginary eigenvalues, or equivalently if $\exp(u)$ is hyperbolic (Definition 5.6). Note in passing that an endomorphism with hyperbolic flow is invertible since it has no zero eigenvalue.

7.9.3. Let u be an arbitrary endomorphism of E. We apply the construction of Proposition 3.4, grouping together the eigenvalues of u whose real part is < 0 (resp. > 0). Let $E_s(u)$ and $E_u(u)$ denote the invariant subspaces of E so obtained. We have $E_s(u) \cap E_u(u) = \{0\}$. For the subspaces $E_s(u)$ and $E_u(u)$ to be complementary it is necessary and sufficient that u have hyperbolic flow. The restriction of u to $E_s(u)$ (resp. $E_u(u)$) has as its eigenvalues those eigenvalues of u which have real part < 0 (resp. > 0). We say that $E_s(u)$ and $E_u(u)$ are (respectively) the *stable subspace* and the *unstable subspace* of u. We have

$$E_u(-u) = E_s(u) = E_c(\exp(u)), \quad E_s(-u) = E_u(u) = E_e(\exp(u)).$$

The exponential flow of u is contracting on $E_s(u)$ and expanding on $E_u(u)$. More precisely, for all $x \in E$ we have the following equivalences:

$$(x \in E_s(u)) \Longleftrightarrow \left(\lim_{t \to +\infty} \exp(tu) \cdot x = 0 \right)$$

$$(x \in E_u(u)) \Longleftrightarrow \left(\lim_{t \to -\infty} \exp(tu) \cdot x = 0 \right).$$

Let us look again at the system

$$\dot{x} = \mathrm{grad}(f)(x) = a \cdot x$$

from 7.7. As the eigenvalues of a are all real the flow is hyperbolic when f is nondegenerate. The dimension of $E_s(a)$ is the index (4.4.1) of the form F. To say that the flow is contracting means that f has maximum index, and is therefore negative definite. It is because of this example that some authors call the dimension of its stable subspace the *index* of a linear vector field.

Let us take this opportunity to correct a common error: the subspace $E_s(a)$, with dimension equal to the index of a and on which f is negative definite, is not intrinsically attached to f, but to the operator a; to fix it we have to be given not only f but also the scalar product that is used.

In the second example of 7.7, namely the Lagrangian system

$$\ddot{x} + \mathrm{grad}(f)(x) = 0,$$

the flow is hyperbolic when f is negative definite; the interesting case is precisely the 'opposite' one in which f is positive definite, which implies that all the eigenvalues of the linear system are purely imaginary and nonzero.

Proposition 7.9.4. *In* $\mathrm{End}(E)$ *the endomorphisms u with hyperbolic exponential flow and for which* $\dim(E_s(u))$ *and* $\dim(E_u(u))$ *take given values form an open set, while the endomorphisms with hyperbolic flow form an open and dense set.*

The fact that the subsets in question are open is a consequence of Proposition 2.6. Moreover, for every endomorphism u, the endomorphism $u - \alpha 1$, which is arbitrarily close to u, has hyperbolic flow when the real number α is not equal to the real part of any eigenvalue of u. $\qquad\square$

7.9.5. Suppose that *the flow of u is hyperbolic* and let $\|x\|$ be a norm on E. Choose constants α and β such that $0 < \alpha < |\Re(\lambda)| < \beta$ for every eigenvalue λ of u. Applying Corollary 6.9 to the restrictions of u or $-u$ to $E_s(u)$ or $E_u(u)$ we see that there exists a constant γ such that

$$\gamma^{-1} e^{-\beta t} \|y\| \le \| \exp(tu) \cdot y \| \le \gamma e^{-\alpha t} \|y\|, \qquad y \in E_s(u), \qquad t \ge 0; \quad (7.9.4)$$

$$\gamma^{-1} e^{\alpha t} \|z\| \le \| \exp(tu) \cdot z \| \le \gamma e^{\beta t} \|z\|, \qquad z \in E_u(u), \qquad t \ge 0; \quad (7.9.5)$$

$$\gamma^{-1}e^{-\alpha t}\|y\| \le \|\exp(tu)\cdot y\| \le \gamma e^{-\beta t}\|y\|, \qquad y \in E_s(u), \qquad t \le 0; \qquad (7.9.6)$$
$$\gamma^{-1}e^{\beta t}\|z\| \le \|\exp(tu)\cdot z\| \le \gamma e^{\alpha t}\|z\|, \qquad z \in E_u(u), \qquad t \le 0. \qquad (7.9.7)$$

Let $x = y + z$ be a point of E, with $y \in E_s(u)$ and $z \in E_u(u)$, so that $\exp(tu)\cdot x = \exp(tu)\cdot y + \exp(tu)\cdot z$. If x belongs to $E_s(u)$ or to $E_u(u)$ then the orbit of x is contained in this subspace and has 0 as a unique limit point (future if $x \in E_s(u)$, past if $x \in E_u(u)$). Otherwise, this orbit meets neither $E_s(u)$ nor $E_u(u)$, is closed in E and is asymptotic to $E_u(u)$ in the future and to $E_s(u)$ in the past (Fig. 7.1a).

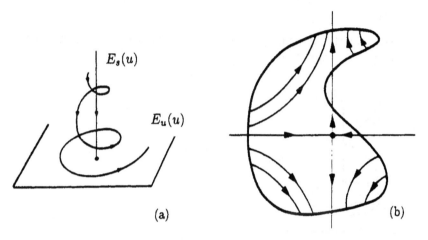

Fig. 7.1. Behaviour of orbits for a hyperbolic linear flow

7.9.6. With the same assumptions, let U be an open set in E containing 0 (Fig. 7.1b). The orbits of the field induced on U are of four types:

a) the singular point 0;
b) the orbits which have 0 as future limit point; their union together with $\{0\}$ forms a neighbourhood of 0 in $E_s(u)$;
c) the points which have 0 as past limit point; their union together with $\{0\}$ forms a neighbourhood of 0 in $E_u(u)$;
d) the orbits which do not have 0 in their closure; they are closed in U.

7.9.7. Now consider an arbitrary endomorphism $u \in \text{End}(E)$. Here E decomposes into the direct sum of $E_s(u)$, $E_u(u)$ and the invariant subspace $E_0(u)$ corresponding to the purely imaginary eigenvalues. If we group together $E_s(u)$ and $E_u(u)$ we obtain an invariant subspace $E_h(u)$, complementary to $E_0(u)$, which is the largest one on which the restriction of the flow of u is hyperbolic. The dynamics on $E_h(u)$ have been described above; on the other hand the dynamics on $E_0(u)$ can be infinitely more complicated.

Suppose for simplicity that all the eigenvalues of u have real part ≤ 0 so that $E_u(u) = \{0\}$ and $E_h(u) = E_s(u)$. Let α be a constant > 0 such that $(\lambda \in \mathrm{Sp}(u)$ and $\Re(\lambda) < 0)$ implies $(\Re(\lambda) < -\alpha)$. Fix $\beta \in (0, \alpha)$. If we apply Corollary 6.9 to $E_s(u)$ and $E_0(u)$ we see that there exists γ with the properties:

$$\| \exp(tu) \cdot y \| \le \gamma e^{-\alpha t} \| y \|, \qquad y \in E_s(u), \qquad t \ge 0,$$
$$\| \exp(tu) \cdot w \| \ge \gamma^{-1} e^{(\beta - \alpha)t} \| w \|, \quad w \in E_0(u), \qquad t \ge 0.$$

For every point $x = y + w$ of E which is not in $E_s(u)$ we have $\exp(tu) \cdot x = \exp(tu) \cdot y + \exp(tu) \cdot w$, with

$$\frac{\| \exp(tu) \cdot y \|}{\| \exp(tu) \cdot w \|} \le \gamma^2 \frac{\| y \|}{\| w \|} e^{-\beta t}, \qquad t \ge 0,$$

which shows that the dynamics of u reduce for many practical purposes to the dynamics of its restriction to $E_0(u)$ (see Fig. 7.2).

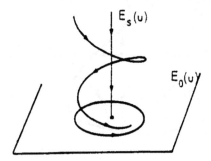

Fig. 7.2 The significant dynamics take place on $E_0(u)$

7.10 Topological Classification of Linear Vector Fields

Later we shall use a result that is a little more precise than the estimate (6.9). Let u be an endomorphism of E. The Lie derivative with respect to the linear vector field u is the operator L_u such that $L_u r(x) = dr(x) \cdot (u \cdot x)$.

Proposition 7.10.1. *Suppose the flow of u is contracting and let $q(x)$ be a positive definite quadratic form on E. Then there exists a unique quadratic form $r(x)$ such that $L_u r = -q$; it is positive definite.*

Let Q and R be the symmetric bilinear forms associated to q and r respectively; the relation above can then be written $2R(x, u(x)) = -Q(x, x)$. Taking an orthonormal basis for Q, letting A denote the matrix for u and B the matrix for the bilinear form R, and writing ${}^t M$ for the transpose of the matrix M, we obtain the matrix condition $2 {}^t X B A X = - {}^t X X$ for all column-vectors X, or equivalently $BA + {}^t AB = -I$. Let f denote the endomorphism of the vector space $M_n(\mathbf{R})$ given by $f(B) = BA + {}^t AB$. To say

$f(B) = 0$ means that $-{}^t\!AB = BA$. However, since we have $\mathrm{Sp}(A) = \mathrm{Sp}(u)$ and $\mathrm{Sp}(-{}^t\!A) = -\mathrm{Sp}(u)$, and since $\mathrm{Sp}(u)$ and $-\mathrm{Sp}(u)$ are contained in the *disjoint* subspaces $\{\Re(\lambda) < 0\}$ and $\{\Re(\lambda) > 0\}$, the matrices A and $-{}^t\!A$ have disjoint spectra. Applying Lemma 3.2, we deduce from this that $B = 0$. Thus f is injective, therefore bijective, and there exists a unique matrix B with $BA + {}^t\!AB = -I$. Transposing, we obtain ${}^t\!BA + {}^t\!A{}^t\!B = -I$ and we have ${}^t\!B = B$ by uniqueness. We have thus proved that there exists a unique quadratic form $r(x)$ with $L_u r = -q$. It remains to prove $r(x) > 0$ for all nonzero x in E. Now, since $L_u r < 0$ on $E \setminus \{0\}$, the function $t \mapsto r(\exp(tu)\cdot x)$ is strictly decreasing. It tends to zero as t tends to infinity (here we use the fact that the flow of u is contracting), and is therefore > 0 for all t and in particular for $t = 0$. $\square$

Corollary 7.10.2. *Let $\alpha > 0$ be such that $\Re(\lambda) < -\alpha$ for all $\lambda \in \mathrm{Sp}(u)$. Then there is a euclidean scalar product $(x \mid y)$ on E such that $(x \mid u\!\cdot\! x) \leq -\alpha(x \mid x)$ for all $x \in E$.*

Apply the above to the endomorphism $u + \alpha 1$ and an arbitrary positive definite quadratic form q, and take the scalar product associated to r. $\square$

The model for linear contracting flows is $x \mapsto e^{-t}x$. It is the integral flow for the vector field $-1 : x \mapsto -x$. We shall see next that a contracting linear flow can be converted into this model flow by a homeomorphism of E.

Proposition 7.10.3. *Let E be a finite-dimensional vector space and let u be an endomorphism of E all of whose eigenvalues have real part < 0. Then there exists a homeomorphism h from E into itself such that $h(0) = 0$ and such that $\exp(tu) \cdot h(x) = h(e^{-t}x)$ for all $x \in E$ and all $t \in \mathbf{R}$.*

As we have just seen, there exists a global euclidean scalar product $(x \mid y)$ on E such that $(x \mid u \cdot x) < 0$ for all nonzero x. Put $q(x) = (x \mid x)$. For all nonzero $x \in E$ the map $t \mapsto q(\exp(tu) \cdot x)$ is strictly decreasing since its derivative is $2(\exp(tu) \cdot x \mid u \cdot (\exp(tu) \cdot x)) < 0$. Moreover, since E is contracted by u and expanded by $-u$ this map tends to zero as $t \to +\infty$, and tends to $+\infty$ as $t \to -\infty$. Therefore there exists a unique $t \in \mathbf{R}$ such that $q(\exp(-tu) \cdot x) = 1$. Let S denote the sphere $\{y \mid q(y) = 1\}$ and consider the map

$$\phi : \mathbf{R} \times S \to E \setminus \{0\}$$

which associates $\exp(tu) \cdot y$ to (t, y). By construction we have

$$\exp(su) \cdot \phi(t, y) = \phi(s + t, y). \tag{7.10.1}$$

The map ϕ is bijective as we have just seen. It is clear that it is continuous (and even C^∞).

The map inverse to ϕ is continuous (and even C^∞): we deduce this for example from the bounds (9.4) which here give

$$\gamma^{-2} e^{-2\beta t} \le q(\phi(t, y)) \le \gamma^2 e^{-2\alpha t}. \tag{7.10.2}$$

If $x = \phi(t, y)$ and $x' = \phi(t', y')$ are close then $q(x)$ and $q(x')$ are close and (10.2) implies that t and t' are close. It then follows that $y = \exp(-tu) \cdot x$ and $y' = \exp(-t'u) \cdot x'$ are close.

If we apply the above to the case when $u = -1$ we obtain the map $\psi : (t, y) \mapsto e^{-t}y$. Let $h : E \backslash \{0\} \to E \backslash \{0\}$ denote the composed map $\phi \circ \psi^{-1}$. It is a homeomorphism from $E \backslash \{0\}$ to itself. Moreover, by the inequalities (10.2) the assertions "$\phi(t, y)$ tends to 0" and "$t \to +\infty$" are equivalent. As this holds for ψ also, it follows that h extends to a homeomorphism from E to itself, sending 0 to 0. Let $t \in \mathbf{R}$ and $x \in E \backslash \{0\}$. There exists $s \in \mathbf{R}$ and $y \in S$ with $x = \psi(s, y) = e^{-s}y$ and by construction we have

$$\exp(tu) \cdot h(x) = \exp(tu) \cdot \phi(s, y) = \phi(s + t, y) = h(e^{-s-t}y) = h(e^{-t}x).$$

This completes the proof. $\square$

Note that the homeomorphism h constructed in this way is a C^∞ diffeomorphism away from the origin. It is not differentiable at the origin (except when u is multiplication by a scalar – see below).

7.10.4. In order to exploit this result it is convenient to introduce a definition. Let E and E' be two finite-dimensional vector spaces and let $u \in \mathrm{End}(E)$ and $u' \in \mathrm{End}(E')$. We shall compare the integral flows of u and u'. If $h : E \to E'$ is a vector space isomorphism such that $u' \circ h = h \circ u$ (we say then that u and u' are *linearly conjugate*) we immediately have $\exp(tu') \circ h = h \circ \exp(tu)$ for all $t \in \mathbf{R}$. Conversely, the latter relation yields $u' \circ h = h \circ u$ by differentiation with respect to t at $t = 0$.

Suppose more generally that we are given a map h from E (or only a neighbourhood of 0 in E) into E' such that $\exp(tu') \cdot h(x) = h(\exp(tu) \cdot x)$ for all $t \in \mathbf{R}$ and all $x \in E$ (or only for t sufficiently small and x sufficiently close to 0). Suppose h is C^1 at 0 and let $v \in \mathrm{End}(E)$ denote its derivative at 0. On differentiating the relation above we obtain $\exp(tu') \circ v = v \circ \exp(tu)$, from which $u' \circ v = v \circ u$ follows as above. Thus if h is a local diffeomorphism of class C^1 then u and u' are linearly conjugate. In order to weaken the notion of linear conjugacy we therefore have to suppress all assumptions of differentiability. This leads us to make the following definition:

Definition 7.10.5. The flows $(\exp(tu))$ and $(\exp(tu'))$ are said to be *topologically conjugate* if there exists a homeomorphism $h : E \to E'$ such that $h(0) = 0$ and $\exp(tu') \circ h = h \circ \exp(tu)$ for all $t \in \mathbf{R}$.

In this way we have defined what is clearly an equivalence relation on $\mathrm{End}(E)$ and which we now intend to study. We start with a few simple remarks.

a) As we have seen, the flows $(\exp(tu))$ and $(\exp(tu'))$ are topologically conjugate if the endomorphisms u and u' are linearly conjugate, but the converse is false.

b) Proposition 10.3 above means that every contracting linear flow is topologically equivalent to the flow $(x \mapsto e^{-t}x)$ (scalar multiplications).

c) With the notation of Definition 10.5, the map h transforms the orbits of the first flow into those of the second. Hence two linear flows that are topologically conjugate are orbitally equivalent in the sense defined in 6.11.5, and *a fortiori* orbitally equivalent in a neighbourhood of the origin.

We can now give the general criterion for topological conjugacy:

Theorem 7.10.6. (Topological classification of hyperbolic linear flows.) *Let E and E' be two finite-dimensional vector spaces and let $u \in \text{End}(E)$ and $u' \in \text{End}(E')$ be endomorphisms with hyperbolic exponential flow (that is u and u' have no eigenvalue with zero real part). The following conditions are equivalent:*

a) *the exponential flows of u and u' are topologically cojugate: there exists a homeomorphism $h : E \to E'$ such that $h(0) = 0$ and $\exp(tu') = h \circ \exp(tu) \circ h^{-1}$ for all $t \in \mathbf{R}$;*

b) *the exponential flows of u and u' are orbitally equivalent;*

c) *the exponential flows of u and u' are orbitally equivalent in a neighbourhood of the origin;*

d) *we have $\dim(E_s(u)) = \dim(E_s'(u'))$ and $\dim(E_u(u)) = \dim(E_u'(u'))$.*

The implications a)$\Rightarrow$b)$\Rightarrow$c) are immediate. It is a question of proving d)$\Rightarrow$a) and c)$\Rightarrow$d).

First suppose that the equalities in d) hold. Let $m = \dim(E_s(u)) = \dim(E_s'(u'))$. By Proposition 10.3 there exist homeomorphisms $g_s : \mathbf{R}^m \to E_s(u)$ and $g_s' : \mathbf{R}^m \to E_s'(u')$ such that $g_s(0) = 0$ and $g_s'(0) = 0$ and such that for $x \in \mathbf{R}^m$ and $t \in \mathbf{R}$ we have $\exp(tu) \cdot g_s(x) = g_s(e^{-t}x)$ and $\exp(tu') \cdot g_s'(x) = g_s'(e^{-t}x)$. Then let $h_s = g_s' \circ g_s^{-1}$; it is a homeomorphism from $E_s(u)$ onto $E_s'(u')$ such that $h_s(0) = 0$ and such that for all $y \in E_s(u)$ and all $t \in \mathbf{R}$ we have $\exp(tu') \cdot h_s(y) = h_s(\exp(tu) \cdot y)$. Likewise we construct a homeomorphism $h_u : E_u(u) \to E_u'(u')$ having analogous properties. However, the subspaces $E_s(u)$ and $E_u(u)$ are complementary in E and similarly for E'. We then define a homeomorphism $h : E \to E'$ with the required property by letting $h(y + z) = h_s(y) + h_u(z)$ for all $y \in E_s(u)$ and all $z \in E_u(u)$. This proves a).

Now let us assume c). There exists an open set U in E containing the origin, and open set U' in E' containing the origin, and a homeomorphism $h : U \to U'$ transforming the orbits of the field $x \mapsto u \cdot x$ in U into the orbits of the field $y \mapsto u' \cdot y$ in U'. For $x \in U$ it is equivalent to say $x = 0$ or $u \cdot x = 0$ (as u has no zero eigenvalue). The unique singular point of the first field is therefore 0, and h takes it to the unique singular point of the second field. Thus $h(0) = 0$. Let U_s (resp. U_s') denote the union of orbits of the first (resp. second) field which have the origin as future limit point. The homeomorphism h induces a homeomorphism of U_s onto U_s'. Now, we saw in 9.6 that U_s is a neighbourhood of the origin in the vector space $E_s(u)$, and U_s' is a neighbourhood of the origin in the vector space $E_s'(u')$. We then apply the theorem of *invariance of dimension* (1.6.4) which states that these two vector spaces have the same dimension. Arguing similarly for the unstable subspaces, we obtain d).   □

From this theorem we deduce a criterion for 'linear structural stability':

Corollary 7.10.7. *Let E be a finite-dimensional vector space and let $u \in$ End(E). The following conditions are equivalent:*

(i)   *the exponential flow of u is hyperbolic: no eigenvalue of u is purely imaginary;*

(ii) *for all $u' \in$ End(E) sufficiently close to u the exponential flows of u and u' are topologically conjugate (resp. orbitally equivalent).*

To simplify matters, call the pair of integers $\big(\dim(E_s(w)), \dim(E_u(w))\big)$ the *numerical type* of the endomorphism $w \in$ End(E). If u satisfies (i) then every endomorphism u' sufficiently close to u has the same numerical type as u (Proposition 9.4) and therefore also has hyperbolic flow, and we apply the theorem to obtain (ii). Conversely, if u has any purely imaginary eigenvalues there exist endomorphisms arbitrarily close to u having hyperbolic flows with different numerical types, and therefore (by the theorem) not mutually orbitally equivalent, which contradicts (ii).   □

The general philosophy illustrated in Sect. 4.11, for example, certainly applies here: the 'structurally stable' linear flows form an open dense subset of End(E), being the disjoint union of $\dim(E) + 1$ open conjugacy classes corresponding to the different possible numerical types $(p, \dim(E) - p)$ with $0 \leq p \leq \dim(E)$. Endomorphisms of all numerical types can be realized by taking diagonal matrices whose diagonal elements are 1 or -1.

7.10.8. Let $u \in$ End(E). As in 9.7, we split the space E into the direct sum of the subspaces $E_s(u), E_u(u)$ and $E_0(u)$, and let u_0 denote the restriction of u to this last subspace. Let us call the triple (m_s, m_u, m_0) with $m_s = \dim(E_s(u))$, $m_u = \dim(E_u(u))$ and $m_0 = \dim(E_0(u))$ the *numerical type* of u. The Classification Theorem 10.6 can be generalized as follows: in order for the two endomorphisms $u \in$ End(E) and $u' \in$ End(E') to have flows that are topologically conjugate (resp. orbitally equivalent), it is necessary and sufficient that they have the same numerical type *and* that the endomorphisms u_0 and u_0' have topologically equivalent (resp. orbitally equivalent) flows.

This reduces the topogical classification of general linear flows to that of flows $(\exp(tu))$ where all the eigenvalues of u are purely imaginary. It can be proved that for such endomorphisms topological conjugacy implies (and is therefore equivalent to) the existence of a linear conjugacy between one of the endomorphisms and a multiple of the other: if two linear vector fields $u \in \mathrm{End}(E)$ and $u' \in \mathrm{End}(E')$ with purely imaginary eigenvalues have orbitally equivalent flows then there exists a real number $\lambda > 0$ and a vector space isomorphism $h : E \to E'$ with $\lambda u' = h \circ u \circ h^{-1}$. This is a theorem of N.Kuiper (1973).

Here we shall look just at the easy case of dimension 2. Let $u \in \mathrm{End}(E)$ with $\dim(E) = 2$ be an endomorphism with purely imaginary eigenvalues. The linear classification is then as follows (see Sect. 12 below): if u is invertible there exists a basis of E in which the matrix of u is $\left(\begin{smallmatrix} 0 & -\tau \\ \tau & 0 \end{smallmatrix} \right)$ with $\tau > 0$; if u is not invertible and not zero then there exists a basis in which the matrix of u is $\left(\begin{smallmatrix} 0 & 1 \\ 0 & 0 \end{smallmatrix} \right)$; finally, if u is zero ... then it is zero! The three cases can immediately be distinguished 'orbitally' using the fixed points of the flow (the origin in the first case, the whole space in the third case, and something else in the second). It remains, in the first case, to recover 'topologically' the invariant τ. The flow of u in this case consists of rotations, and we retrieve τ as $2\pi/T$ where T is the common period of the nonzero orbits.

7.11 Topological Classification of Automorphisms

In fact the results of the previous section have analogues for automorphisms that are not necessarily in the image of the exponential. Let E and E' be two finite-dimensional vector spaces and let $v \in \mathrm{End}(E)$ and $v' \in \mathrm{End}(E')$. We shall say that v and v' are *topologically conjugate* if there exists a homeomorphism $h : E \to E'$ such that $v' \circ h = h \circ v$. We give without proof a criterion for 'linear structural stability' that directly generalizes Corollary 10.7.

Proposition 7.11.1. *Let E be a finite-dimensional vector space and let v be an automorphism of E. The following conditions are equivalent:*

(i)   *v is hyperbolic (Definition 5.6);*
(ii) *every automorphism of E sufficiently close to E is topologically conjugate to v.*   □

Here, also without proof, is a criterion for topological conjugacy of hyperbolic automorphisms that generalizes Theorem 10.6. To state it conveniently we first introduce a piece of terminology. Let $v \in \mathrm{Aut}(E)$ be a hyperbolic automorphism. We shall say that the *numerical type* of v is the quadruple $(m_c(v), s_c(v), m_e(v), s_e(v))$ defined as follows:

$$m_c(v) = \dim(E_c(v)) = \sum_{|\lambda|<1} m(\lambda) \in [0, \dim(E)],$$

$$s_c(v) = sign(\det(v|_{E_c(v)})) = \prod_{|\lambda|<1} sign(\lambda)^{m(\lambda)} \in \{1, -1\},$$

$$m_e(v) = \dim(E_e(v)) = \sum_{|\lambda|>1} m(\lambda) \in [0, \dim(E)],$$

$$s_e(v) = sign(\det(v|_{E_e(v)})) = \prod_{|\lambda|>1} sign(\lambda)^{m(\lambda)} \in \{1, -1\}.$$

In these formulae $m(\lambda)$ denotes the multiplicity of the complex number λ as an eigenvalue of v, and $sign(x) \in \{1, -1\}$ denotes the sign of a nonzero real number x. In the two products involving signs we restrict ourselves to real λ (or make the convention that the sign of a non-real number is 1).

We have $m_c(v) + m_e(v) = \dim(E)$ and $s_c(v)s_e(v) = sign(\det(v))$.

Theorem 7.11.2. *In order for two hyperbolic automorphisms $v \in \mathrm{Aut}(E)$ and $v' \in \mathrm{Aut}(E')$ to be topologically conjugate it is necessary and sufficient that they have the same numerical type.*   □

There are $4n$ possible numerical types for a hyperbolic automorphism of a vector space of dimension n. To see this, note that for the two extreme cases where one of the m is equal to 0 we have a choice of sign for the other, while for the $n - 1$ intermediate cases we have two independent sign choices. Each of these $4n$ cases can be realized by diagonal matrices each of whose diagonal terms is equal to $1/2$, $-1/2$, 2 or -2. For example, we can take m_c times $1/2$ if $s_c = +1$, or $(m_c - 1)$ times $1/2$ and one $-1/2$ if $s_c = -1$, and likewise m_e times 2 if $s_e = +1$ or $(m_e - 1)$ times 2 and one -2 if $s_e = -1$. This shows that in order for a topological conjugacy class to intersect the image of the exponential it is necessary and sufficient that the two signs equal $+1$; but note that the fact of belonging to the image of the exponential is not necessarily preserved under topological conjugacy when $n > 1$ (see Fig. 7.7 for example). Also, the sign of the determinant of v is the product $s_c(v)s_e(v)$; among the $4n$ conjugacy classes of hyperbolic automorphisms in dimension n there are $2n$ which preserve orientation and $2n$ which reverse orientation.

7.11.3. Let $u \in \mathrm{Aut}(E)$. Split E into the direct sum of the invariant subspace $E_h(u) = E_c(u) + E_e(u)$, which is the largest subspace on which the restriction of u is hyperbolic, and the invariant subspace $E_1(u)$ corresponding to the eigenvalues of u which belong to the unit circle. The Classification Theorem 11.2 can be generalized along the lines of the result indicated in 10.8 with a final additional difficulty: the topological classification of automorphisms having spectrum on the unit circle is different from the linear classification.

7.12 Classification of Linear Flows in Dimension 2

7.12.1. Let E be a finite-dimensional vector space of dimension 2 and let u be an endomorphism of E. The characteristic polynomial of u may be written

$$P_u(T) = T^2 - \mathrm{Tr}(u)u + \det(u).$$

Its discriminant is $\mathrm{Tr}(u)^2 - 4\det(u)$. The various possibilities are as follows (see Fig. 7.3 which indicates in each case the position of the eigenvalues in the complex plane):

A) $\det(u) < 0$. Here the eigenvalues of u are real and of opposite sign, and the flow is hyperbolic. We call this a *saddle*.
B) $\det(u) > 0$ and $\mathrm{Tr}(u)^2 \geq 4\det(u)$. Here the eigenvalues of u are real and of the same sign, which is the sign of $\mathrm{Tr}(u)$. We call this a *node*, which is *attracting* if $\mathrm{Tr}(u) < 0$ and *repelling* if $\mathrm{Tr}(u) > 0$. We say the node is *proper* if u is a multiple of the identity, and *improper* otherwise.
C) $\mathrm{Tr}(u) \neq 0$ and $\mathrm{Tr}(u)^2 < 4\det(u)$. Here the eigenvalues of u are complex, with real part $\mathrm{Tr}(u)/2$ and nonzero imaginary part. We call this a *focus*, which is *attracting* if $\mathrm{Tr}(u) < 0$ and *repelling* if $\mathrm{Tr}(u) > 0$.
D) $\det(u) > 0$ and $\mathrm{Tr}(u) = 0$. The eigenvalues of u are purely imaginary. We call this a *centre*.
E) $\det(u) = 0$. In this case u is not invertible. If $\mathrm{Tr}(u)$ is nonzero this is sometimes called a *saddle-node*.

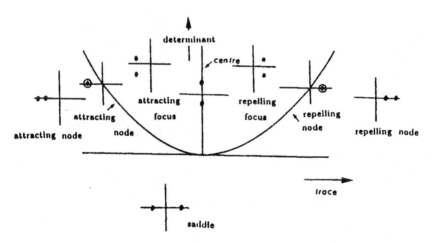

Fig. 7.3 Types of linear flow in $\mathbf{R}^2$ according to trace and determinant

In this list, the structurally stable cases are A), B) and C). The topological classification ignores the distinction between nodes and foci and knows only three distinct forms. These are the three cases: attracting, repelling and mixed, with respective models $(x,y) \mapsto (-x,-y)$, $(x,y) \mapsto (x,y)$ and

$(x, y) \mapsto (-x, y)$, which can be respectively characterized by the conditions $(\det(u) > 0$ and $\mathrm{Tr}(u) < 0)$, $(\det(u) > 0$ and $\mathrm{Tr}(u) > 0)$, and $(\det(u) < 0)$ (see Fig. 7.4).

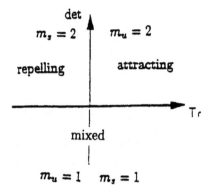

Fig. 7.4 Topological classification of structurally stable linear flows in $\mathbf{R}^2$ according to trace and determinant

We shall now study each of these possibilities in greater detail. They are illustrated in Figs. 7.5 and 7.6.

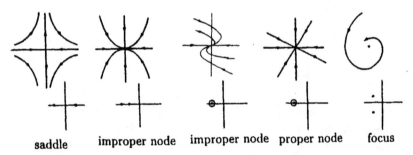

saddle   improper node   improper node   proper node   focus

Fig. 7.5 Types of structurally stable linear flows in $\mathbf{R}^2$

7.12.2. Saddles. Let $\lambda < 0 < \mu$ be the two eigenvalues of u. Then in a suitable basis we have $u \cdot (x, y) = (\lambda x, \mu y)$ and

$$\exp(tu) \cdot (x, y) = (e^{\lambda t} x, e^{\mu t} y).$$

The x-axis is the stable subspace, and the y-axis is the unstable subspace. The orbits are of three types: First the orbit $\{0\}$, then the four semi-axes, and finally all the others. These last are closed in E and are asymptotic to an unstable semi-axis in the future and a stable semi-axis in the past. They look approximately like hyperbolae, which is the origin of the term 'hyperbolic flow' (for $\lambda = -\mu$ they are genuine hyperbolae).

7.12.3. Attracting nodes. Let $\lambda \leq \mu < 0$ be the two eigenvalues. There are three possible cases.

If $\lambda < \mu$ there exists a basis for E in which we have $u \cdot (x, y) = (\lambda x, \mu y)$ and

$$\exp(tu) \cdot (x, y) = (e^{\lambda t} x, e^{\mu t} y).$$

If $\lambda = \mu$ but $u \neq \lambda 1$ then in a suitable basis we have $u \cdot (x, y) = (\lambda x + y, \lambda y)$ and

$$\exp(tu) \cdot (x, y) = (e^{\lambda t}(x + ty), e^{\lambda t} y).$$

If $u = \lambda 1$ we have $\exp(u) = e^{\lambda t} 1$.

In the first case there are four straight line orbits, namely the semi-axes, and all the orbits except $\{0\}$ and the semi-axes $x = 0$ are tangent to $y = 0$ at the origin. In the second case two of the exceptional straight line orbits disappear. Finally, in the case of a proper node the orbits are $\{0\}$ and all the rays from the origin.

Repelling nodes. The situation is analogous, with the direction along the orbits reversed.

7.12.4. Attracting foci. Let $\sigma \pm i\tau$, with $\sigma < 0$ and $\tau > 0$, denote the two eigenvalues. In a suitable basis we have $u \cdot (x, y) = (\sigma x - \tau y, \tau x + \sigma y)$ and

$$\exp(tu) \cdot (x, y) = e^{\sigma t}(x \cos(\tau t) - y \sin(\tau t), x \sin(\tau t) + y \cos(\tau t)).$$

The orbits other than $\{0\}$ spiral around the origin. They all have 0 as future limit point.

Repelling foci. The situation is analogous, with the direction along the orbits reversed.

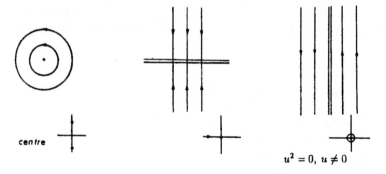

Fig. 7.6. A centre and degenerate linear flows in $\mathbf{R}^2$

7.12.5. Centres. Let $\pm i\tau$, $\tau > 0$ be the eigenvalues. In a suitable basis we have $u \cdot (x, y) = (-\tau y, \tau x)$ and

$$\exp(tu) \cdot (x, y) = (x\cos(\tau t) - y\sin(\tau t), x\sin(\tau t) + y\cos(\tau t)).$$

The orbits are $\{0\}$ and the circles centred at 0, which are periodic orbits all with the same period $2\pi/\tau$.

7.12.6. Degenerate cases.

If $\det(u) = 0$ and $\mathrm{Tr}(u) \neq 0$ the eigenvalues of u are 0 and $\lambda \neq 0$. In a suitable basis we have $u \cdot (x, y) = (0, \lambda y)$ and

$$\exp(tu) \cdot (x, y) = (x, e^{\lambda t} y).$$

All the points on the x-axis are fixed by the flow, and the other orbits are half-lines $x = a$, $y > 0$ and $x = a$, $y < 0$, $a \in \mathbf{R}$.

If $\det(u) = 0$, $\mathrm{Tr}(u) = 0$ and $u \neq 0$ then in a suitable basis we have $u \cdot (x, y) = (0, x)$, and so $\exp(tu) \cdot (x, y) = (x, y + tz)$. The points $(0, b)$ are fixed by the flow; the other orbits are the lines $x = a$, traversed with constant speed a.

Finally, the case $u = 0$, $\exp(tu) = 1$ needs no comment.

7.12.7. Note that in every case the orbits distinct from 0 which have 0 as a limit point are tangent at 0 to an eigendirection of u corresponding to a nonzero eigenvalue.

7.12.8. It is instructive to insert the classification of these flows into the broader classification of automorphisms of E. By Theorem 11.2, the hyperbolic automorphisms of vector spaces of dimension 2 form eight topological conjugacy classes. A list of these is set out below, giving to each one its numerical type and a representative of the class. Here $diag(\alpha, \beta)$ denotes the endomorphism $(x, y) \mapsto (\alpha x, \beta y)$ of $\mathbf{R}^2$, or in other words the diagonal matrix $\begin{pmatrix} \alpha & 0 \\ 0 & \beta \end{pmatrix}$.

> type $(2^+, 0)$, contracting: $diag(1/2, 1/2)$;
> type $(2^-, 0)$, contracting: $diag(-1/2, 1/2)$;
> type $(0, 2^+)$, expanding: $diag(2, 2)$;
> type $(0, 2^-)$, expanding: $diag(-2, 2)$;
> type $(1^+, 1^+)$, mixed: $diag(1/2, 2)$;
> type $(1^-, 1^+)$, mixed: $diag(-1/2, 2)$;
> type $(1^+, 1^-)$, mixed: $diag(1/2, -2)$;
> type $(1^-, 1^-)$, mixed: $diag(-1/2, -2)$.

Let us associate to every $v \in \mathrm{Aut}(E)$ the point $(\mathrm{Tr}(v), \det(v))$ in $\mathbf{R}^2$. In the plane, the eight regions corresponding to the eight equivalence classes above are separated by the following lines: the straight line $\{\det(v) = 0\}$, the two straight lines $\{P_v(\pm 1) = 0\}$, that is $\{\mathrm{Tr}(v) = \pm(1 + \det(v))\}$, and the straight line segment $\{\det(v) = 1,\ -2 < \mathrm{Tr}(v) < 2\}$ corresponding to the presence of complex eigenvalues with absolute value 1. Adding in the parabola $\{\mathrm{Tr}(v)^2 = 4\det(v)\}$ and the type of endomorphism of which v is (possibly) the exponential, we obtain Fig. 7.7.

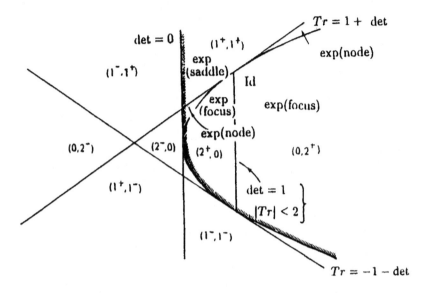

Fig. 7.7 Classification of linear flows as part of the classification of linear automorphisms of $\mathbf{R}^2$

8 Singular Points of Vector Fields

8.1 Introduction

This chapter is concerned with just one question: how valid is the operation of linearizing a differential system in the neighbourhood of an equilibrium point? This translates into our technical language as follows. Consider a vector field X on a phase space V, and a point $a \in V$ at which $X(a) = 0$ (recall that such a point is traditionally called a *singular point* of X). By differentiation at a we associate to these a linear vector field L, which we naturally call the *linearization* of X at a, whose phase space is, no less naturally, the tangent space T to V at a. We aim to compare the integral flow of X (in a neighbourhood of a) with that of L (in a neighbourhood of 0).

Thus we are led to the following question: how can the two situations (V, a, X) and $(T, 0, L)$ be compared? Here we discover again a theme we have met before. The most natural and simplest answer would be that under reasonable assumptions (for example, if L is invertible) there should exist a local diffeomorphism between V and T which sends a to 0 and transforms X into L. In other words, there should exist a system of local coordinates on V in a neighbourhood of a on which X becomes linear. We would then have a very nice 'Linearization Theorem' analogous to the 'Straightening-out Theorem' 6.8.2 valid when $X(a)$ is nonzero. After this we could dream of generalization in the spirit of the 'sufficient jets' of Chapt. 5 and try to prove a theorem of the following kind: *In a suitably chosen system of local coordinates every vector field X reduces to its principal part up to a certain order*, for example to a constant field when $X(a)$ is nonzero (straightening out), to a linear field when $X(a)$ is zero and with invertible derivative (linearization), and so on.

Unfortunately there is no chance of this, as easy counterexamples show. The most striking example is without a doubt the differential system

$$\dot{x} = 2x + y^2, \quad \dot{y} = y,$$

which cannot be written in the form

$$\dot{u} = 2u, \quad \dot{v} = v$$

in *any* system of local coordinates (u, v) of class C^2. The 'reason' for this, as we shall see, is that one of the eigenvalues of the matrix $\left(\begin{smallmatrix} 2 & 0 \\ 0 & 1 \end{smallmatrix}\right)$ of the linearization is twice the other one, and the nonlinear term $(x, y) \mapsto (y^2, 0)$ is cunningly adapted to this relationship. It is what is technically called a *resonance*, borrowing the word from another context.

From now on there are, as usual, two ways forward ... and, as usual, we shall follow both of them at the start. On the one hand we can weaken the conclusion sought and be content with a 'topological' comparison between X and L (in fact between their integral flows, since if we do not have derivatives then we cannot carry out a change of variables in the differential equation). On the other hand, we can demand a C^∞ linearization, but we then have to arm ourselves against resonance phenomena. A third approach is possible, engaging fully in the game of 'sufficient jets'. Thus in the case of the example above we could prove that every differential system whose linearization is $L(u, v) = (2u, v)$ can be reduced to one of the two forms $X_1(x, y) = (2x, y)$ and $X_2(x, y) = (2x + y^2, y)$, which are not mutually equivalent. However, this theme would take us too far afield.

We will therefore content ourselves here with the first two approaches, stating the topological linearization theorem of Hartman (Sect. 7) and the C^∞ linearization theorem of Sternberg (Sect. 11), with some small embellishments in dimension 2 (Sects. 11,12).

The text of the chapter contains two additions to this programme. The first is a digression in Sect. 6 on Lyapunov theory which directly generalizes the technique used in Sect. 5 to linearize topologically those singular points that are attracting. The second, more essential for our purposes, concerns the following fact stated in Sect. 8. Although we cannot hope for any differentiability in the operation of linearizing an arbitrary singular point, the subsets (topological submanifolds) which correspond to the stable and unstable subspaces of the linearized system turn out to be differentiable submanifolds (of the same class as the original vector field), called the *stable and unstable manifolds* of the singular point. Once again we regret that the specialists use the adjective *stable* (and also *unstable*) with varying meanings, and sometimes in the same sentence. However, this usage is so universal[46] that it did not seem possible to escape it here.

8.2 The Classification Problem

Let E be a finite-dimensional vector space with V a submanifold of E and X a vector field on V of class C^r, $r \in [1, \infty]$. If $a \in V$ is such that $X(a) \neq 0$ we saw in 6.8.2 that there exists a C^r diffeomorphism of an open neighbourhood of a in V onto an open set in a vector space which transforms X into a constant vector field. Here we intend to study what happens in the neighbourhood

---

[46] Arnol'd says 'incoming manifold' and 'outgoing manifold', which is more intuitive and less easily lends itself to confusion.

of *singular points* of X, that is to say (6.2.2) points $a \in V$ with $X(a) = 0$. In order to be able to ask precise questions we introduce some appropriate terminology.

We consider triplets (X, a, V) where V is a submanifold of a finite-dimensional vector space, X is a vector field (of class at least C^1) on V, and a is a point of V.

Definition 8.2.1. Two triplets (V, a, X) and (V', a', X') as above are said to be C^r-*conjugate*, with $r \in [0, \infty]$, if there exists an open set U of V containing a, an open set U' of V' containing a' and a C^r diffeomorphism $h : U \to U'$ such that $h(a) = a'$ and $h_*(X|_U) = X'|_{U'}$, that is to say

$$T_x h \cdot X(x) = X'(h(x)) \tag{8.2.1}$$

for every point x of V.

If the integral flows of X and X' are denoted by Φ_t and Φ'_t respectively, then (see 6.10.5) condition (2.1) implies

$$h \circ \Phi_t \circ h^{-1} = \Phi'_t. \tag{8.2.2}$$

Conversely, differentiation of (2.2) at $t = 0$ yields (2.1). Note, however, that while it is necessary for h to be differentiable in order to formulate (2.1), the identity (2.2) is still meaningful for any bijection h without assumptions of differentiability. This allows us to make the following definition:

Definition 8.2.2. The triplets (X, a, V) and (X', a', V') are C^0-*conjugate* or *topologically conjugate* if it is possible to find an open set U of V containing a, an open set U' containing a' and a homeomorphism $h : U \to U'$ with $h(a) = a'$ that satisfies (2.2).

Let us be precise about what this means from the point of view of the sets in the definition: for $x \in U$ and $t \in \mathbf{R}$ the condition "$\Phi_t(x)$ is defined and belongs to U" is equivalent to the condition "$\Phi'_t(h(x))$ is defined and belongs to U'", and then we have $\Phi'_t(h(x)) = h(\Phi_t(x))$. In particular, h transforms the local phase portrait of X in a neighbourhood of a to the local phase portrait of X' in a neighbourhood of a'. In the terminology introduced in 6.11.5, *topological equivalence implies orbital equivalence*. Recall that (X, a, V) and (X', a', V') are *orbitally equivalent* if there exists a local homeomorphism $h : U \to U'$ as above which maps the orbits of X in U to the orbits of X' in U', respecting the sense of the flow.

8.2.3. Clearly C^r-conjugacy is an equivalence relation, becoming finer as r becomes larger. Here are some examples:

a) All triplets (V, a, X) where X is C^r, where $\dim_a V$ is a fixed integer $n \geq 0$ and where $X(a)$ is nonzero are C^r-conjugate to the triplet type

$(\mathbf{R}^n, 0, e_1)$ where e_1 is the constant vector field $x \mapsto (1, 0, \ldots, 0)$: this is the statement we recalled above. Next we have to classify the cases where a is a singular point of X. To simplify the language, we shall often say 'singularity' instead of 'triplet (V, a, X) with $X(a) = 0$'.

b) If (V, a, X) and (V', a', X') are orbitally equivalent (for example, if they are topologically conjugate) and if $X(a) = 0$ then $X'(a') = 0$.

c) All singularities $(E, 0, u)$ for which E is a vector space of a given dimension n and u is an endomorphism of E whose eigenvalues all have negative real part are topologically conjugate to the singularity $(\mathbf{R}^n, 0, -Id)$: this is Proposition 7.10.3.

d) The topological classification of linear vector fields without any purely imaginary eigenvalues follows from Theorem 7.10.6: the topological conjugacy class of $(E, 0, u)$ is given by the two integers $\dim(E_s(u))$ and $\dim(E_u(u))$.

e) The C^1 classification of linear vector fields is the same as the linear classification (Jordan form, ...); this is essentially what we have seen in 7.10.4.

8.2.4. To end this section on definitions we give a few more that are quite classical:

A singularity is *stable in the Lyapunov sense* if for every neighbourhood U_1 of a there exists a neighbourhood U_2 of a having the following property: for every $x \in U_2$ and all $t \geq 0$ the point $\Phi_t(x)$ exists and belongs to U_1.

We say that a is *asymptotically stable in the Lyapunov sense* if a is stable and if, moreover, for all x sufficiently close to a we have that $\Phi_t(x) \to a$ as $t \to +\infty$.

We shall avoid using this terminology as it conflicts with the modern usage of the word 'stable' that we have already met. Instead of *asymptotically stable in the Lyapunov sense* we prefer to say *attracting*, and we call the *domain of attraction* of a the set of those $x \in V$ such that $\Phi_t(x)$ tends to a as $t \to +\infty$.

The domain of attraction A of an attracting point a is an open set. By definition it contains an open neighbourhood U of a; for every $x \in A$ there exists $t \in \mathbf{R}$ such that $\Phi_t(x) \in U$ and then $\Phi_t^{-1}(U)$ is an open set contained in A and containing a. Moreover, A is invariant under the flow (6.10.6): for every $x \in A$ and every $t \in \mathbf{R}$ for which $\Phi_t(x)$ is defined (which holds in any case for $t \geq 0$) we have $\Phi_t(x) \in A$.

8.2.5. These definitions involve only the integral flow of X. They could be translated as follows.

The point a is stable in the Lyapunov sense if for every neighbourhood U_1 of a there exists a neighbourhood U_2 of a having the following property: the future semi-orbit of every element x of U_2 is contained in U_1.

The point a is attracting if it is stable in the Lyapunov sense and if, moreover, for every x sufficiently close to a, the point a is the unique limit point of the future semi-orbit of x. The domain of attraction of a is the set of those $x \in V$ whose future semi-orbit has a as its limit point.

Consequently the fact that a singular point is Lyapunov stable or attracting is preserved under orbital equivalence.

Likewise we say that the singular point a is *repelling* if is it an attracting point for the vector field $-X$. It is best to avoid using the word 'unstable' since different authors take it to mean either 'not stable' or 'stable for the opposite vector field'. As above, we can translate the property of being repelling into purely topological terms; it is therefore preserved under orbital equivalence.

8.2.6. Finally, we often say that a is *exponentially attracting* if it is attracting and if, for some arbitrary norm $\|x\|$ fixed on E, there exist positive constants α, β and γ such that for $\|x - a\| < \gamma$ we have $\|\Phi_t(x) - a\| \leq \beta\|x - a\|e^{-\alpha t}$ for $t \geq 0$.

Similarly we define *exponentially repelling* points. The fact of being exponentially attracting or repelling is preserved under C^1 conjugacy. With this terminology we can translate 7.9.1 as follows:

Proposition 8.2.7. *Let u be an endomorphism of the vector space E. The following conditions are equivalent:*

(i)   the eigenvalues of u all have real part < 0;
(ii)   0 is an attracting singular point for the vector field u;
(iii) 0 is an exponentially attracting singular point for the vector field u.

Moreover, the domain of attraction of 0 is then the whole space E.   □

Now we shall look at the first technique that comes to mind for studying a singularity, namely *linearization*.

8.3 Linearization of a Vector Field in the Neighbourhood of a Singular Point

Let E be a finite-dimensional vector space with V a submanifold of E and X a vector field on V of class C^r, $r \in [1, \infty]$. Let $a \in V$ be a singular point of X. We shall define an endomorphism $\Lambda_a X$ of the tangent space $T_a V$, called the *linearization* of X at a.

8.3.1. Suppose first of all that V is an open subset of E. Then X is a C^1 map from V to E ; it has a derivative at a that we denote by $\Lambda_a X$. We have $\Lambda_a X \in \text{End}(E)$. If we choose a basis for E and denote the components of the vector field X by $X_1, \ldots, X_n$ then the matrix of the endomorphism $\Lambda_a X$

is the Jacobian matrix $(\partial X_j/\partial x_i)(a)$. Since $X(a)$ is zero, for x close to a we have

$$X(x) = (\Lambda_a X) \cdot (x - a) + o(x - a). \tag{8.3.1}$$

This operation of linearization *commutes with local diffeomorphisms*. To be more precise, let E' be another vector space and let h be a local diffeomorphism at a between E and E'. Let $a' = h(a) \in E'$ and let X' denote the vector field h_*X which is the image of X under h. Then $X'(a') = 0$ and the endomorphism $\Lambda_{a'}X'$ of E' is the transform of the endomorphism $\Lambda_a X$ of E by the isomorphism $T_a h : E \to E'$: we have

$$\Lambda_{a'}X' \circ T_a h = T_a h \circ \Lambda_a X. \tag{8.3.2}$$

Indeed, for x close to a we have $X'(h(x)) = (T_x h) \cdot X(x)$; differentiating at $x = a$ and using the fact that $X(a) = 0$ we obtain the stated formula.

Note in passing that if $X(a)$ were nonzero there would be a supplementary term involving the second derivatives of h. This is what makes it impossible to define intrinsically a derivative vector field of a vector field on a manifold.

In particular, if $(x_1, \ldots, x_n)$ is a system of local coordinates on E at a and if $X_1, \ldots, X_n$ denote the corresponding components of the vector field X (recall $X_i = L_X x_i$) then the matrix of $\Lambda_a X$ in the basis of the $(\partial/\partial x_i)(a)$ is formed from the $(\partial X_j/\partial x_i)(a)$ just as in the case of linear coordinates.

8.3.2. Let us turn now to the general case where V is an arbitrary submanifold of E. We can define $\Lambda_a X$ by taking a system of local coordinates on V at a and applying the formulae above. The fact that this definition is independent of the chosen system of local coordinates follows from the invariance under local diffeomorphisms that we have already verified. We can also argue as follows. Consider X as a map from V into E and differentiate it at a. We obtain a linear map $T_a X : T_a V \to E$. However, the fact that $X(a)$ is zero implies that the image of $T_a X$ is contained in $T_a V$. To see this, let f be a C^1 function defined on a neighbourhood of a in E and vanishing on V. We have $f'(x) \cdot X(x) = 0$ for all x, hence by differentiation

$$f''(a) \cdot (X(a), \eta) + f'(a) \cdot ((T_a X) \cdot \eta) = 0, \quad \eta \in T_a V. \tag{8.3.3}$$

But since $X(a) = 0$ we deduce that $\mathrm{Im}(T_a X) \subset \mathrm{Ker}(f'(a))$. As this holds for all f we do indeed have $\mathrm{Im}(T_a X) \subset T_a V$. Then $\Lambda_a X$ is defined as the endomorphism of $T_a V$ obtained by restricting $T_a X$.

Let (Φ_t) denote the integral flow of X. Since $X(a) = 0$ we have $\Phi_t(a) = a$ for all t and we can consider the tangent map to Φ_t at a: it is an endomorphism of $T_a V$.

Proposition 8.3.3.
$$T_a \Phi_t = \exp(t\Lambda_a X).$$

It suffices to treat the case where V is an open subset of E. By definition we have $X \circ \Phi_t = d\Phi_t/dt$. Differentiating this we obtain

$$\Lambda_a X \circ T_a \Phi_t = T_a \frac{d}{dt}\Phi_t = \frac{d}{dt}T_a\Phi_t \; ;$$

moreover $T_a\Phi_0 = T_a Id_V = Id_E$. The map $u : t \mapsto T_a\Phi_t$ is therefore a solution in $\mathrm{End}(E)$ to the linear differential equation $du/dt = \Lambda_a X \circ u$ with the initial condition $u(0) = 1$. This implies the proposition. $\qquad\qquad\square$

In the case of an open subset of a vector space we derive from the above a limited expansion analogous to (3.1):

$$\Phi_t(x) = a + \exp(t\Lambda_a X) \cdot (x - a) + o(x - a) \qquad (8.3.4)$$

in which t is *fixed* or varies in a bounded subset.

8.3.4. Now for an example. Take V to be an open subset of E; suppose we are given a euclidean scalar product $(\xi \mid \eta)$ on E, and take X to be the gradient of a function f on V, namely the vector field on V defined by

$$(\mathrm{grad}(f)(x) \mid \eta) = f'(x) \cdot \eta, \quad x \in V, \; \eta \in E.$$

To say that a is a singular point of $\mathrm{grad}(f)$ means that it is a critical point of f. Let $L \in \mathrm{End}(E)$ be the linearization of $\mathrm{grad}(f)$ at a. Differentiating the above relation we obtain $(L \cdot \xi \mid \eta) = f''(a)(\xi, \eta)$. Thus L is the (symmetric) endomorphism of E associated to the Hessian form of f at a (or, which amounts to the same thing, the gradient of this quadratic form). To say that a is a nondegenerate critical point (4.4.4) of f means that L is invertible. This makes it natural to introduce the following definition.

Definition 8.3.5. The singularity (V, a, X) is *nondegenerate* if the endomorphism $\Lambda_a X$ is invertible.

Nondegenerate singular points are isolated points in the set of all singular points. This set consists of the solutions to the equation $X(x) = 0$, and the derivative of the left hand side of the equation at the point a is precisely $\Lambda_a X$, and is therefore invertible since a is nondegenerate.

It is important to distinguish a particular type of nondegenerate singular point:

Definition 8.3.6. The singularity (V, a, X) is *hyperbolic* if the endomorphism $\Lambda_a X$ has no purely imaginary eigenvalue.

Note as an example that a nondegenerate singularity of a gradient field is always hyperbolic, and its numerical type (see Sect. 8.7 below) is the signature of the Hessian form at that point.

8.3.7. Just as for the linear case in the previous chapter, we can generalize the above definitions to arbitrary local diffeomorphisms not necessarily belonging to an integral flow. Let Φ be a local diffeomorphism from a manifold V to itself and let a be a fixed point of Φ. We have $\Phi(a) = a$ and can consider the endomorphism of the vector space $T_a V$.

Definition 8.3.8. The point a is said to be a *nondegenerate fixed point* of Φ if the endomorphism $T_a\Phi$ has no eigenvalue equal to 1. It is said to be a *hyperbolic fixed point* of Φ if the endomorphism is hyperbolic, that is to say (7.5.6) it has no eigenvalue with modulus 1.

Now we return to the case of a vector field X with integral flow (Φ_t) and having a singular point a. By 3.3 and 7.6.6 the eigenvalues of $T_a\Phi_t$ are the exponentials of the eigenvalues of $t\Lambda_a X$. Hence the following conditions are equivalent:

(i) a is a nondegenerate (hyperbolic) singular point of the vector field X,
(ii) there exists $t \in \mathbf{R}$ such that a is a nondegenerate (hyperbolic) fixed point of the local diffeomorphism Φ_t,
(iii) for almost all $t \in \mathbf{R}$ (for all $t \neq 0$ in $\mathbf{R}$) a is a nondegenerate (hyperbolic) fixed point of the local diffeomorphism Φ_t.

8.4 Difficulties with Linearization

We begin by showing why difficulties are to be expected. For simplicity we work in a finite-dimensional vector space E. It is true that a vector field X (of class C^1) such that $X(0) = 0$ is, in a neighbourhood of 0, close to its linearization L, and by the fundamental theorems of differential equations this implies that the integral flows (Φ_t) of X and $(\exp(tL))$ of L are close. More precisely, for x close to 0 and for fixed t, or more generally for *bounded* t, the points $\Phi_t(x)$ and $\exp(tL) \cdot x$ are close. Unfortunately, as we have already observed, the local study of the flow of X close to a singular point involves unbounded values of t.

8.4.1. Naturally we may nevertheless hope that the local phase portrait of X in a neighbourhood of a might resemble the local phase portrait of the linear vector field $\Lambda_a X$ in a neighbourhood of 0. But it all depends on what we mean by 'resemble'.

Let us return first of all to the invariance of the operation of linearization under local diffeomorphisms as shown above. This can be stated as follows. Let (V, a, X) and (V', a', X') be two singularities that are C^r-conjugate, with $r \geq 1$. Let h be a local diffeomorphism of class C^r between V and V' satisfying the condition (2.2) of Definition 2.1. Then we have

$$\Lambda_{a'} X' = (T_a h) \circ (\Lambda_a X) \circ (T_a h)^{-1}.$$

Consequently the endomorphisms $\Lambda_a X$ and $\Lambda_{a'} X'$ are linearly conjugate. In particular, a' is nondegenerate or hyperbolic if a is. If we could show that every nondegenerate singularity (V, a, X) were C^r-equivalent to its linearization $(T_a V, 0, \Lambda_a X)$ then the differentiable classification of singularities would be reduced to the linear classification of endomorphisms. Unfortunately this is nothing like the case. We now give two counterexamples to indicate the limits of what we can expect. In both cases we take $V = \mathbf{R}^2$ and the singular point considered is $a = (0, 0)$.

8.4.2. The first example is as follows: consider a C^∞ function $\phi : \mathbf{R} \to \mathbf{R}$ such that $\phi(0) = 0$, and the C^∞ vector field X on $\mathbf{R}^2$ given by

$$X(x, y) = (y + \phi(x^2 + y^2)x, \ -x + \phi(x^2 + y^2)y).$$

We immediately have $L \cdot (x, y) = (y, -x)$ and the orbits of L are circles centred at 0.

On the other hand, the phase portrait of the field X depends on the function ϕ. Moving to polar coordinates (ρ, θ), an immediate calculation gives $L_X \rho = \rho\phi(\rho^2)$ and $L_X \theta = 1$. The orbits are therefore the integral curves of the equation $d\rho/d\theta = \rho\phi(\rho^2)$. For every $\rho > 0$ with $\phi(\rho^2) = 0$ the circle of radius ρ is an orbit of X. Suppose we are given α and β with $\alpha < \beta$ such that $\phi(\alpha^2) = \phi(\beta^2) = 0$ and such that ϕ does not vanish in the interval (α, β). Then in the annulus $\alpha^2 < x^2 + y^2 < \beta^2$ each orbit of X has as its limit sets the closed orbits that form the boundary of the annulus. If $\phi > 0$ in I, the outer circle is the future limit set and the inner circle is the past limit set. If $\phi < 0$ in I the outer circle is the past limit set and the inner circle is the future limit set.

However, since we can manufacture a function ϕ whose zero set is an arbitrary closed subset of $\mathbf{R}^+ = [0, \infty)$ we see that there are at least as many orbitally inequivalent possible phase portraits as there are closed subsets of $\mathbf{R}^+$ (or germs at 0 of these subsets, at any rate)! The situation is therefore desperate. In fact there is nothing too surprising in this phenomenon. The vector field L already fails to be structurally stable as a linear vector field: arbitrarily close to L in $M_2(\mathbf{R})$ there are attracting linear vector fields and repelling linear vector fields. There is no reason to hope that a nonlinear perturbation of L of second order behaves any better than a linear perturbation of first order.

By slightly generalizing the above we easily obtain the following negative result: *for every nonhyperbolic endomorphism L the set of orbital equivalence classes of singular points with linearization L is infinite* (indeed uncountable, and so 'unclassifiable').

8.4.3. The second example is more astonishing, as this time the linearized vector field is attracting, and therefore 'linearly structurally stable' (7.10.7). Consider the vector field X on $\mathbf{R}^2$ given by

$$X(x, y) = (2x + y^2, y).$$

The linearized field L has as matrix $\left(\begin{smallmatrix} 2 & 0 \\ 0 & 1 \end{smallmatrix}\right)$. The integral flows of L and X are

$$\Phi_t^L(x, y) = (e^{2t}x, e^t y), \quad \Phi_t^X(x, y) = (e^{2t}(x + ty^2), e^t y).$$

The phase portraits 'look like' each other and we would say there ought to be a differentiable equivalence between X and L. Therefore we try to construct two functions $f(x, y)$ and $g(x, y)$, with nonvanishing Jacobian at 0, for which

$$\Phi_t^L\big(f(x, y), g(x, y)\big) = \big(f(\Phi_t^X(x, y), g(\Phi_t^X(x, y))\big).$$

In particular this gives

$$e^{2t} f(x, y) = f(e^{2t}(x + ty^2), e^t y).$$

Now, since f is C^2 this latter relation, even for a single nonzero value of t, implies that the two partial derivatives of f at 0 are zero (exercise: differentiate twice with respect to y). Hence X and its linearization L are *not* C^2-*conjugate*. We shall see later (Theorem 7.2) that they are nevertheless topologicaly conjugate.

In this example, X and L are in fact C^1-conjugate, but this is a phenomenon which is peculiar to both dimension 2 and to the attracting case (see 11.4). As an exercise, find an analogous example in $\mathbf{R}^3$ where there is no C^1-conjugacy between X and its linearization even though the latter is hyperbolic and therefore topologically stable among linear vector fields.

For the time being we remain in the 'topological' context; we shall return to the more subtle problems of differentiability later.

8.5 Singularities with Attracting Linearization

Let E be a finite-dimensional vector space with V a submanifold of E and X a C^1 vector field on V; let a be a singular point of X.

Theorem 8.5.1. (Poincaré-Lyapunov) *If 0 is an (exponentially) attracting singular point for the linearized field $\Lambda_a X$ then a is an exponentially attracting singular point for X.*

By 2.7 the hypothesis means that all the eigenvalues of $\Lambda_a X$ have real part < 0. Fix a constant $\alpha > 0$ with

$$\Re(\lambda) < -\alpha \tag{8.5.1}$$

for all eigenvalues λ of $\Lambda_a X$. Let (Φ_t) denote the integral flow of X. We shall prove the following proposition that makes the theorem precise:

Proposition 8.5.2. *Under the above hypotheses, there exists a euclidean norm $\|x\|$ on E and a number $\varepsilon > 0$ such that for every $x \in V$ with $\|x-a\| < \varepsilon$ we have $\Phi_t(x)$ defined for all $t \geq 0$ and*

$$\|\Phi_t(x) - a\| \leq \|x - a\|e^{-\alpha t}. \tag{8.5.2}$$

To simplify the notation let $T = T_a V$ and $u = \Lambda_a X$. Since the inequalities (5.1) are strict, we may replace α in them by $\alpha + \eta$ where η is a sufficiently small constant > 0. Applying 7.10.2, we can construct a euclidean scalar product on T for which

$$(u \cdot z \mid z) \leq -(\alpha + \eta)(z \mid z), \quad z \in T. \tag{8.5.3}$$

Extend this scalar product to a euclidean scalar product on the whole of E, and fix a sufficiently small constant θ. We saw above that u is the linear tangent map at a to the map X from V into E. It follows from this (2.7.5) that we can find $\varepsilon > 0$ such that, for all $x \in V$ with $\|x - a\| \leq \varepsilon$, there exists $z \in T$ with

$$\|x - a - z\| \leq \theta\|x - a\|, \quad \|X(x) - u \cdot z\| \leq \theta\|x - a\|. \tag{8.5.4}$$

Take $x \in V$ with $\|x-a\| \leq \varepsilon$. If θ has been chosen small enough, the relations (5.3) and (5.4) imply

$$(X(x) \mid x - a) \leq -\alpha\|x - a\|^2. \tag{8.5.5}$$

Now fix ε sufficiently small so that the subset $\{\, x \in V \mid \|x-a\| \leq \varepsilon \,\}$ is closed in E and therefore compact. Let U be the open set $\{\, x \in V \mid \|x - a\| < \varepsilon \,\}$. For simplicity, write $f(x) = \|x - a\|$ for $x \in U$. Then f is a C^∞ function on $U \setminus \{a\}$ and we have $L_X f(x) = (X(x) \mid x-a)/\|x-a\|$. Hence the inequalities (5.5) can be re-written

$$L_X f \leq -\alpha f, \quad x \in U. \tag{8.5.6}$$

In particular we have $L_X f \leq 0$ on U. Take $x \in U$. First we shall prove that $\Phi_t(x)$ is defined and belongs to U for all $t \geq 0$, by applying the *principle of a priori bounds* from 6.9.2 to the restriction of the vector field X to U. Thus let $J = [0, A]$ be an interval such that $\Phi_t(x)$ is defined and belongs to U for all $t \in J$. Let $\gamma(t) = \Phi_t(x)$ for $t \in J$. We then have $d/dt(f \circ \gamma) = (L_X f) \circ \gamma \leq 0$, and $f \circ \gamma$ is decreasing. In particular we have $\|\gamma(t) - a\| = f(\gamma(t)) \leq f(0) = \|x-a\|$. The principle of a priori bounds therefore applies, and $\gamma(t)$ is defined for all $t \geq 0$. The inequalities (5.6) then give $d(f \circ \gamma)/dt \leq -\alpha f \circ \gamma$. But this implies that the function $e^{\alpha t} f(\gamma(t))$ is decreasing and therefore everywhere

less than $f(x)$. We have thus proved Proposition 5.2 and also Theorem 5.1.
□

Note that the appearance of a norm on E in 5.2 is a pure convenience (see the next section; in any case, it is only the restriction of the function $\|x - a\|$ to a neighbourhood of a in V which counts). As in 7.10.3, Theorem 5.1 can be given in another version as follows:

Theorem 8.5.3. *If 0 is an attracting point for $\Lambda_a X$, then (V, a, X) is topologically conjugate to $(\mathbf{R}^n, 0, -Id)$, $n = \dim_a(V)$.*

By taking a local parametrization on V we can first reduce to the case where V is an open subset of E. Applying the above, fix an arbitrary δ with $0 < \delta < \varepsilon$ and let $S = \{ x \mid f(x) = \delta \}$ and $B = \{ x \mid 0 < f(x) \le \delta \}$. Consider then the map $(t, y) \mapsto \Phi_t(y)$ from $\mathbf{R}^+ \times S$ to B, and argue just as in the proof of 7.10.3. The only point that has to be altered is the proof of surjectivity: it must be proved that for all $x \in B$ there exists $t \le 0$ with $\Phi_t(x) \in S$. However, if this were false we would see from the principle of *a priori* bounds that $\Phi_t(x)$ exists and belongs to B for all $t \le 0$. However, applying (5.2) to $-t$ and $\Phi_t(x)$ then gives a contradiction.
□

Applying 5.3 to the reversed vector field we obtain:

Corollary 8.5.4. *If 0 is a repelling point for $\Lambda_a X$ then (V, a, X) is topologically conjugate to $(\mathbf{R}^n, 0, Id)$, $n = \dim_a(V)$.*
□

It follows from 5.3 (or 5.4) that two singularities with attracting (or repelling) linearizations on manifolds of the same dimension are topologically conjugate. Hence if 0 is an attracting (or repelling) point for $\Lambda_a X$ then (V, a, X) and $(T_a V, 0, \Lambda_a X)$ are topologically conjugate.

8.6 Lyapunov Theory

The results in the previous section made use of the auxiliary function $f(x) = \|x - a\|$. Lyapunov theory extends these results to the case of more general functions. Two types of property of f were used: those that relate it to the topology ("x tends to a" is equivalent to "$f(x)$ tends to 0") and those which involve the vector field under consideration. We begin by studying the first aspect.

8.6.1. Let E be a finite-dimensional vector space with V a submanifold of E and a a point of V (although in fact what we are about to do is not tied to this particular situation). Consider a function f defined and continuous on a neighbourhood of a in V and satisfying the following property:

a) $f(a) = 0$ and $f(x) > 0$ for $x \neq a$.

We shall see that f can replace $\|x - a\|$ in the topological arguments of the previous section. Note that in the vast majority of cases where such a function is used, the properties we are about to establish are in any case self-evident.

8.6.2. Let K be a compact neighbourhood of a in V on which f is defined and continuous. For all $\varepsilon > 0$ let U_ε denote the set of those $x \in K$ for which $f(x) < \varepsilon$. The $\{U_\varepsilon\}$ are a decreasing family of neighbourhoods of a and form a so-called 'fundamental system of neighbourhoods of a ': every open set of V containing a contains one of the U_ε. To see this, let W be such an open set. The set of those $x \in K$ which do not belong to W is closed in K and therefore compact, and the function f is > 0 everywhere on it. It thus has a greatest lower bound $\varepsilon > 0$ and all $x \in U_\varepsilon$ belong to W.

Note also that for ε sufficiently small, U_ε is an open set in V (and not just in K). Indeed, let U be an open neighbourhood of a contained in K. If U_ε is contained in U (which is true for ε sufficiently small, by the above) then U_ε is defined by the condition $f < \varepsilon$ in U and is therefore open.

When f is defined on the whole of V and is proper (2.10.1) we can dispense with introducing the compact set K. We define U_ε as the set of all $x \in V$ with $f(x) < \varepsilon$. The U_ε are then open in V and form a fundamental system of neighbourhoods of a.

8.6.3. Suppose now that f is of class C^1 (except perhaps at the point a itself), and moreover suppose that we are given a C^1 vector field X on a neighbourhood of a in V. We have the Lie derivative $L_X f$ which is defined on a neighbourhood of a except perhaps at the point a itself.

Assume that f satisfies the condition a) and also one of the following three properties of increasing strength:

b1) $L_X f \leq 0$,
b2) $L_X f < 0$ for $x \neq a$,
b3) there exists $\alpha > 0$ with $L_X f \leq -\alpha f$.

Here are some examples:

1)   If f satisfies one of the conditions a), b1), b2) or b3) then so does f^μ for all $\mu > 0$. This is clear for condition a); for the three b) conditions it follows immediately from the identity $L_X f^\mu = \mu f^{\mu-1} L_X f$.

2)   For any norm $x \mapsto \|x\|$ on E, the function $f(x) = \|x - a\|$ satisfies the property a). If $\|x\|$ comes from a scalar product $(x \mid y)$ then f is certainly C^1 away from a; we have $L_X f(x) = (X(x) \mid x - a)/f(x)$ and the conditions b1), b2) and b3) are respectively equivalent to the three following conditions: $(X(x) \mid x - a) \leq 0$, $(X(x) \mid x - a) < 0$ for $x \neq a$, and $(X(x) \mid x - a) \leq -\alpha\|x - a\|^2$.

3)   If the linearization of X at a is attracting then there exists a euclidean norm on E such that $\|x - a\|$ satisfies b3); this is just as we saw in 5.2.

4)   Consider the vector field

$$X = 2y(z - 1)\frac{\partial}{\partial x} - x(z - 1)\frac{\partial}{\partial y} - z^3\frac{\partial}{\partial z}$$

on $\mathbf{R}^3$ and let f be the function $x^2 + 2y^2 + z^2$ which clearly satisfies a). Then $L_X f = -2z^4$, and f satisfies b1).

We retain the previous notation; in particular, f satisfies condition a).

Proposition 8.6.4.

a) If f satisfies the condition b1) then the singular point a is stable in the Lyapunov sense (see 2.4).

b) If f satisfies the condition b2) then the singular point a is attracting.

c) Suppose that f satisfies the condition b3) and let α be the constant that appears in this condition. Let (Φ_t) denote the integral flow of X. Then for x sufficiently close to a, the flow (Φ_t) is defined for all $t \geq 0$ and we have

$$f(\Phi_t(x)) \leq e^{-\alpha t} f(x), \quad t \geq 0. \tag{8.6.1}$$

Let K be a compact neighbourhood of a in V small enough so that X and f are defined on K. For all $\varepsilon > 0$ let U_ε denote the set of $x \in K$ such that $f(x) < \varepsilon$. Let $\varepsilon > 0$ be small enough so that U_ε is open in V. We shall prove the following statements which imply the theorem:

1)   If f satisfies b1) then for all $x \in U_\varepsilon$ and all $t \geq 0$ the flow Φ_t exists and belongs to U_ε.

2)   If f satisfies b2) then U_ε is contained in the domain of attraction of a.

3)   If f satisfies b3) then (6.1) is satisfied when x belongs to U_ε.

Let $x \in U_\varepsilon$. Suppose f satisfies b1). We apply the principle of *a priori bounds* from 6.9.2 to the restriction to U_ε of the vector field X. We have $f(x) < \varepsilon$; let L denote the set of those $y \in K$ such that $f(y) \leq f(x)$. This is a subset of U_ε which is closed in K and therefore compact. Let $\gamma(t) = \Phi_t(x)$ and let $T > 0$ be such that $\gamma(t)$ is defined and belongs to U_ε for every element t of the interval $J = [0, T]$. For all $t \in J$ we have

$$\frac{d}{dt}(f(\gamma(t))) = L_X f(\gamma(t)) \leq 0,$$

and the function $f \circ \gamma$ is decreasing on J. Hence we have that $f(\gamma(t)) \leq f(\gamma(0)) = f(x)$ for all $t \in J$, and so $\gamma(t) \in L$. It follows that $\gamma(J)$ is contained in the compact set L. The said principle therefore applies and shows that $\gamma(t)$ is defined for all $t \geq 0$ and belongs to U_ε. This proves 1).

Now suppose that f satisfies b2). Then the function $f \circ \gamma$ is strictly decreasing on $\mathbf{R}^+$. It remains to show that it tends to 0 as t tends to $+\infty$. Suppose not: then there exists $\eta > 0$ with $f(\gamma(t)) \geq \eta$ for all $t \geq 0$. But this implies that the image of γ is contained in the compact set $L - U_\varepsilon$. On this compact set the function $L_X f$ is continuous and everywhere < 0 and therefore less than some constant $c < 0$. Thus the function $f \circ \gamma$ on $\mathbf{R}^+$ is everywhere > 0 and with derivative everywhere $\leq c < 0$, which is absurd.

Finally, suppose f satisfies the condition b3). Then we have $\frac{d}{dt}(f \circ \gamma) \leq -\alpha f \circ \gamma$, or in other words $\frac{d}{dt}(e^{\alpha t} f \circ \gamma(t)) \leq 0$, from which (6.1) follows immediately. $\square$

8.6.5. A function satisfying conditions a) and b1) is traditionally called a *Lyapunov function*, while a function satisfying a) and b2) is a *strong Lyapunov function*. The use of such functions is called *Lyapunov's direct method*, the adjective 'direct' referring to the fact that it is not necessary to solve the differential equation in advance, since all that is necessary is to calculate a Lie derivative. Clearly, the difficulty is in finding a good function to use.

8.7 The Theorems of Grobman and Hartman

The local topological classification of diffeomorphisms in the neighbourhood of a hyperbolic fixed point and of vector fields in the neighbourhood of a hyperbolic singular point is given by the following two parallel theorems that we state without proof[47]. The second theorem generalizes Theorem 5.3 and its Corollary 5.4 to the 'mixed' case.

Let E be a finite-dimensional vector space with V a submanifold of E and a a point of V.

Theorem 8.7.1. *Let Φ be a local diffeomorphism (of class C^1) of V having a hyperbolic fixed point at a. Then Φ (in a neighbourhood of a) and $T_a\Phi$ (in a neighbourhood of 0) are topologically conjugate: there exists an open subset U of V containing a, an open subset Ω of T_aV containing the origin and a homeomorphism $h : \Omega \to U$ with $\Phi = h \circ T_a\Phi \circ h^{-1}$.* $\square$

Naturally, this latter relation is to be understood in terms of maps not defined everywhere, as in Definition 2.2.

Theorem 8.7.2. *Let X be a C^1 vector field on V having a hyperbolic singular point at a. Then (V, a, X) and $(T_aV, 0, \Lambda_a X)$ are topologically conjugate.* $\square$

---

[47]Theorem 7.1 is due to the American Philip HARTMAN (1960). Theorem 7.2 is due independently to HARTMAN and to the Russian mathematician D.GROBMAN (1959). The best proof is that due to Charles PUGH, for which see [IR] pp. 113-119, [PM] pp. 60-67 or [RO] §5.7.

From now on we shall concentrate on the vector field case, leaving the generalization to local diffeomorphisms as an exercise. To state the following corollary, let us define the *numerical type* of the hyperbolic singular point a to be the pair (m_s, m_u) where m_s (resp. m_u) is the dimension of the stable (resp. unstable) subspace of the linearization, that is the number – counting multiplicities – of those eigenvalues whose real part is < 0 (resp. > 0). There are thus $n + 1$ possible numerical types in dimension n.

Corollary 8.7.3. *For two hyperbolic singularities (V, a, X) and (V', a', X') the following conditions are equivalent:*

(i) (V, a, X) and (V', a', X') are topologically conjugate,
(ii) (V, a, X) and (V', a', X') are orbitally equivalent,
(iii) (V, a, X) and (V', a', X') have the same numerical type.

By the theorem, (V, a, X) and (V', a', X') are respectively topologically conjugate to their linearizations; hence each of the conditions (i) and (ii) is equivalent to the analogous condition for the linearizations. The same holds by definition for condition (iii). The result then follows from 7.10.6. $\square$

8.8 Stable and Unstable Manifolds of a Hyperbolic Singularity

Let E be a finite-dimensional vector space with V a submanifold of E and X a vector field on V of class C^r, $r \in [1, +\infty]$, with integral flow (Φ_t). Suppose (V, a, X) is a hyperbolic singularity.

Definition 8.8.1. The *stable manifold* of the hyperbolic singularity (V, a, X) is the set $V^s(a)$ consisting of those $x \in V$ for which $\Phi_t(x)$ is defined for all $t \geq 0$ and tends to a as $t \to +\infty$.

This definition can be translated into a purely 'topological' vocabulary, showing that it is invariant under orbital equivalence: in order for x to belong to $V^s(a)$ it is necessary and sufficient that the future semi-orbit of x should have a as a unique limit point. Equivalently, $V^s(a)$ can be defined as the union of the orbits of X that have a as a unique future limit point. This shows that $V^s(a)$ is a union of orbits of X. The *unstable manifold* $V^u(a)$ is defined by replacing t by $-t$ (or X by $-X$, or 'future' by 'past') in the previous definitions. The use of the word *manifold* in these definitions will be justified below: we shall see in 8.5 that $V^s(a)$ is an 'immersed submanifold of class C^r'.

Two examples are already familiar. First, if a is attracting then $V^s(a)$ is the basin of attraction of a; conversely, to say that a is attracting (resp.

repelling) is equivalent to saying that $V^s(a)$ (resp. $V^u(a)$) is open in V. The other case is the linear case: if X is a linear vector field on the vector space V then $V^s(0)$ and $V^u(0)$ are the stable and unstable subspaces introduced in 7.9.3 (see 7.9.5).

8.8.2. To simplify notation write $T = T_a V$ and $L = \Lambda_a X$. By definition the endomorphism $L \in \text{End}(T)$ has hyperbolic flow. Applying 7.9.3, we decompose the vector space T into the sum of two complementary subspaces $T^s = T_s(L)$ and $T^u = T_u(L)$, which we call the *stable tangent subspace* and the *unstable tangent subspace* of X at a, respectively. Recall that the *numerical type* of the hyperbolic singular point a is the pair of integers $(\dim(T^s), \dim(T^u))$ whose sum is $\dim(T) = \dim_a V$. If we replace X by $-X$ then L is replaced by $-L$ and the roles of T^s and T^u are exchanged. We shall therefore be content to study the 'stable part' and leave it to the reader to make the translations necessary for the 'unstable part'.

As the Hartman-Grobman Theorem allows us to do, let us fix an open set Ω in T containing 0, an open subset U in V containing a, and a homeomorphism $h : \Omega \to U$ such that

$$\Phi_t \circ h = h \circ \exp(tL).$$

Recall that this equation means precisely the following: let $y \in \Omega$ and $t \in \mathbf{R}$; then for $\Phi_t(h(y))$ to be defined it is necessary and sufficient that $\exp(tL) \cdot y$ belong to Ω, and then we have $\Phi_t(h(y)) = h(\exp(tL) \cdot y)$. For simplicity we identify T with $T^s \times T^u$ and suppose that Ω has been chosen to be of the form $\Omega^s \times \Omega^u$ where Ω^u is an open set in T^u containing 0 and Ω^s is an open subset of T^s containing 0 and such that the condition $(y \in \Omega^s$ and $t \geq 0)$ implies $(\exp(tL) \cdot y \in \Omega^s)$ (for this it suffices to take a ball in a suitable norm, cf. 7.10.2).

Having laboriously carried out these preliminaries, we now return to the subject. For the linear field L on T we saw in 7.9.5 that the stable manifold $T^s(0)$ is the stable subspace T^s and that, for every $y \in T$ which does not belong to T^s, the point $\exp(tL) \cdot y$ leaves every compact set as t tends to $+\infty$. In view of the condition imposed on Ω^s, we have $\Omega^s(0) = \Omega^s$. Transporting by the homeomorphism h we immediately see that $h(\Omega^s) = V^s(a) \cap U$; thus $V^s(a)$ is a 'submanifold of class C^0' in a neighbourhood of a. Since every point of $V^s(a)$ can be brought by the flow into the connected set $V^s(a) \cap U$, it follows that $V^s(a)$ is connected. In passing, we have in fact obtained more:

Proposition 8.8.3. *Let V' be an open set in V containing a. If V' is sufficiently small then every element of V' whose future semi-orbit is contained in V' belongs to the stable manifold $V^s(a)$.* $\qquad\square$

Now let x be an arbitrary point of the stable manifold $V^s(a)$. For all sufficiently large $u > 0$ we have $\Phi_u(x) \in U$ and therefore $\Phi_u(x) \in h(\Omega^s) = V^s(a) \cap U$. This shows first of all that $V^s(a)$ is the union of the $\Phi_u^{-1}(V^s(a) \cap U)$. Let $g(x) = \exp(-uL) \cdot (h^{-1}(\Phi_u(x)) \in T^s$. Then $g(x)$ does not depend on the choice of the element u such that x belongs to $\Phi_u^{-1}(V^s(a) \cap U)$. Indeed, if $u, v \in \mathbf{R}$ with $v \geq u$ are such that $\Phi_u(x)$ and $\Phi_v(x)$ belong to U and we put $w = v - u$ then we have

$$\exp(-vL) \cdot (h^{-1}(\Phi_v(x)) = \exp(-uL) \cdot ((\exp(-wL) \circ h^{-1} \circ \Phi_w)(\Phi_u(x))),$$

and by construction $\exp(-wL) \circ h^{-1} \circ \Phi_w$ is the identity. Thus we have obtained a well-defined map g from $V^s(a)$ into T^s.

We shall see that g is a *bijection* from $V^s(a)$ onto an *open set* in T^s. For this we shall explicitly construct an inverse map f. For every $u \in \mathbf{R}$ let U_u denote the open subset of U consisting of those z such that $\Phi_{-u}(z)$ is defined, and consider the open subset

$$W_u = T^s \cap \exp(-uL)(h^{-1}(U_u))$$

of T^s. For $y \in W_u$ let

$$f_u(y) = \Phi_{-u}(h(\exp(uL) \cdot y)) \in V,$$

which defines a map from W_u into V. This map is the restriction of a composition of three homeomorphisms. It is therefore a continuous map which induces a homeomorphism of W_u onto its image. This latter set consists of those $x \in V$ such that $\Phi_u(x)$ is defined and belongs to $h(\Omega^s)$, that is to say $V^s(a) \cap U$, and it is therefore the open subset $\Phi_u^{-1}(V^s(a) \cap U)$ of $V^s(a)$ considered previously. Moreover, it may be verified as above that for $v \geq u$ we also have $y \in W_v$ and $f_v(y) = f_u(y)$. Thus if W denotes the open subset of T^s that is the union of the W_u, we have defined a continuous map $f : W \to V$ which is by construction the inverse of g. Note that $W_0 = \Omega^s$ and $f_0(y) = h(y)$ for $y \in W_0$. Assembling these results we obtain:

Proposition 8.8.4.

a)   The set W is an open, connected subset of the stable tangent space T^s.
b)   The map $f : W \to V^s(a)$ is bijective, it extends the homeomorphism $\Omega^s \to V^s(a) \cap U$ induced by h, and it is a local homeomorphism in the neighbourhood of every point of W.
c)   For all $t \geq 0$ and all $x \in W$ we have $f(\exp(tL) \circ x) = \Phi_t(f(x))$.   □

There are a few further remarks. First of all, f is a bijective 'immersion of class C^0', but it is not always a homeomorphism (see Fig. 8.1). Next, the properties given in the proposition characterize W and f once the homeomorphism h (or merely its germ at 0) is fixed. Finally, suppose the vector field X is complete. We then have $U_u = U$ for all u, so $W_u = \exp(-uL)(\Omega^s)$.

Since the flow of $-X$ is expanding, it follows that every bounded open subset of T^s is contained in one of the W_u. Thus W is equal to the whole of T^s and the restriction of f to every bounded open subset of W is a homeomorphism onto its image.

In general the homeomorphism h is not of class C^r (examples indicated in 4.3 show that we can not even insist that it be C^1). Hence f is no longer *a priori* of class C^r. This shows the importance of the following result, which we shall accept without proof:

Theorem 8.8.5. (Stable Manifold Theorem.) *The stable manifold $V^s(a)$ is an immersed submanifold of class C^r, whose tangent space at the point a is the stable subspace T^s.*   □

It is useful to make a some remarks here to avoid confusion. Let U be an open set in V containing a. We can consider the restriction of the vector field X to U and construct the stable manifold $U^s(a)$ which is an open subset of $V^s(a)$. If U is small enough, $U^s(a)$ is a submanifold (immersed) of class C^r, tangent at a to T^s. Beware, however, that $U^s(a)$ can be strictly smaller than the intersection of U and $V^s(a)$. Likewise, for U sufficiently small, the submanifolds $U^s(a)$ and $U^u(a)$ intersect transversely at a. On the other hand, this is not necessarily the case for $V^s(a)$ and $V^u(a)$; they can for example be the same set. See Fig. 8.1 for examples of these different phenomena.

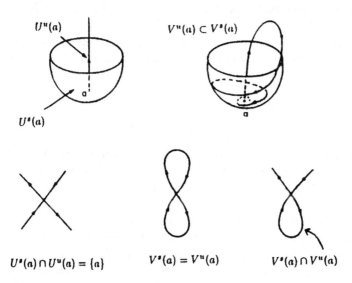

Fig. 8.1. Some possiblities for the stable and unstable manifolds of a fixed point

8.9 Differentiable Linearization: Statement of the Problem

Now we come to the problem of differentiable linearization. Here too we shall restrict ourselves to the vector field case, leaving it to the reader to formulate the analogous results for local diffeomorphisms. Placing ourselves in the context of the Grobman-Hartman Theorem 7.2, we suppose that X is of class C^r with $r \in [1, +\infty]$, and we wish to know if it is possible to choose the local homeomorphism h to be of class C^r, or at least of class C^s with $s \in [1, r]$. For linguistic convenience we make a definition:

Definition 8.9.1. A singularity is said to be C^r-linearizable if it is C^r-conjugate to its linearization.

Using this we can state a minor additional result:

Proposition 8.9.2. *The following conditions are equivalent:*

(i) (V, a, X) is C^r-equivalent to a linear vector field $(E, 0, L)$,
(ii) (V, a, X) is C^r-linearizable,
(iii) there exists a C^r map h defined on a neighbourhood of 0 in $T_a V$ and with values in V such that $h(0) = a$, $T_0 h = Id_E$ and $h_*(\Lambda_a X) = X$.

The implications $(iii) \Rightarrow (ii) \Rightarrow (i)$ are clear. Suppose (i) is satisfied, and let u be a local diffeomorphism defined on a neighbourhood of 0 in E and with values in V such that $u(0) = a$ and $u_*(L) = X$. By (3.2) we have:

$$(\Lambda_a X) \circ (T_a u) = (T_a u) \circ L,$$

and it is enough to take $h = u \circ (T_a u)^{-1}$ to obtain (iii). □

Thus we are seeking criteria for linearizability of a singularity (V, a, X) and, to be more precise, criteria that can be verified from the linearization $(T_a V, 0, \Lambda_a X)$. This leads naturally to a definition of the following kind:

Definition 8.9.3. Let E be a finite-dimensional vector space and let L be an endomorphism of E. We say that L is C^r-rigid if it satisfies the following condition: for every C^r vector field X defined on a neighbourhood of 0 in E such that $X(0) = 0$ and $X'(0) = L$ there exists a C^r map h defined on a neighbourhood of 0 in E and with values in E such that $h(0) = 0$, $h'(0) = Id_E$ and $h_*(L) = X$, that is to say

$$h'(x) \cdot (L \cdot x) = X(h(x)), \qquad (8.9.1)$$

for all $x \in E$ sufficiently close to 0.

In this relation , $h'(x)$ denotes the derivative of h at a point x, and it is therefore an element of $End(E)$. Immediately we have the following almost tautological proposition:

Proposition 8.9.4. *If X is of class C^r and $\Lambda_a X \in End(T_a V)$ is C^r-rigid then (V, a, X) is C^r-linearizable.* $\qquad\square$

The counterexample in 4.3 can be regarded as a demonstration that the matrix $\begin{pmatrix} 2 & 0 \\ 0 & 1 \end{pmatrix}$ is not C^2-rigid. With an analogous 'topological' definition that is easy to formulate, the Grobman-Hartman Theorem can be stated as: every endomorphism with no purely imaginary eigenvalue is topologically rigid. Moreover, the converse is true, as already noted in 4.2. Hence a C^1-rigid endomorphism has no purely imaginary eigenvalue.

8.10 Differentiable Linearization: Resonances

We continue with the notation of 9.3. Given an endomorphism $L \in End(E)$ the problem is to solve equation (9.1). In fact it is equivalent, but a little easier from the notational point of view, to find the inverse of the diffeomorphism h, that is (changing notation) to solve the equation

$$h'(x) \cdot X(x) = L \cdot h(x), \qquad (8.10.1)$$

with $h(0) = 0$ and $h'(0) = Id_E$.

8.10.1. To analyze where the difficulties come from, we begin by trying to determine the Taylor expansion of h. For $k = 1, 2, \ldots$ let $P_k(E)$ denote the vector space of homogeneous polynomial maps of degree k from E to E. For example, we have $P_1(E) = End(E)$. Fix an integer $r > 0$ and write the Taylor expansions of X and h up to order r

$$X = L + u_2 + \cdots + u_r + R^r(X), \qquad (8.10.2)$$

$$h = Id_E + h_2 + \cdots + h_r + R^r(h), \qquad (8.10.3)$$

where u_i and h_i belong to $P_i(E)$ for all $i \leq r$. The relation (10.1) can then be written

$$(L \cdot x + u_2(x) + \cdots) + h'_2(x) \cdot (L \cdot x + u_2(x) + \cdots) + \cdots = L \cdot x + L \cdot h_2(x) + \cdots,$$

which gives

$$L \cdot h_2(x) - h'_2(x) \cdot (L \cdot x) = u_2(x),$$

$$L \cdot h_3(x) - h'_3(x) \cdot (L \cdot x) = u_3(x) + h'_2(x) \cdot u_2(x),$$

and so on. For each k let a_k denote the map from $P_k = P_k(E)$ into itself such that $a_k(p)(x) = L \cdot p(x) - p'(x) \cdot (L \cdot x)$ for all $p \in P_k$ (to clarify the meaning of the symbols: for x fixed, L and $p'(x)$ are elements of $\text{End}(E)$ that are applied respectively to the elements $p(x)$ and $L \cdot x$ of E).

Since L is linear, the map a_k is linear and is therefore an *endomorphism of the vector space* P_k. The relations above can be written in the form

$$a_2(h_2) = u_2, \tag{8.10.4}$$

$$a_3(h_3) = u_3 + h_2' \cdot u_2, \tag{8.10.5}$$

and at order k

$$a_k(h_k) = u_k + F_{k-1}(h, u), \tag{8.10.6}$$

where $F_{k-1}(h, u)$ is an expression involving the h_i' and u_i only for $i < k$ (and which vanishes when these are zero).

8.10.2. We can already extract some partial information from these calculations:

a) If for an integer $k > 1$ the endomorphism a_k is not surjective, take an element u_k of P_k which is not in the image of a_k. *There is then no diffeomorphism of class C^k which transforms L into $L + u_k$.* Hence in order for L to be C^r-rigid it is necessary that a_k be surjective for $2 \le k \le r$.

b) Conversely, if a_k is surjective for all $k \le r$ we can successively solve the equations (10.6) for $k = 2, \ldots, r$ and obtain a polynomial diffeomorphism $h = \text{Id}_E + h_2 + \cdots + h_r$ which transforms the field X into a field whose Taylor expansion to order r reduces to L. We can therefore 'linearize X up to order r'. Note that the surjectivity of the endomorphism a_k implies that it is bijective, and that the Taylor expansion of the linearizing diffeomorphism h is therefore uniquely determined up to order r.

c) If we assume that the Taylor series for X is convergent and that all the a_k are surjective then it is possible to show that the Taylor expansion of h so obtained is also convergent. We thus obtain the theorems of Poincaré and Siegel on analytic linearization. See [A4], for example.

8.10.3. We are therefore led to study the endomorphism a_k of the vector space $P_k(E)$ for $k \ge 2$. First suppose that the endomorphism L is *diagonalizable* and let $(e_i)_{i=1,\ldots,n}$ be a basis for E consisting of eigenvectors of L. Let λ_i be the eigenvalue associated to e_i, so that

$$L \cdot (x_1 e_1 + \cdots + x_n e_n) = \lambda_1 x_1 e_1 + \cdots + \lambda_n x_n e_n.$$

Fix an integer $k \ge 2$. For every multi-index $\alpha = (\alpha_1, \ldots, \alpha_n)$ of total degree $\alpha_1 + \cdots + \alpha_n = k$ and each $i \in \{1, \ldots, n\}$ define an element $p_{\alpha,i}$ of P_k by

$$p_{\alpha,i}(x_1 e_1 + \cdots + x_n e_n) = x_1^{\alpha_1} \cdots x_n^{\alpha_n} e_i.$$

The $p_{\alpha,i}$ so obtained clearly form a basis for P_k. We shall see that each of the $p_{\alpha,i}$ is in fact an eigenvector of a_k. Fix α and i and write $p = p_{\alpha,i}$. First of all, we have $L \cdot p(x) = \lambda_i p(x)$ for all x. Moreover, the partial derivatives of p at the point $x = x_1 e_1 + \cdots + x_n e_n$ satisfy the identities $x_j \partial p(x) / \partial x_j = \alpha_j p(x)$, which implies that $p'(x) \cdot L(x)$ is equal to $\mu p(x)$ with $\mu = \alpha_1 \lambda_1 + \cdots + \alpha_n \lambda_n$. Specifically we have

$$\alpha_k(p_{\alpha,i}) = \big(\lambda_i - (\alpha_1 \lambda_1 + \cdots + \alpha_n \lambda_n)\big) p_{\alpha,i}.$$

Therefore, assuming that L is diagonalizable, we have proved that the endomorphism a_k is diagonalizable and we have exhibited its set of eigenvalues. To say that a_k is surjective means that none of its eigenvalues is zero. This points to the following general definition.

Definition 8.10.4. Let $(\lambda_1, \ldots, \lambda_n)$ be the list of eigenvalues (real or complex) of the endomorphism $L \in \mathrm{End}(E)$, each repeated a number of times equal to its multiplicity. Let k be an integer ≥ 2. We say that L has a *resonance of order k* if there exists an integer $i \in \{1, \ldots, n\}$ and a multi-index $(\alpha_1, \ldots, \alpha_n)$ with $\alpha_1 + \cdots + \alpha_n = k$ and

$$\lambda_i = \alpha_1 \lambda_1 + \cdots + \alpha_n \lambda_n.$$

With this terminology we have therefore proved (for L diagonalizable) that the presence of a resonance of order $k \leq r$ is equivalent to the non-surjectivity of a_k, and this therefore implies that L is not C^r-rigid. This remains true without the assumption on L (exercise) and so we obtain:

Proposition 8.10.5. *For the endomorphism L to be C^r-rigid it is necessary and sufficient that it have no resonance of order k with $2 \leq k \leq r$.* □

Suppose that L is diagonalizable and has a resonance of order $k \geq 2$, namely $\lambda_i = \alpha_1 \lambda_1 + \cdots + \alpha_n \lambda_n$. We have seen that a_k is diagonalizable and that the element $p_{\alpha,i}$ of P_k is an eigenvector for the eigenvalue 0. Hence $p_{\alpha,i}$ does not belong to the image of a_k, the vector fields L and $X = L + p_{\alpha,i}$ are not C^k-conjugate and X is not C^k-linearizable. In the eigenbasis that we are using we have

$$X_j(x_1, \ldots, x_n) = \lambda_j x_j, \quad j \neq i,$$
$$X_i(x_1, \ldots, x_n) = \lambda_i x_i + x_1^{\alpha_1} \cdots x_n^{\alpha_n}.$$

This is described by saying that the presence in X of a 'resonant monomial' of degree k prevents the C^k-linearization. This was precisely how the counterexample in 4.3 was constructed: there the resonance is $\lambda_1 = 2\lambda_2$ and the resonant monomial is $(x, y) \mapsto (y^2, 0)$.

We shall return to consider dimension 2 in some detail in Sect. 12. Meanwhile, here are some obvious cases of resonances:

a) If L is not injective it has resonances of order 2 of the form $\lambda = 1.0 + 1.\lambda$ and is therefore not C^2-rigid.

b) If L has two opposite eigenvalues it has resonances of order 3 of type $\lambda = 2.\lambda + 1.(-\lambda)$, and is therefore not C^3-rigid.

8.11 Differentiable Linearization: the Theorems of Sternberg and Hartman

The calculations in the previous section have given us a necessary condition for linearizability: the absence of resonance (Proposition 10.5). Clearly, what are important for us are sufficient conditions. First let us take the C^∞ case. Proposition 10.5 then has the following converse, a difficult result that we shall assume.

Theorem 8.11.1. *For an endomorphism L to be C^∞-rigid it is necessary and sufficient that it have no resonance of order ≥ 2.*   □

Similarly:

Theorem 8.11.2. (Sternberg's Linearization Theorem.) *If X is of class C^∞ and if the linearized endomorphism $\Lambda_a X$ has no resonance of order ≥ 2 then the singularity (V, a, X) is C^∞-linearizable.*   □

The extension of Sternberg's theorem to finite differentiability classes is quite subtle. For example, in Theorem 11.2 if we assume only that X is of class C^r with r sufficiently large (and that there is no resonance), it is possible to prove the existence of a linearization of class C^s, where s depends on r and tends to infinity with r (see [NE], for example).

It is useful to know two statements about C^1 linearization, due to Hartman and contained in the following general criterion:

Theorem 8.11.3. *If L has no purely imaginary eigenvalue and if it has no resonance of order 2 of the form $\lambda_i = \lambda_1 + \lambda_2$ with $\Re(\lambda_1) < 0$ and $\Re(\lambda_2) > 0$, then it is C^1-rigid.*   □

The latter condition is automatically satisfied when the eigenvalues of L all have real part < 0 or all have real part > 0. It is also satisfied when E has dimension 2, since the resonance relation requires three eigenvalues to be distinct. Thus on removing the intervening verbiage we obtain:

Corollary 8.11.4. (Hartman) *Let E be a finite-dimensional vector space, with V a submanifold of E, and let X be a vector field on V, of class C^1, having a singular point a. Let $L = \Lambda_a X \in \operatorname{End}(T_a V)$ be the linearization of X at a. Assume one of the following:*

a) *all the eigenvalues of L have real part < 0,*

b)   *all the eigenvalues of L have real part > 0,*
c) $\dim_a V = 2$ *and no eigenvalue of L is purely imaginary.*

Then (V, a, X) is C^1-linearizable: there exists an open set Ω of $T_a V$ containing 0, an open subset U of V containing a and a diffeomorphism $h : \Omega \to U$ of class C^1 such that $h(0) = a$, $T_a h = Id$ and $h_*(L) = X$.   □

Note that a) (or b)) implies and sharpens Theorem 5.3 (or 5.4). The corollary shows that to find a counterexample to C^1-linearization for a hyperbolic singular point we need to go up to dimension 3, taking for example the resonance $1 = 2 + (-1)$ (exercise).

8.12 Linearization in Dimension 2

Suppose $\dim_a V = 2$, and to fix the ideas assume that X is of class C^∞. Denote the linearization of X at the singular point a by L, an endomorphism of the 2-dimensional vector space $T_a V$. Suppose that a is nondegenerate, that is L is invertible. First take the hyperbolic case, so that the eigenvalues λ and μ of L are not purely imaginary. A resonance may be written either as $\lambda = a\lambda + b\mu$ or as $\mu = a\lambda + b\mu$ with $a + b = k \geq 2$. We immediately deduce that the resonant cases are as follows: λ/μ is an integer > 1, or μ/λ is an integer > 1, or λ/μ is rational and < 0. In all these cases λ and μ are real. Having noted all this, now return to the classification given in Sect. 7.12. Observe in passing that the trace of L is the value at the point a of the *divergence* of X.

8.12.1. The simplest case is the *focus*, which is characterized by the condition $0 \neq \operatorname{Tr}(L)^2 < 4\det(L)$. The eigenvalues of L are neither real nor purely imaginary. There is no resonance possible: by Theorem 11.2 we therefore have C^∞ linearization. Thus there exists a system of local coordinates on V centred at a in which X is linear, and hence in which X and its integral flow can be expressed by the formulae given in 7.12.4, namely

$$X(x, y) = (\sigma x - \tau y, \tau x + \sigma y),$$
$$\Phi_t(x, y) = e^{\sigma t}(x\cos(\tau t) - y\sin(\tau t), x\sin(\tau t) + y\cos(\tau t)).$$

8.12.2. The next case in order of complexity is the *node*, characterized by the condition $0 < \det(L) \leq \operatorname{Tr}(L)^2/4$. The eigenvalues of L are then real and of the same sign, and the only possible resonances are of the form $\lambda_1 = k\lambda_2$ and $\lambda_2 = k\lambda_1$ with k an integer ≥ 2. If there is no resonance there is C^∞ linearization and there exists a system of local coordinates on V centred at a in which X is linear, and hence in which X and its integral flow can be expressed by the formulae given in 7.12.3. Suppose there is resonance of order $k \geq 2$. The eigenvalues of L are thus of the form $\lambda_1 = k\lambda$ and $\lambda_2 = \lambda$. By a polynomial change of coordinates of degree $\leq k$ we can then bring the Taylor expansion of X up to order k into the form

$$X(x,y) = (k\lambda x + \alpha y^k, \lambda y) + \cdots .$$

If α is nonzero, a suitable rescaling in the x variable reduces us to the case where $\alpha = \lambda$. For ε taking values 0 or 1, let X^ε denote the vector field

$$X^\varepsilon(x,y) = (kx + \varepsilon y^k, y) \qquad (8.12.1)$$

corresponding to the differential system

$$\dot{x} = kx + \varepsilon y^k, \quad \dot{y} = y$$

with integral flow

$$\Phi_t^\varepsilon(x,y) = (e^{kt}(x + \varepsilon t y^k), e^t y). \qquad (8.2.2)$$

We have therefore just seen that if the eigenvalues of L are λ and $k\lambda$ then there exists a system of local coordinates (x,y) on V centred at a in which the Taylor expansion of X up to order k is of the form $\lambda X^\varepsilon + \cdots$, for $\varepsilon = 0$ or $\varepsilon = 1$. In fact a more precise study enables us to prove that X is C^∞-conjugate to λX^ε: we can choose a system of local coordinates such that there is no remainder in the above expansion.

To summarize, the C^∞ classification of nodes is as follows: apart from the linearizable nodes, for each resonance possibility there is a single class of non-linearizable nodes, all C^∞-conjugate.

8.12.3. Now we turn to the case of *saddle points*, characterized by the condition $\det(L) < 0$. The eigenvalues of L are then real and of opposite sign, so the numerical type of the singular point is $(1,1)$. From the general theory we have stable and unstable manifolds which are 1-dimensional and of class C^∞. The stable manifold $V^s(a)$ is the union of $\{a\}$ and two orbits of X, each of which is tangent at a to one of the semi-axes defined by the line T^s. A similar statement holds for $V^u(a)$. The orbits so obtained are called the *separatrices* of the saddle point a. There are two, three or four of them. By Corollary 11.4 and Theorem 11.2 there is always C^1 linearization, and there is C^∞ linearization if the ratio of the two eigenvalues of L is irrational. Now assume the opposite case: let λ and μ denote the eigenvalues of L, and let p and q be coprime integers > 0 such that $p\lambda + q\mu = 0$. The resonances are therefore $\lambda = (1+ps)\lambda + qs\mu$ and $\mu = ps\lambda + (1+qs)\mu$, where s runs through the integers ≥ 1, and there are infinitely many resonant monomials. This makes precise analysis much more difficult than in the case of nodes. In any case, what can be deduced from 10.2-b) is that if we fix an integer r then we can choose a system of local coordinates (x,y) centred at a in which X takes the form

$$\dot{x} = x\left(q\theta + \sum_{s=1}^{r} \alpha_s (x^p y^q)^s + \cdots \right)$$

$$\dot{y} = y\left(-p\theta + \sum_{s=1}^{r} \beta_s (x^p y^q)^s + \cdots \right),$$

with the remaining terms being of order $> (r+1)(p+q)$.

8.12.4. Now let us turn to the nonhyperbolic case where L has a *centre*. This case is characterized by the conditions $\text{Tr}(L) = 0$ and $\det(L) < 0$. The eigenvalues of L are purely imaginary. In a suitable system of local coordinates the vector field can be expressed as

$$X(x,y) = (-\omega y + a(x,y),\ \omega(x,y) + b(x,y)) \tag{8.12.3}$$

where $\omega > 0$ and the functions a and b are of order ≥ 2 at the origin. We shall attack the problem in two different ways, first by the technique already used above and then by a direct approach.

8.12.5. With the notation of 10.1 we have

$$L = \begin{pmatrix} 0 & -\omega \\ \omega & 0 \end{pmatrix} ;$$

the vector space P_k consists of the maps

$$h\begin{pmatrix} x \\ y \end{pmatrix} = \begin{pmatrix} A(x,y) \\ B(x,y) \end{pmatrix} \tag{8.12.4}$$

where A and B are homogeneous polynomials of degree k, and the endomorphism a_k is

$$a_k(h) = \begin{pmatrix} 0 & -\omega \\ \omega & 0 \end{pmatrix} \begin{pmatrix} A \\ B \end{pmatrix} - \begin{pmatrix} A'_x & A'_y \\ B'_x & B'_y \end{pmatrix} \begin{pmatrix} -\omega y \\ \omega x \end{pmatrix}$$
$$= \omega \begin{pmatrix} -B - x A'_y + y A'_x \\ A - x B'_y + y B'_x \end{pmatrix} . \tag{8.12.5}$$

Calculations show (exercise) that the kernel of a_k consists of the elements

$$h\begin{pmatrix} x \\ y \end{pmatrix} = (x^2 + y^2)^m \begin{pmatrix} \alpha x - \beta y \\ \beta x + \alpha y \end{pmatrix}$$

with $k = 2m + 1$, and the kernel and the image of a_k are complementary subspaces. It then follows, by arguments as in 10.1, that for each integer m we can choose a system of local coordinates (x, y) (of class C^m), centred at a, in which X may be written

$$\dot{x} = -(\omega + P(x^2 + y^2))y + Q(x^2 + y^2)x + \cdots$$
$$\dot{y} = (\omega + P(x^2 + y^2))x + Q(x^2 + y^2)y + \cdots$$

where P and Q are polynomials of degree $\leq m$ without constant term, and where the remainder terms are of order $> 2m + 2$.

8.12.6. Now let us return to the formula (12.3) and move to polar coordinates, putting $x = \rho \cos \theta$ and $y = \rho \sin \theta$. By Hadamard's Lemma (Proposition 4.2.3), the function $a(x, y)$ can be written in the form $ux^2 + vxy + wy^2$ where u, v and w are C^∞ functions, or alternatively as $\rho^2 c(\rho, \theta)$ with c of class C^∞. Likewise we write $b(x, y) = \rho^2 d(\rho, \theta)$. A direct calculation then gives

$$L_X \rho = \rho^2 (c \cos \theta + d \sin \theta), \quad L_X \theta = w + \rho(d \cos \theta - c \sin \theta).$$

The second equation shows that $L_X \theta$ is bounded below by a number > 0 when ρ is small enough. It follows that along every integral curve sufficiently close to the singular point, the polar angle θ increases indefinitely and the corresponding curves are the curves $\rho = \phi(\theta)$ obtained by solving the differential equation $d\rho/d\theta = L_X \rho / L_X \theta$ which is of the form $d\rho/d\theta = \rho^2 f(\rho, \theta)$. Fix a ray $\theta = \theta_0$; for sufficiently small $r > 0$ let $h(r)$ denote the value for $\theta = \theta_0 + 2\pi$ of that solution of the above differential equation that takes the value r at $\theta = \theta_0$. Clearly the function h is continuous and always > 0 (an orbit cannot pass through the singular point 0) and a little thought shows that it is strictly increasing (two orbits cannot cross; see Fig. 8.2a).

Let $r > 0$ be fixed, and let Ω be the orbit which cuts the fixed ray at the point r. The successive intersections of Ω with this ray are the points $\ldots, h^{-1}(r), r, h(r), h \circ h(r), \ldots$. Some important consequences of this are the following:

a) If $h(r) = r$ then Ω is periodic.
b) If $h(r) < r$ then the sequence $r, h(r), h \circ h(r), \ldots$ is strictly decreasing; let r_0 be its limit. If $r_0 = 0$ then Ω has 0 as future limit point; if $r_0 > 0$ then Ω is (future) asymptotic to the orbit Ω_0 of r_0, which is periodic. In both cases the orbit of every point of the segment (r_0, r) has the same destiny as Ω.
c) If $h(r) > r$ a similar argument applies to the sequence $r, h^{-1}(r), \ldots$.

8.12.7. In view of the above, there are three possible situations:

a) $h(r) = r$ for all sufficently small r. Here all the orbits sufficently close to r are closed and we have a *centre*.
b) $h(r) \neq r$ for all sufficiently small r. Then all the orbits sufficiently close to 0 have 0 as a limit point, and 0 is a *focus* which is attracting or repelling according to whether $h(r)$ is always $< r$ or always $> r$ for small r.
c) In the remaining cases there are infinitely many closed orbits. Around zero there are annular regions alternating between those in which all the orbits are closed (such a region may possibly reduce to a single orbit) and those bounded by two closed orbits between which all orbits are asymptotic to the two boundary orbits. We say that 0 is a *centre-focus* (see Fig. 8.2b). It is a singular point that is Lyapunov stable.

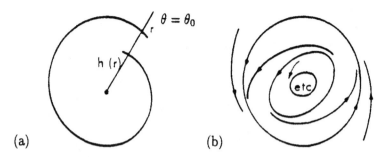

Fig. 8.2. If the linearized flow has a centre the nonlinear flow may have spirals or worse

Poincaré showed that an analytic vector field cannot have a centre-focus.

Note that when 0 is a focus it is not in general exponentially attracting or repelling. In fact, given the form of the differential equation, we see that typically the orbits are spirals of the form $\rho = 1/\theta$.

8.13 Some Historical Landmarks

The founders of the 'qualitative' analysis of differential systems are Henri Poincaré (1854-1912) and Alexander Lyapunov (1854-1918). We should also mention Ivar Bendixson (1861-1936) and George Birkhoff (1884-1944). Essentially it was celestial mechanics that motivated the early work, and this manifested itself simultaneously in the general framework (Hamiltonian mechanics), in the basic assumptions (everything has analytic coefficients) and in the methods (perturbation series). It was in this context that Poincaré established the first versions of the theorems in this chapter: for example, he proved the existence of separatrices of an analytic saddle point (1879) and the possibility of linearizing analytically an attracting analytic singular point (under the appropriate non-resonance assumptions).

It was only much later (1957-1962) that linearization and existence of stable and unstable manifolds – for a vector field and for a local diffeomorphism – were studied in the differentiable and topological contexts (Sternberg, Grobman, Hartman).

Likewise it was Poincaré who introduced the 'global' approach to differential systems, who identified the essential elements of phase portraits in the plane (singular points, closed orbits, limit cycles, separatrices ...) and who drew attention to the need to give high priority to studying those properties that we now call *generic* (particularly as regards 'multipliers' and their resonances).

For further study of the subjects treated in this chapter see one of the many texts on dynamical systems such as [AA],[GH],[IR],[KH], [PM] or [RO].

9 Closed Orbits – Structural Stability

9.1 Introduction

After singular points (that is, equilibrium positions) the most significant elements of the phase portrait of a vector field are the periodic or closed orbits (known also under the poetic name of *nonlinear oscillations*). They are studied by a method that goes back to Poincaré. This consists of choosing a point a of the closed orbit Ω, taking a small piece of hypersurface W through a and transverse to Ω, and for each $x \in W$ considering the first point $p(x)$ at which the orbit of x cuts W again (we say that p is the *Poincaré map*, or the *first-return map*). The fact that W was chosen transverse to Ω implies that p is well defined in a neighbourhood of the point a in W (clearly $p(a) = a$) and that p is a local diffeomorphism. The destiny of x is reflected in the successive intersections with $W : \ldots, p^{-1}(p^{-1}(x)), p^{-1}(x), x, p(x), p(p(x)), \ldots$ and knowledge of p enables us to reconstruct the dynamics in a neighbourhood of Ω, at least if we decide to disregard the time between two successive intersections, which amounts to working to within *orbital equivalence* (6.11.5, 6.11.6).

The theory developed in the previous chapter for singular points can be repeated for closed orbits: we replace the study of the integral flow of the field X in the neighbourhood of a singular point by the study of the 'discrete flow' formed by the iteration of p in a neighbourhood of the fixed point a. The role of the linearization $\Lambda_a X$ (or rather its exponential $\exp(\Lambda_a X)$) is played by the tangent map $T_a p$ of p at a, which is an automorphism of the vector space $T_a W$ tangent to W at a, this being a hyperplane (in the tangent space $T_a V$ of the ambient manifold V) complementary to the straight line in which $X(a)$ lies. Thus the notion of a *hyperbolic orbit* arises naturally (these are the ones for which $T_a p$ has no eigenvalue on the unit circle), and among these are the two extreme cases of attracting orbits (the spectrum of $T_a p$ lies in the interior of the unit disc) and repelling orbits (the spectrum of $T_a p$ lies outside the unit disc). Just as for singular points, there are linearization theorems that give the cases when p is equivalent to $T_a p$, with the same troubles concerning differentiablity (resonance among eigenvalues). As we are already working only to within orbital equivalence, there is no need for us to worry too much about differentiability problems and we shall be content with the purely topological result (Hartman). In dimension $n + 1$ (therefore with $\dim(W) = n$) we obtain $4n$ classes of hyperbolic orbits up to orbital equivalence (although here there is an annoying business to do with sign – the number reduces to $2n$ when the manifold is orientable). Moreover, the Stable Manifold Theorem that we have seen for singular points extends to diffeomorphisms and therefore to closed orbits. A hyperbolic closed orbit Ω has a *stable manifold* and an *unstable manifold* consisting of the points whose orbit (future or past, respectively) is asymptotic to Ω. These two (immersed) submanifolds meet transversely along Ω and often elsewhere. In the 'generic' case this additional intersection is the union of orbits asymptotic to Ω both in the future and in the past: the famous *homoclinic* orbits of Poincaré.

This study of closed orbits is one theme of this chapter. The other theme is that of *structural stability*. This has to do with characterizing those vector fields whose phase portrait remains unchanged (topologically) when the field is replace by a nearby one. By definition these form an open set in the space (yet to be defined) of all vector fields, and 'naturally' it is to be hoped that this set is dense. The situation turns out to be reasonable in dimension ≤ 2 (by Peixoto's Theorem 9.3 the structurally stable fields are dense and their phase portraits are characterized by simple topological properties). After that, things become much more complicated: structurally stable systems are no longer dense and moreover there exist some extremely bizarre ones! Nevertheless, it is possible to describe on the one hand *generic* properties which hold for a dense set of vector fields (the *Kupka-Smale Theorem*), and on the other hand a class of nice structurally stable fields (the so-called *Morse-Smale* fields).

As we shall see in the next chapter, this bad state of the fundamental scenario makes the study of bifurcations of phase portraits extremely difficult. The 'bifurcation-free' situation is still not fully understood, in contrast to that for fields derived from a potential where the theory of catastrophes is firmly anchored in the stability and deformation theorems of Mather.

The plan of the chapter is as follows. In Sects. 2 and 3 the Poincaré map is constructed and its main properties are verified. Attracting orbits are studied from an elementary viewpoint in Sect. 4, while Sects. 5 and 6 are devoted to hyperbolic orbits. Questions related to structural stability occupy Sects. 7 (generalities), 8 (the Kupka-Smale Theorem) and 9 (Morse-Smale fields). In Sect. 10 we give a brief historical survey, from Poincaré to Smale.

9.2 The Poincaré Map

Let E be a finite-dimensional vector space, with V a submanifold of E and X a vector field on V of class C^r, $r \in [1, \infty]$.

9.2.1. Let W be a submanifold of V and let a be a point of W. We say that W is *transverse to X at a* (or is a *transversal* to X at a) if a is not a singular point of the vector field X and the tangent subspace T_aW is complementary to the line $\mathbf{R}X(a)$ in T_aV, which implies in particular that W has codimension 1 in V at the point a. If W is transverse to X at a then it is transverse at nearby points. We say that W is *transverse* to X if it is transverse at all of its points. Through each nonsingular point there exist transversals to X.

Let (Φ_t) denote the integral flow of X. Let $\gamma : [0, T] \to V$ be an integral curve of X, joining the points $a = \gamma(0)$ and $b = \gamma(T)$ in V. We thus have $b = \Phi_T(a)$ and $a = \Phi_{-T}(b)$. Suppose a (and therefore b) is nonsingular, and fix a submanifold W_b of V transverse to X at b.

Lemma 9.2.2. *There exists an open interval J of $\mathbf{R}$ containing T, an open set U in V containing a and a C^r map $\tau : U \to J$ such that for each $x \in U$ the value $\tau(x)$ is the unique element of J such that $\Phi_{\tau(x)}(x)$ belongs to W_b.*

This is a slightly more elaborate form of the Straightening-out Theorem 6.8.2 which is proved by the same method. Let f be the map which associates to each point (t, y) of $\mathbf{R} \times W_b$ sufficiently close to $(0, b)$ the point $f(t, y) = \Phi_t(y)$ of V. We have $f(0, x) = x$ for all x, and in particular $f(0, b) = b$. A direct calculation as in Theorem 6.8.2 shows that the tangent map $T_{(0,b)}f$ is bijective and f is a local diffeomorphism at b. Therefore there exists a real number $\epsilon > 0$ and an open subset A of W_b containing b such that f induces a C^r diffeomorphism from $[-\epsilon, \epsilon] \times A$ onto an open subset U' of V containing b. We can choose ϵ and A small enough so that $W_b \cap U'$ consists just of A, and so that Φ_{-T} is defined on the whole of U'. Let $U = \Phi_{-T}(U')$ and $J = [T - \epsilon, T + \epsilon]$. The map $(t, y) \mapsto \Phi_{-t}(y)$ induces a diffeomorphism from $J \times A$ to U. For t and u in J and y in A the condition $(\Phi_u(\Phi_t(y)) \in W_b)$ is equivalent to $(u = t)$. The result follows, taking $\tau(\Phi_{-t}(y)) = t$. $\qquad\square$

9.2.3. For $x \in U$ let $g(x) = \Phi_{\tau(x)}(x) \in W_b$. By construction, $g(x)$ is the unique point of the orbit of x which belongs to W_b *and which can be written as $\Phi_t(x)$ for a time t that belongs to J* (there can certainly be other points of the orbit of x in W_b). The map g is C^r and the map $x \mapsto (\tau(x), g(x))$ is a diffeomorphism from U to $J \times A$, inverse to the diffeomorphism f. We have $g(a) = b$ and $\tau(a) = T$.

9.2.4. Now suppose we have also chosen a transversal W_a to X at a. Let $W'_a = U \cap W_a$; this is an open subset of W_a containing a, and the restriction of g is a C^r map $p : W'_a \to W_b$ such that $p(a) = b$. Applying the same construction to the integral curve $t \mapsto \gamma(T - t)$ of the vector field $-X$ we obtain an open subset W'_b of W_b containing b and a C^r map $q : W'_b \to W_a$ such that $q(b) = a$. We immediately verify that for $x \in W'_a$ sufficiently close to a we have $q(p(x)) = x$, and likewise $p(q(y)) = y$ for $y \in W'_b$ sufficiently close to b. Reducing W'_a and W'_b if necessary we may therefore suppose that p is a diffeomorphism from W'_a onto W'_b. It is called the (germ of) the *Poincaré* *diffeomorphism* or *Poincaré map* associated to the integral curve γ (or to the point a and the time T) and to the transversals W_a and W_b.

Note that no assumptions have been made about the integral curve γ. We could for example have $T = 0$ (therefore $b = a$); we could also have $b = a$ but $T \neq 0$ (the orbit of a is therefore periodic and T is a multiple of the period); we could also have $b \neq a$ but on a periodic orbit and with several circuits between a and b, and so on.

Lemma 9.2.5. *Suppose one of the following two assumptions holds:*

a) γ *is injective,*
b) *the restriction of γ to $[0, T)$ is injective, and $a = b$ and $W_a = W_b$.*

Then, for each $x \in W'_a$, if W'_a and W'_b are chosen sufficiently small, $\tau(x)$ is the smallest $t > 0$ such that $\Phi_t(x)$ belongs to W'_b.

In other words, $p(x)$ is the *first point* of the orbit of x which comes after x and belongs to W'_b.

If the conclusion were false we would be able to find a sequence (x_i) of elements of W'_a tending to a, and a sequence of real numbers t_i with $0 < t_i < \tau(x_i)$ and $\Phi_{t_i}(x_i) \in W'_b$ for all i. Since the sequence $\tau(x_i)$ tends to $\tau(a) = T$ by extracting a sub-sequence if necessary we can assume that the t_i converge to an element θ of $[0, T]$. Taking the limit, we obtain $\Phi_\theta(a) \in W'_b$ and therefore, if W'_b has been chosen small enough, $\Phi_\theta(a) = b$. Under assumption a) this implies $\theta = T$; under assumption b) it implies $\theta = T$ or $\theta = 0$; in the latter case we obtain $\theta = T$ by passing to the reverse vector field. We may therefore assume $\theta = T$; but then for i sufficiently large t_i and $\tau(x_i)$ both belong to J, which contradicts Lemma 2.2.   □

9.3 Characteristic Multipliers of a Closed Orbit

9.3.1. Consider a periodic orbit Ω of X that is not just one point, together with a point a of Ω and a transversal W to Ω at a (see Fig. 9.1). Let T denote the period of Ω, and let $p : W' \to W''$ be the (or a) Poincaré diffeomorphism associated to a, to the time T and to the transversal W. Recall that W' and W'' are open subsets of W containing a, and that p is a diffeomorphism with fixed point a such that $p(x)$ belongs to the orbit $\Omega(x)$ of x for all $x \in W'$. Moreover, from Sect. 2 above, $p(x)$ is distinguished among the points of $\Omega(x) \cap W'_a$ by either one of the two following properties: it can be written $\Phi_t(x)$ for t sufficiently close to T; it is the first point after x on $\Omega(x)$ (note that this characterization is preserved under orbital equivalence). Since a is a fixed point of p we can consider the derivative map $T_a p \in \text{End}(T_a W)$. Its eigenvalues are classically called the *characteristic multipliers* of the periodic orbit Ω. In fact we shall see below that they depend only on Ω and not on the auxiliary choices of a, W and W'.

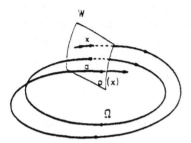

Fig. 9.1. Poincaré map p for the periodic orbit Ω

9.3.2. The eigenvalues can also be calculated as follows. The map Φ_T takes a to a. Its derivative at a is an endomorphism of the vector space $T_a V$ which can be decomposed into the direct sum of the line $\mathbf{R}X(a)$ and the hyperplane $T_a W$. We shall calculate the components of $T_a \Phi_T$ in this decomposition. First we have $\Phi_T(\Phi_t(a)) = \Phi_t(\Phi_T(a)) = \Phi_t(a)$; differentiating with respect to t at $t = 0$ we obtain

$$T_a \Phi_T \cdot X(a) = X(a).$$

Moreover, for $y \in W$ sufficiently close to a we have $\Phi_T(y) = \Phi_{T-\tau(y)}(p(y))$. Also $T - \tau(a) = 0$ and $p(a) = a$; differentiating the previous expression with respect to y at $y = 0$ we obtain

$$T_a \Phi_T \cdot \eta = T_a p \cdot \eta - \langle T_a \tau, \eta \rangle X(a), \quad \eta \in T_a W.$$

Specifically, the matrix of $T_a \Phi_T$ is triangular, with diagonal blocks that are 1 and $T_a p$ and hence:

Proposition 9.3.3. *The spectrum of $T_a p$ is obtained by deleting the eigenvalue 1 once from the spectrum of $T_a \Phi_T$.* $\qquad\square$

This implies that this spectrum, like that of $T_a \Phi_T$, is independent of the choice of $a \in \Omega$. To see this, write $b = \Phi_\theta(a)$ and let $\phi = T_a \Phi_\theta$. Differentiating the relation $\Phi_T = \Phi_\theta \circ \Phi_T \circ (\Phi_\theta)^{-1}$ we obtain $T_b \Phi_T = \phi \circ T_a \Phi_T \circ \phi^{-1}$, so that $T_a \Phi_T$ and $T_b \Phi_T$ are linearly conjugate.

This is perhaps the moment to make the following remark. The very strong parallelism between the theory for singular points and for closed orbits ought not to make us forget an essential concrete difference: to determine $T_a p$ or $T_a \Phi_T$ requires the *solution of the differential equation* (or at least a suitable linear equation with variable coefficients), while the analogous objects for singular points can be determined without any integration.

9.3.4. A manifold is said to be *orientable* if an orientation of each of its tangent spaces can be chosen in a continuous way. Every compact hypersurface is orientable; every curve is orientable. There exist nonorientable manifolds of dimension 2. Assume that V is orientable, and let each T_xV be given an orientation that depends continuously on x. Then each of the tangent maps $T_x\Phi_t : T_xV \to T_{\Phi_t(x)}V$ preserves orientation (this is true for $t = 0$ and remains so for all t by continuity). In particular, the endomorphism $T_a\Phi_T$ of T_aV preserves orientation, which means that its determinant is > 0. From Proposition 3.3 we therefore deduce

Proposition 9.3.5. *If V is orientable the automorphism $T_a p$ has determinant > 0.*   □

9.3.6. Now we give a classical example. Let F be a finite-dimensional vector space, with M a submanifold of F and T a real number > 0. Consider a time-dependent vector field on M, periodic with period T, which means a map $Y : M \times \mathbf{R} \to F$ of class C^1, say, such that $Y(y,t) \in T_yM$ and $Y(y,t+T) = Y(y,t)$ for all $(y,t) \in M \times \mathbf{R}$; note in passing that this applies in particular to a usual (time-independent) vector field.

Let $\omega : \mathbf{R} \to M$ be an integral curve of Y (so we have $\dot{\omega}(t) = Y(\omega(t),t)$ for all t) such that $\omega(t + T) = \omega(t)$ for all t; in particular this applies to a constant integral curve corresponding to a *singular point a* of Y (a point of M such that $Y(a,t) = 0$ for all t). We manufacture an extended phase space V and a vector field X on V as follows. Consider the circle $\mathbf{S}^1$ parametrized by an angle θ with values in $\mathbf{R}$ modulo 2π. Take V to be the product $\mathbf{S}^1 \times M$, for example embedded in $\mathbf{R}^2 \times F$, and let X be the field which associates to the point $(\theta, y) \in \mathbf{S}^1 \times M$ the vector $(\partial/\partial\theta, Y(y, 2\pi\theta/T))$. To each integral curve γ of Y there corresponds the integral curve $t \mapsto (t \bmod 2\pi, \gamma(2\pi t/T))$ of X and conversely. In particular, ω gives a periodic orbit Ω and, for all $\theta \in \mathbf{S}^1$, the manifold $\{\theta\} \times M$ is a transversal to Ω at $(\theta, \omega(\theta))$. The corresponding Poincaré map can be identified with the map from M to M (defined in a neighbourhood of $\omega(\theta)$) which takes $y \in M$ to the value at $t = \theta + T$ of the integral curve starting at y at time θ.

9.4 Attracting Closed Orbits

Theorem 9.4.1. *Let Ω be a periodic orbit of X whose characteristic multipliers all have absolute value < 1. Then there exists an open set U in V containing Ω such that every element $x \in U$ has the following two properties:*

a) *$\Phi_t(x)$ is defined for all $t \geq 0$;*
b) *there exists a point $b(x)$ of Ω such that the distance from $\Phi_t(x)$ to $\Phi_t(b(x))$ tends to 0 as $t \to +\infty$ (see Fig. 9.2).*

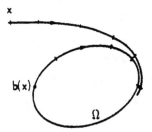

Fig. 9.2. Illustration of limiting phase

In the condition b) we use an arbitrary metric, for example defined by a norm on the ambient space E. We could express this condition without involving a metric (exercise). Moreover, $b(x)$ is clearly uniquely determined by x. We can see, either directly or by completing the proof which follows below, that the map $b : U \to \Omega$ is of class C^r. Theorem 4.1 is sometimes called the Limiting Phase Theorem.

The idea of the proof is simple. We choose a transversal W to X at a point a of Ω and let $p : W' \to W''$ be an associated Poincaré map. For $k = 0, 1, \ldots$ let p^k denote the kth iterate of p (defined in a neighbourhood of a). First take the case when x belongs to W. As the eigenvalues of the derivative endomorphism $T_a p \in \text{End}(T_a W)$ have absolute value < 1, the fixed point a of p is attracting and the successive images $p^k(x)$ tend to a. Then all that remains is to allow for the time the orbit of x takes in going between successive intersections with W.

Now we turn to the actual proof. Let $\alpha < 1$ be a real number such that $|\lambda| < \alpha$ for all $\lambda \in \text{Sp}(T_a p)$. By 7.5.1 there exists a norm on the vector space $T_a W$ for which $\|T_a p\| < \alpha$. Extend this norm to a norm on the space E. Since p is C^1 there exists $r > 0$ such that

$$\left(y \in W \text{ and } \|y - a\| < r \right) \Rightarrow \left(y \in W' \text{ and } \|p(y) - a\| \leq \alpha \|y - a\| \right). \quad (9.4.1)$$

Set $W(r) = \{ y \in W \mid \|y - a\| < r \}$. Let T be the period of Ω and let $A > T$ be arbitrary. If r is small enough, the map $(t, y) \mapsto \Phi_t(y)$ from $(0, A) \times W(r)$ to V is a local diffeomorphism at all points, and its image is therefore an open subset U of V that contains Ω. Every point x of U is of the form $\Phi_\theta(y)$ with $y \in W(r)$ and it suffices to show the required properties hold for y; then we shall take $b(x) = \Phi_\theta(b(y))$.

It follows from (4.1) that $p^k(y)$ is defined for all $y \in W(r)$ and for every integer $k \geq 0$, and that

$$\|p^k(y) - a\| \leq r\alpha^k, \quad y \in W(r), \qquad k \in \mathbf{N}. \quad (9.4.2)$$

In particular, $p^k(y)$ tends to a as $k \to \infty$.

The function $y \mapsto \tau(y)$ of 2.2 is of class C^1 and $\tau(a)$ is equal to T. Therefore if τ has been chosen small enough there exists a constant β with

$$|\tau(y) - T| \le \beta\|y - a\|, \quad y \in W(r). \tag{9.4.3}$$

Moreover, we may assume that r has been chosen so small that $T - \beta r > 0$, and then we have $T' \le \tau(y) \le T''$ for all $y \in W(r)$, with $T' = T - \beta r > 0$ and $T'' = T + \beta r$. By construction, $p^k(y)$ is the point $\Phi_{\tau(k,y)}(y)$ where

$$\tau(k, y) = \tau(y) + \tau(p(y)) + \cdots + \tau(p^{k-1}(y)). \tag{9.4.4}$$

Since $\tau(k, y)$ is bounded below by kT', it increases without bound as $k \to \infty$ and so $\Phi_t(y)$ is indeed defined for all $t \ge 0$. We have therefore proved a).

Now consider the series whose general term is $\tau(p^k(y)) - T$. Since

$$|\tau(p^k(y)) - T| \le \beta\|p^k(y) - a\| \le \beta r \alpha^k, \tag{9.4.5}$$

it is absolutely convergent and it follows from (4.4) that we can write

$$\tau(k, y) = kT - s(y) + \epsilon_k(y), \tag{9.4.6}$$

where $s(y)$ depends only on y (it is the negative of the sum of the series) and where $\epsilon_k(y)$ tends to 0 as $k \to \infty$. We then put $b(y) = \Phi_{s(y)}(a)$ and will verify condition b). By construction, we have

$$\Phi_{\tau(k,y)}(y) = p^k(y) \quad \text{and} \quad \Phi_{\tau(k,y)}(b(y)) = \Phi_{\epsilon_k(y)}(a).$$

As $k \to \infty$ these two points tend to a which implies

$$\lim_{k \to \infty} \|\Phi_{\tau(k,y)}(y) - \Phi_{\tau(k,y)}(b(y))\| = 0.$$

It remains only to pass to an arbitrary t, not necessarily of the form $\tau(k, y)$. Now the difference between $\tau(k, y)$ and $\tau(k + 1, y)$ is bounded above by T'' and therefore there exists, for all $t \ge 0$, an integer $k \ge 0$ with $\tau(k, y) \le t \le \tau(k + 1, y) + T''$. The uniform continuity of the Φ_t for $0 \le t \le T''$ then gives the result.   □

Changing X to $-X$ or t to $-t$ we obtain an analogous result for orbits all of whose characteristic multipliers have absolute value > 1. Moreover, it follows from (4.1) that Ω is the only closed orbit of X which meets U.

9.5 Classification of Closed Orbits and Classification of Diffeomorphisms

9.5.1. We continue with the notation of Sect. 3. Thus consider a periodic orbit Ω of X, a point a of Ω and a transversal W to Ω at a. We associate to this situation (the germ of) a Poincaré diffeomorphism $p : W' \to W''$. Note first of all that up to conjugacy p and $T_a p$ depend only on Ω and not on the auxiliary choices of a and W. To see this, consider another point a_1 of Ω

(possibly equal to a) and a transversal W_1 to Ω at a_1, and let $p_1 : W_1' \to W_1''$ be a Poincaré diffeomorphism associated to a_1, time T and W_1. Let T' be the smallest real number ≥ 0 such that $a' = \Phi_{T'}(a)$ and let r be a Poincaré diffeomorphism associated to a, T' and the pair of transversals (W, W_1). Then $r \circ p$ and $p_1 \circ r$ are two Poincaré diffeomorphisms associated to a, $T + T'$ and the transversals W and W_1. They therefore coincide in a neighbourhood of a and we have $p_1 = r \circ p \circ r^{-1}$ and $T_{a_1}p_1 = T_a r \circ T_a p \circ (T_a r)^{-1}$.

Let W and W_1 be two manifolds, with p and p_1 local diffeomorphisms of W and W_1, and let a and a_1 be points of W and W_1 respectively with $p(a) = a$ and $p_1(a_1) = a_1$. We say that (W, a, p) and (W_1, a_1, p_1) are *topologically conjugate* if there exists a homeomorphism r from a neighbourhood of a in W to a neighbourhood of a_1 in W_1 such that $r(a) = a_1$ and $p_1 \circ r = r \circ p$. With this terminology, *two Poincaré maps relative to the same orbit are topologically conjugate* (they are even 'C^r-conjugate', as we have just seen) *and their tangent endomorphisms are linearly conjugate*. We also recover the fact that the spectrum of the tangent endomorphism to the Poincaré map depends only on Ω.

9.5.2. Conversely, knowledge of the (conjugacy class of the) Poincaré map allows us to reconstruct the flow of X in a neighbourhood of Ω up to orbital equivalence.

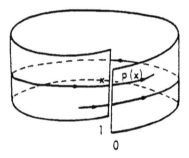

Fig. 9.3. Reconstructing the flow from a Poincaré map

To see this, suppose the map $p : W' \to W''$ is known. Consider the product $[0, 1] \times W'$, and for each $x \in W' \cap W''$ identify the points $(1, x)$ and $(0, p(x))$; removing some superfluous points we obtain a manifold $C(p)$ (of class C^r) equipped with an obvious vector field for which the curves $[0, 1] \times \{x\}$ are the integral curves (see Fig. 9.3). On taking each point (t, x) to $\Phi_{tr(x)}(x)$ we realise a homeomorphism from $C(p)$ to a neighbourhood of Ω in V taking orbits to orbits.

Note that this construction shows that *every local diffeomorphism (germ) having a fixed point is a Poincaré map*.

From this construction we can also deduce the following:

Proposition 9.5.3. *Let X_1 and X_2 be two C^1 vector fields (on manifolds V_1 and V_2). Let Ω_1, Ω_2 be periodic orbits of X_1, X_2 respectively, with Poincaré maps p_1, p_2. If the local diffeomorphisms p_1 and p_2 are topologically conjugate then the vector fields X_1 and X_2 are orbitally eqivalent in a neighbourhood of Ω_1 and Ω_2.*   □

9.5.4. Independently of this technical construction, it is easy to translate properties of the Poincaré map into properties of the flow in a neighbourhood of Ω. Here are some examples:

a) If $x \in W'$ is a fixed point of p, the orbit $\Omega(x)$ of x is periodic, with period close to T.
b) If $x \in W'$ is a periodic point of p with period k (by which we mean that $p(x), p^2(x), \ldots, p^{k-1}(x)$ are defined and distinct from x and that $p^k(x) = x$), then the orbit $\Omega(x)$ of x – which is also the common orbit of the points $p^i(x)$ – is periodic with period close to kT.
c) If $x \in W$ is such that $p^k(x)$ is defined for all $k \geq 0$ and tends to a as $k \to \infty$, then $\Phi_t(x)$ is defined for all $t \geq 0$ and $\Omega(x)$ is future asymptotic to Ω.
d) Replacing $p : W' \to W''$ by $p^{-1} : W'' \to W'$ we obtain the analogous statement for the past.

9.6 Hyperbolic Closed Orbits

Definition 9.6.1. A periodic orbit Ω is said to be *hyperbolic* if none of its characteristic multipliers belongs to the unit circle.

A hyperbolic periodic orbit Ω is said to be *attracting* (*repelling*) if all its characteristic multipliers lie inside (outside) the unit circle.

9.6.2. Let $a \in \Omega$. By Proposition 3.3, to say that Ω is hyperbolic amounts to saying that no eigenvalue of the endomorphism $T_a\Phi_T$ of T_aV belongs to the unit circle, with the exception of 1, which is of multiplicity 1 and is associated to the eigenvector $X(a)$. Following the construction of 7.3.5, we can therefore decompose T_aV into the direct sum of the line $\mathbf{R}X(a)$ tangent to Ω at a and contracting and expanding subspaces for $T_a\Phi_T$. Let T_a^s (T_a^u) denote the subspace of T_aV which is the direct sum of $\mathbf{R}X(a)$ and the contracting (expanding) subspace for $T_a\Phi_T$. We therefore have

$$T_a^s + T_a^u = T_aV, \quad T_a^s \cap T_a^u = \mathbf{R}X(a) = T_a\Omega.$$

These constructions do not depend on the chosen point a in the following sense: if $b = \Phi_\theta(a)$ is another point on the orbit Ω then the isomorphism $T_a\Phi_\theta$ from T_aV to T_bV, which as we have already seen transforms $T_a\Phi_T$ into $T_b\Phi_T$, takes T_a^s to T_b^s, T_a^u to T_b^u and $T_a\Omega$ to $T_b\Omega$.

If W is a transversal to Ω at a and if $p : W' \to W''$ is an associated Poincaré map then to say that Ω is hyperbolic means also that a is a hyperbolic fixed point (Definition 8.3.8) of the diffeomorphism p. The triangular structure of $T_a \Phi_T$ in the decomposition of $T_a V$ as the direct sum of $\mathbf{R} X(a)$ and $T_a W$ shows that the contracting (expanding) subspace of $T_a p$ is complementary to $\mathbf{R} X(a)$ in T_a^s (T_a^u).

9.6.3. If we use Proposition 5.3, Hartman's Topological Linearization Theorem 8.7.1 and the Linear Classification Theorem 7.11.2 we see that *the orbital equivalence class of X in a neighbourhood of Ω is given by the numerical type of $T_a p$.* The latter can take $4n$ distinct values, where $n = \dim(\Omega) = \dim_a V - 1$. If $m(\lambda)$ denotes the multiplicity of the characteristic multiplier λ, this numerical type is the quadruple (m_s, s_s, m_u, s_u) defined by (see 7.11.2)

$$m_s = \sum_{|\lambda|<1} m(\lambda) \in [0, n-1], \quad s_s = \prod_{|\lambda|<1} sign(\lambda)^{m(\lambda)} \in \{-1, 1\},$$

$$m_u = \sum_{|\lambda|>1} m(\lambda) \in [0, n-1], \quad s_u = \prod_{|\lambda|>1} sign(\lambda)^{m(\lambda)} \in \{-1, 1\}.$$

As we saw in Proposition 3.5, only the $2n$ types with $s_s = s_u$ can occur in an orientable manifold. For example, in dimension 2 (hence with $n = 1$) we find 4 types of hyperbolic orbit; two are attracting, two repelling. The four 'model' Poincaré maps are the scalar multiplications in $\mathbf{R}$ by $1/2$, $-1/2$, 2 and -2. If V is orientable only the two types corresponding to $1/2$ and 2 can occur.

9.6.4. Fix a norm on the ambient space E and let d denote the associated metric $(d(x, y) = \|x - y\|)$. Let

$$d(x, \Omega) = \inf_{a \in \Omega} d(x, a).$$

Since Ω is compact this lower bound is attained and there exists $a \in \Omega$ with $d(x, a) = d(x, \Omega)$. In particular, the condition $d(x, \Omega) = 0$ is equivalent to $x \in \Omega$.

The *stable manifold of the closed orbit* Ω is the set $V^s(\Omega)$ consisting of those $x \in V$ which satisfy the following conditions:

a) $\Phi_t(x)$ is defined for all $t \geq 0$,
b) $d(\Phi_t(x), \Omega)$ tends to 0 as $t \to +\infty$.

This is the same (exercise) as saying that the orbit of x has Ω as future limit set.

For $a \in \Omega$, let $V^s(a)$ denote the set of those $x \in V$ satisfying a) and such that $d(\Phi_t(x), \Phi_t(a))$ tends to 0 as $t \to +\infty$. Clearly we have $\Omega \subset V^s(\Omega)$ and $a \in V^s(a) \subset V^s(\Omega)$. With this notation the hypothesis of Theorem 4.1 can be written $T_a^s = T_a V$ and the conclusion can be stated as follows: U is equal to $U^s(\Omega)$ and is the union of the $U^s(a)$. The following theorem extends these results to the general case:

Theorem 9.6.5. (Stable Manifold Theorem for Closed Orbits.) *Suppose the periodic orbit Ω is hyperbolic.*

a) *Let U be a sufficiently small open set in V containing Ω. Then $U^s(\Omega)$ consists of the points x such that $\Phi_t(x)$ belongs to U for all $t \geq 0$. It is a C^r submanifold containing Ω. For all $a \in \Omega$ we have that $U^s(a)$ is a C^r submanifold with $T_a U^s(\Omega) = T_a^s$, and $T_a U^s(a)$ is complementary to $T_a \Omega$ in T_a^s.*

b) *$V^s(\Omega)$ is a C^r immersed submanifold which is the union of the $V^s(a)$ for $a \in \Omega$, these also being C^r immersed submanifolds.* □

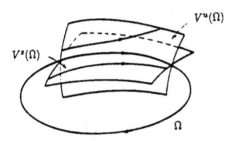

Fig. 9.4. Stable and unstable manifolds for a periodic orbit Ω

By changing t to $-t$ in the above definitions, or by reversing the vector field, we define the *unstable manifolds* $V^u(\Omega)$ and $V^u(a)$ and we have an analogous theorem. For U small enough, Ω is the transverse intersection of the manifolds $U^s(\Omega)$ and $U^u(\Omega)$. See Fig. 9.4.

9.7 Local Structural Stability

The results that follow make essential use of a natural topology on the space of C^r vector fields on a compact manifold V. For simplicity we shall restrict ourselves to the case $r = 1$. All the assertions remain valid for class $C^r, r > 1$.

9.7.1. Fix the vector space E and the submanifold V which we shall assume to be *compact*. Let $\Xi(V)$ denote the vector space of C^1 vector fields on V. There are many ways to define the C^1 topology on $\Xi(V)$; here is one of them. An element X of $\Xi(V)$ is in particular a C^1 map from V to E, which therefore has tangent maps $T_a X : T_a V \to E$. Fix a norm on E and give each $T_a V$ the induced norm. For each $a \in V$ we therefore have the norm $\|X(a)\|$ of the element $X(a)$ of E and the norm $\|T_a X\|$ of the linear map $T_a X$. Let

$$\|X\| = \|X\|_{C^1} = \sup_{a \in V}(\|X(a)\| + \|T_a X\|).$$

In this way we make $\Xi(V)$ into a complete normed space. Changing the norm initially chosen on E would replace the norm on $\Xi(V)$ by an equivalent norm.

9.7.2. Let L be an open set (or a submanifold) of a finite-dimensional vector space. A *family of vector fields on V parametrized by L* is a map $\lambda \mapsto X_\lambda$ from L into the space of vector fields on V. For $\lambda \in L$ and $x \in V$ let $\overline{X}(\lambda, x)$ denote the value of the vector field X_λ at the point x, so that $\overline{X}(\lambda, x) = X_\lambda(x)$. This then defines a map $\overline{X}$ from $L \times V$ into E such that $\overline{X}(\lambda, x)$ belongs to $T_x V$ for every pair (λ, x) or, what amounts to the same thing, a differential equation of the form

$$\dot{\lambda} = 0, \quad \dot{x} = \overline{X}(\lambda, x)$$

on the product $L \times V$. We shall say that the family X is of class C^1 if the map $\overline{X}$ is of class C^1. Then $X(\lambda)$ belongs to $\Xi(V)$ for all λ and we verify immediately that the map X from L into $\Xi(V)$ is continuous. Thus from a general topological statement about the space $\Xi(V)$ (for example: vector fields of a certain type form an open set in $\Xi(V)$) we shall often be able, by taking inverse images, to deduce an analogous statement about the family X (in the example above: the set of those λ for which $X(\lambda)$ is of that type forms an open set in L). The introduction of the function space $\Xi(V)$ and its topology can to a certain extent be regarded merely as a convenience, allowing us to state results about a family of vector fields without having to carry around the parameter space explicitly throughout the text.

The space $\Xi(V)$, being a complete normed space, is one to which Baire's Theorem (Sect. 3.2) can be applied: every subset of $\Xi(V)$ which contains the intersection of a countable family of open dense sets is itself dense. As in Definition 3.2.5, we shall say that such a subset is *residual* and that a property of a vector field is *generic* if it holds for a residual subset of $\Xi(V)$. A vector field is said to be structurally stable if its phase portrait is invariant under small deformations. To be precise:

Definition 9.7.3. The vector field $X \in \Xi(V)$ is *structurally stable* if there exists an open set L in $\Xi(V)$ containing X such that every element of L is orbitally equivalent to X.

In other words, the structurally stable vector fields are those whose orbital equivalence class is open. By construction they form an open subset of $\Xi(V)$ which is stable under orbital equivalence. One of the major problems is to know whether this subset is dense, which is otherwise expressed by asking if a generic vector field is structurally stable. Independently of whether this assertion is true or not (and we shall see later how we need to view it) the search for properties that characterize structurally stable vector fields is closely related to the search for generic properties of vector fields: if P is a generic property of vector fields then every structurally stable field X is orbitally equivalent to a field possessing the property P.

If we look at the phase portrait of a vector field the first features that we notice are the singular points and the periodic orbits, and it is through these that we usually begin the study. In fact in our case where V is compact every orbit that is closed (in the sense of 'closed subset of V') is either a singular point or is periodic (Proposition 6.11.3). Therefore we shall not be afraid to say *closed orbit* for 'periodic orbit' and we shall regard singular points as closed orbits of period zero.

9.7.4. Let us begin with an example. Take V to be the circle S^1 parametrized by an angle θ. Every vector field X on V can be identified with a function f by $X(\theta) = f(\theta)\partial/\partial\theta$. The presence of periodic orbits can be decided immediately: if f does not vanish then V is a periodic orbit of X, while if f has any zeros then X has no periodic orbit. The singular points of X are the zeros of f. The linearization of X at the singular point a is the scalar $f'(a)$. To say that a is a hyperbolic singular point of X therefore means that $f(a) = 0$ and $f'(a) \neq 0$; the point is repelling or attracting according to the sign of $f'(a)$.

A simple application of the Transversality Theorem shows that C^1 functions that have no common zero with their derivative form an open dense set. The set S of C^1 vector fields on V all of whose singular points are hyperbolic is therefore an open dense subset of $\Xi(V)$. Suppose that X belongs to S. It then has an even number $2n$ (possibly zero) of singular points. If $n = 0$ then V is a closed orbit. If $n > 0$ there are n attracting singular points and n repelling singular points, alternating around the circle. The domain of attraction of an attracting point is an open interval bounded by the immediately neighbouring repelling points; likewise for the repelling points. The phase portrait of X therefore depends only on the integer n. The vector fields $X \in S$ having $2n$ singular points form an open subset S_n of S (apply the Implicit Function Theorem to the equation $f(a) = 0$, as in Proposition 7.7 below), and all the elements of S_n are orbitally equivalent. Therefore all the elements of S are structurally stable (and moreover it can be verified that there are no others). Hence the structurally stable vector fields form a dense subset of $\Xi(V)$.

Now we return to the general case. Let X be a vector field on V, with a a singular point of X and L the linearization of X at a. If a is not hyperbolic, we can find endomorphisms in $\mathrm{End}(T_a V)$ as close to L as we wish having hyperbolic flows of different types. It is not difficult to deduce from this that we can find vector fields, as close to X as we wish in $\Xi(V)$, having hyperbolic singular points at a of different types and which are therefore not mutually orbitally equivalent. This precludes X from being structurally stable. Thus we are able to obtain the following result:

Lemma 9.7.5. *Let X be a structurally stable vector field. Then all the singular points of X are hyperbolic.* □

An 'argument' of the same kind seems to suggest that periodic orbits of a structurally stable vector field are hyperbolic; this is true (and difficult) in the case $r = 1$ which is where we are working, but the question remains open in the general case.

On the other hand, we have some quite elementary results about 'local structural stability' in the neighbourhood of a hyperbolic singular point (or closed orbit). Consider a vector field $X_0 \in \Xi(V)$ and a point $a_0 \in V$.

Proposition 9.7.6. *Suppose a_0 is non-singular point of X_0. Then there exists an open subset U of V containing a_0 and an open subset Λ of $\Xi(V)$ containing X_0 such that every $X \in \Lambda$ has no singular point in U.*

Proposition 9.7.7. *Suppose a_0 is a non-degenerate (resp. hyperbolic) singular point of X_0. Then there exists an open subset U of V containing a_0, an open subset Λ of $\Xi(V)$ containing X_0 and a continuous map $h : \Lambda \to U$ such that for all $X \in \Lambda$ the point $h(X)$ is the unique singular point of X in U. Moreover, this singular point is non-degenerate (resp. hyperbolic of the same type as a_0).*

Proposition 9.7.8. *Suppose the orbit Ω_0 of a_0 for X_0 is periodic and hyperbolic. Then there exists an open subset U of V containing a_0, an open subset Λ of $\Xi(V)$ containing X_0 and a continuous map $h : \Lambda \to U$ such that for all $X \in \Lambda$ the orbit of $h(X)$ for X is periodic and contained in U. Moreover, this orbit is hyperbolic and of the same type as Ω_0.*

Proposition 7.6 is a triviality: the map $u : (X, a) \mapsto X(a)$ from $\Xi(V) \times V$ to E is continuous (no need for the C^1 topology for that!); it is nonzero at the point (X_0, a_0), and therefore in a neighbourhood. To prove Proposition 7.7 we apply the Implicit Function Theorem. The map u is C^1, it vanishes at the point (X_0, a_0); its derivative at this point with respect to the second variable is the linearization of X_0 at a_0 and is therefore invertible by assumption. The Implicit Function Theorem thus gives existence and uniqueness, for X close to X_0, of a solution $a = h(X)$ of the equation $X(a) = 0$. The map h is C^1 and therefore continuous. Finally, the linearization of X at $h(X)$ depends continuously on X and is therefore invertible (resp. hyperbolic and of the same type as that of X_0) when X is sufficiently close to X_0. For Proposition 7.8 we argue similarly (exercise): we fix a transversal W to X_0 at a_0, associate to each X close enough to X_0 the Poincaré map p_X relative to X and W and solve the equation $p_X(x) = x$ by the Implicit Function Theorem.   □

In particular we conclude from Proposition 7.7 that if all the singular points of X_0 are non-degenerate (and in particular if they are all hyperbolic) then they are isolated and therefore finite in number (since V is compact).

It does no harm to note that we have not proved (fortunately, because it is false!) the analogous assertion for periodic orbits. While it is certainly true that in each of the three cases envisaged the vector field X_0 has no periodic orbit entirely contained in the open set U (with the obvious exception of Ω_0 in the third case), there could exist some which meet this open set without being contained in it. Moreover, in the third case it could happen that the Poincaré map p_X possessed periodic points other that $h(X)$, which would give rise to periodic orbits in U distinct from the one corresponding to $h(X)$; note, however, that such an orbit would have a period close to a multiple of the period T of Ω_0, and therefore greater than $2T - \epsilon$. In fact what is true in each of the three cases is that by taking U sufficiently small we could require every periodic orbit meeting U (and distinct from Ω_0 in the third case) to have period greater than an arbitrary constant fixed in advance. It follows from this that if all the closed orbits (singular points and periodic orbits) of X_0 are hyperbolic then X_0 has only a finite number of closed orbits of period $\leq A$ for each chosen A. In particular, there is only a countable number of closed orbits in total.

If we add to the previous results the corresponding results on orbital equivalence (Theorem 6.8.2, Corollary 8.7.3 and 6.3 respectively) we see that in each of the three cases X_0 is 'locally structurally stable' in a neighbourhood of the relevant set ($\{a_0\}$ in the first two cases, Ω_0 in the third).

9.7.9. We end this section with a caution. The definition of structural stability given above refers to the set of all vector fields. In particular, this excludes the possibility of a conservative vector field being structurally stable. More precisely, *a structurally stable vector field possesses no non-constant first integral* (Thom).

Clearly we could restrict the ambient category of vector fields and consider, for example, only Hamiltonian fields. This gives a notion of structurally stable Hamiltonian field. Here we get into a very difficult theory, in which the first result is the famous KAM Theorem, the *invariant torus theorem* of Kolmogorov-Arnol'd-Moser. See for example [AK], [AM] or [BO].

9.8 The Kupka-Smale Theorem

If we pursue the topological analysis of a phase portrait beyond the closed orbits, the next things we see are the stable and unstable manifolds of these orbits. By construction, the stable manifold of two distinct orbits are disjoint (since the future semi-orbit of a point of intersection would be asymptotic to two distinct closed orbits). On the other hand, a stable manifold $V^s(\Omega)$ and an unstable manifold $V^u(\Omega')$ can intersect (if $\Omega = \Omega'$ they meet in any case along Ω but can also meet elsewhere). Suppose this is what happens. We then imagine naturally enough that it ought to be the case that

a) if this intersection is not transverse we could make it transverse by a small modification of X,
b) if this intersection is transverse then it will not be altered topologically by slight changes in X.

This leads us to introduce the following definition:

Definition 9.8.1. The vector field X is said to be of *type* KS (for Kupka-Smale) if:

a) all the closed orbits (singular points and periodic orbits) of X are hyperbolic;
b) for each pair (Ω, Ω') of closed orbits the manifolds $V^s(\Omega)$ and $V^u(\Omega')$ are transverse.

We shall have some comments to make below on the condition b). In any case, the previous discussion lends plausibility to the following two assertions: the fact of being KS is generic, and a structurally stable vector field is KS. While the second statement does pose some problems (it is true in the case $r = 1$ which is where we are, but the question remains open in the general case), the first statement is true without restriction:

Theorem 9.8.2. (Kupka-Smale) *The vector fields of type KS form a residual (and therefore dense) subset of $\Xi(V)$.*   □

Before giving explicit examples, let us go back to the condition b) in the definition. We see immediately that it is automatically satisfied if Ω or Ω' is attracting or repelling, since one of the stable or unstable manifolds either reduces to the corresponding closed orbit or is an open set. This implies that there is nothing to verify in dimension 1 and that the only cases to consider in dimension 2 are those where Ω and Ω' are both saddle points.

9.8.3. Let us look at the general situation in a little more detail. Let Ω and Ω' be two closed orbits, assumed to be hyperbolic; let $A = V^s(\Omega) \cap V^u(\Omega')$. There is nothing to say if A is empty. Also, if $\Omega = \Omega'$ then A contains Ω, and there is nothing to say if A consists of just Ω. Suppose we are in one of the other situations and let x be a point of A (which does not belong to Ω if $\Omega = \Omega'$). Then the orbit $\Omega(x)$ of x is contained in A and is asymptotic to Ω in the future and to Ω' in the past (it follows from this incidentally that x is not singular and hence that $\dim_x(A) > 0$). In the particular case when $\Omega = \Omega'$ (and x does not belong to Ω) we obtain what is called, following Poincaré, a *homoclinic* orbit (see for example the Figures in Sect. 8.8).

This having been said, the transversality condition b) prevents certain situations from arising. Suppose that the intersection A is transverse and non-empty (and not just Ω if $\Omega = \Omega'$). Then we have $\dim(A) > 0$; letting n denote the dimension of V (at the points of A) we have on the one hand

$$\dim(V^s(\Omega)) + \dim(V^u(\Omega')) = n + \dim(A)$$

by transversality, and on the other hand

$$\dim(V^s(\Omega)) + \dim(V^u(\Omega)) = n + \dim(\Omega),$$
$$\dim(V^s(\Omega')) + \dim(V^u(\Omega')) = n + \dim(\Omega').$$

These imply the relations

$$\dim(A) = \dim(V^s(\Omega)) - \dim(V^s(\Omega')) + \dim(\Omega')$$
$$= \dim(V^u(\Omega')) - \dim(V^u(\Omega)) + \dim(\Omega).$$

Evidently this is impossible if Ω and Ω' are singular points of the same type (since $\dim(A) > 0$); in particular there can be no homoclinic orbit for a singular point. In the case when $\Omega = \Omega'$ is a periodic orbit we must have $\dim(A) = 1$ and A is a disjoint union of homoclinic orbits.

9.8.4. Now we look at the simplest cases of Theorem 8.2. For $\dim(V) = 0$ it is really too simple. Take V to have dimension 1 and to be connected. By Proposition 2.11.1 we can identify V with the circle $\mathbf{S}^1$; the condition b) of Definition 8.1 is automatic. Condition a) is the one we have met in 7.4. Therefore Theorem 8.2 reduces in this case to 7.4.

When $\dim(V) = 2$ matters become more subtle. Suppose X satisfies a). There are three possible types of singular point (attracting points, repelling points and saddles) and two types of periodic orbit (attracting and repelling – there being two types of each of these when V is not orientable). As we have seen, condition b) means that there is no orbit connecting two saddles (distinct or otherwise) which is to say that on the one hand the four separatrices of a saddle are distinct and on the other hand two distinct saddles have no separatrix in common.

9.9 Morse-Smale Fields

The classification of structurally stable vector fields in dimension 2 (see later) has led to the following definition (due to Smale, up to terminology):

Definition 9.9.1. The vector field $X \in \Xi(V)$ is said to be a *Morse-Smale* (MS) vector field if it satisfies the following conditions:

a) X has only finitely many closed orbits (singular points or periodic orbits) and they are all hyperbolic;
b) for each pair (Ω, Ω') of closed orbits, the manifolds $V^s(\Omega)$ and $V^u(\Omega')$ are transverse;
c) for every point $x \in V$ whose orbit is not closed, we can find an open set U containing x and a real number T such that $|t| > T$ implies $\Phi_t(U) \cap U = \emptyset$.

A vector field of MS type is therefore a KS field that has only a finite number of periodic orbits and which satisfies the extra condition c). This condition c) prohibits a disagreeable phenomenon: the existence of 'recurrent' non-closed orbits, which we now explain. An orbit Ω of X is said to be *recurrent* if it is contained in one of its limit sets. This means that there exists a point x in Ω and a family t_i of points in $\mathbf{R}$ tending to $+\infty$ or $-\infty$ such that the $\Phi_{t_i}(x)$ tend to x. This condition is then satisfied for every point x of Ω. Closed orbits are clearly recurrent and the condition c) prevents there being any others.

Let us look at the simplest cases. If $\dim(V) = 1$ only the condition a) counts, and we recover the vector fields all of whose singular points are non-degenerate. If $\dim(V) = 2$, for a KS vector field to be MS it is necessary and sufficient that every recurrent orbit be closed (this is hard to prove). On a manifold of dimension 2 the MS fields are therefore those that have the following three properties: their closed orbits are hyperbolic, there is no orbit connecting two saddles, every recurrent orbit is closed. On the sphere $\mathbf{S}^2$ there can exist no non-closed recurrent orbit (this is one of the consequences of the famous Poincaré -Bendixson Theorem) and hence every KS vector field is MS.

That having been said, the link between the notions of MS vector field and structurally stable vector field is given by the following two deep theorems:

Theorem 9.9.2. (Palis-Smale) *The MS vector fields form an open subset of* $\Xi(V)$*. Every MS field is structurally stable.* □

Theorem 9.9.3. (Peixoto) *Suppose V has dimension 2.*

a) *The structurally stable vector fields are precisely the MS vectorfields.*
b) *If V is orientable, these fields form a dense (open) subset of $\Xi(V)$.* □

There a few scattered comments to make. First, the case $\dim(V) = 1$ is elementary: for a vector field on $\mathbf{S}^1$ it is equivalent to be structurally stable, to be KS, to be MS or to have all its singular points nondegenerate, and these fields form an open dense subset of $\Xi(V)$ (Transversality Theorem). Also, part b) of Theorem 9.3 remains true if V is non-orientable, but this is a more delicate result (and known only for genus ≤ 3 in class $r > 1$: see [AA], p.226). Finally, the results obtained for vector fields on the sphere can be applied to vector fields on a disc that are transverse to the boundary.

As soon as $\dim(V) > 2$ everything goes wrong: the structurally stable vector fields do not form a dense subset of $\Xi(V)$ (Smale), and there exist structurally stable vector fields which are not MS, for example because their closed orbits fill a dense subset of V (Anosov).

It is instructive to look a little more closely at the case of gradient fields. Suppose E has a euclidean norm. Let $C^2(V)$ denote the (complete normed) space of C^2 functions on V. To each $\phi \in C^2(V)$ we associate its gradient $X = \mathrm{grad}(\phi) \in \Xi(V)$. We have $L_X\phi(x) = \|X(x)\|^2$, and in particular $\phi(\Phi_t(x)) > \phi(x)$ for every $t > 0$ if x is not singular. This implies first of all that X has no periodic orbit. Moreover, singular points of X are critical points of ϕ; the linearization of X at such a point is the endomorphism associated to the Hessian form (8.3.4) and therefore has all its eigenvalues real. Condition a) of Definition 9.1 thus means that f is a Morse function (Definition 4.4.4). Condition c) is automatically satisfied (exercise). In view of this we have:

Theorem 9.9.4. (Smale) *The set of those $\phi \in C^2(V)$ whose gradient is a MS vector field is open and dense.*   □

Thus *among the gradient fields* the MS fields and therefore also the structurally stable fields form a dense subset. It follows in particular that on all manifolds there exist MS fields and therefore structurally stable fields.

9.10 Structural Stability Through the Ages

It was in his work on celestial mechanics (1880-1890) that Henri Poincaré introduced and systematised the method of transversality. There he discovered the importance of homoclinic orbits as a 'route to chaos'.

The definition of structurally stable systems (under the name of *rough* systems) is due to Andronov and Pontryagin (1937). A sequence of articles on planar systems, from Poincaré (1899) and Bendixson to De Baggis (1952), culminated in 1962 with the classification by Peixoto of all the structurally stable systems in dimension 2. In higher dimensions, the fundamental examples of structurally stable systems of a different type are due to Thom and Anosov (structurally stable diffeomorphisms of the torus T^2 with dense periodic points, 1962) and Smale (the 'horseshoe', 1967). The fact that structurally stable vector fields in dimension n (and structurally stable diffeomorphisms in dimension $n - 1$) are not dense is due to Smale ($n \geq 4$, 1966, later extended to $n \geq 3$ by R.F.Williams), as also is the positive result for gradients (1967-1969). Since then a whole train of work has forged ahead in understanding generic properties of vector fields (and diffeomorphisms) and the characterization of structurally stable phase portraits, with key stability results due to Palis (1970), Robbin (1971), Robinson (1973-76) and eventually Mañé (1987). For details of this story see [RO] or [KH]. Much of the material of this chapter can also be found for example in [AA],[IR] or [PM].

10 Bifurcations of Phase Portraits

10.1 Introduction

The situation that we shall be concerned with in this chapter is the following: we consider a differential system that depends on auxiliary parameters (as in Chapt. 5, we may talk about control parameters, hidden parameters, imperfection parameters, ...) and we wish to understand how the phase portrait changes as the parameters vary. This is the question answered by catastrophe theory when we restrict to dissipative systems governed by a potential, and take as the only significant features of the phase portrait the equilibrium positions and their bifurcations.

It has been known for a considerable time that there are a certain number of typical situations (the 'saddle-node' and so-called 'Hopf' bifurcations) which enjoy a kind of universal character and seem to deserve the name *elementary* bifurcations. It is also known, mainly through numerical experiments, that there are situations of 'evolution towards chaos' which, despite their disordered appearance, seem likewise to play a universal role. Our aim will be to find to what extent the geometric approach (or perhaps rather the geometric 'viewpoint') enables us to understand the more or less inevitable nature of these bifurcations.

As we shall see, this 'top down' approach soon reaches its limits. There are several reasons for this. First of all, the phenomena themselves are very complicated; the considerable mathematical difficulties (loss of differentiability, resonances) force us to weaken significantly the results that we might naïvely hope for. Secondly, despite an explosive development in recent years the theory is still in its infancy and the specialists may not yet have found the right unifying concepts; it is even likely that not enough examples have so far been analyzed to give a fair idea of the generality of the phenomena encountered. Moreover, a context that now seems to be the 'right' one in many situations is that of *nonconservative perturbations of conservative systems* and in this text we do not have the necessary tools available. In any case, the aim of this chapter is only to show how the ideas we have developed previously can be applied and to what extent they are effective.

After having posed in Sect. 2 the general question of knowing just what we may call a bifurcation – which is not so simple – we introduce the basic technique called the *centre manifold* method which allows us to eliminate a 'transversely hyperbolic' part which plays no role in the bifurcation. This technique is described for singular points in Sect. 3 and briefly sketched for closed orbits in Sect. 6. It enables us to describe and justify fairly adequately (without rigorous proof) the simplest cases of bifurcations of codimension 1: the saddle-node and Hopf bifurcations for a singular point in Sects. 4 and 5, the saddle-node and period-doubling bifurcations for closed orbits in Sects. 7 and 8. The considerably more complicated case of Hopf bifurcation for a closed orbit is treated in Sect. 9.

All this concerns only *local* bifurcations, that is those that arise from a lack of hyperbolicity of a singular point or closed orbit (although, as we shall see in Sects. 9 and 11, the 'local/global' boundary is not so clear). As far as truly global bifurcations are concerned, we shall merely give in Sect. 11 a sample of homoclinic bifurcation treated by the Melnikov method. It arises from the study of a local codimension 2 bifurcation that we treat in Sect. 10 as a somewhat complicated example taken directly from [GH], pp.364-371.

10.2 What Do We Mean by Bifurcation?

Let us start from the catastrophe theory model. Suppose we are working in a compact manifold V. Fix a sufficiently large integer r and let $C^r(V)$ denote the space of C^r functions on V. The (structurally) stable functions (see 4.11.3) are characterized by the fact that all their critical points are Morse points and their critical values are pairwise distinct; they form an open dense subset of the space $C^r(V)$. A 'generic' function is therefore stable.

10.2.1. A family (f_μ) of elements of $C^r(V)$, parametrized by a point μ of a manifold M, is by definition a function (C^r, say) on the product $V \times M$, or alternatively a map (suitably regular) from M to $C^r(V)$. We say that this family exhibits a *bifurcation* for the parameter value μ_0 if the function f_{μ_0} has a critical point which is not Morse and is therefore not stable. We aim to describe the *bifurcation set* or *bifurcation locus* consisting of the bifurcation values of the parameter, and more precisely the subset of $V \times M$ consisting of pairs (x, μ) such that f_μ has a degenerate critical point at x (the *equilibrium manifold*). In the theoretical study we always assume that the family we are considering is *generic* (since we can arrange this to be the case by extending M – see the discussion in the Introduction on hidden parameters). The theorems that we have seen in Chapts. 3,4 and 5 then give us the following descriptions, in order of increasing complexity:

a) If M consists of just a point then there is no bifurcation.
b) Suppose M is a curve. Then M can be decomposed into two disjoint subsets M_0 and M_1. When μ belongs to the open dense subset M_0 all the critical points of f_μ are Morse. When μ belongs to the complement M_1, which is discrete (and therefore finite as M is compact), all the

critical points of f_μ except just one are Morse with the latter being a fold (codimension 1 singularity). The bifurcation locus is therefore M_1 and the (discrete) equilibrium manifold projects bijectively onto M_1.

c) If M is a surface it can be divided into three disjoint sets M_0, M_1 and M_2. When μ belongs to the open dense set M_0 all the critical points of f_μ are Morse. When μ belongs to the closed set M_1, which is a union (finite, as M is compact) of curves, all the critical points of f_μ except just one are Morse with the latter being a fold. Finally, when μ belongs to the discrete set M_2, either all the critical points of f_μ except two are Morse, with these two being folds, or all the critical points of f_μ except one are Morse and the latter is a cusp. The bifurcation locus is $M_1 \cup M_2$. The equilibrium locus is a curve. The part of this curve lying above M_1 projects bijectively onto M_1; above M_2 either we have two points or we have one point at which the tangent is vertical. The bifurcation set is therefore an immersed curve with regular part M_1 and with the points of M_2 being either double points or cusp points.

d) If M has dimension 3 the discussion is analogous; the additional feature is a discrete set M_3 of values of μ for which f_μ possesses a swallowtail, and so on.

10.2.2. In general the Transversality Theorem implies that the equilibrium manifold is indeed a submanifold of $V \times M$ of dimension equal to $\dim(M) - 1$. Moreover, the local situation in the neighbourhood of a point on the equilibrium manifold is given by Theorem 5.11.1: if for a value μ_0 of the parameter the element f_{μ_0} of the generic family (f_μ) has a codimension p singularity at x_0 (a Morse point if $p = 0$, a fold if $p = 1$, a cusp if $p = 2$, ...) then, in a neighbourhood of the pair (x_0, μ_0), the deformation f_μ is the sum of a deformation that is 'universal up to a constant' (hence with p parameters) and a 'fixed part'.

10.2.3. Transporting this approach just as it stands to the case of families of vector fields $M \mapsto \Xi(V)$ (notation as in 9.7.1) meets with two major obstacles.

The first difficulty is that strictly speaking there cannot exist any 'universal deformation' of a vector field, even in the simplest case. Suppose for example that the field X has a singularity at the point a that is as well-behaved as possible, let us say hyperbolic (the analogous situation for functions would be a function f having a non-degenerate critical point at a). We know (Proposition 9.7.7) that the situation will remain unchanged (locally) by deformation and therefore we shall have no bifurcation. Hence the universal deformation must be trivial (0 parameter). This is indeed what happens in the case of functions, where the Morse Lemma gives precisely a fixed form for f that depends only on the type of the point a (signature of the Hessian form).

Following this model, therefore, suppose that there exists a fixed form for all the vector fields sufficiently close to X; it is not hard to see that this form has to be the linearization of X at a, which implies that all the vector fields close to X have to be linearizable. But even if X itself is linearizable we cannot prevent certain of its neighbours from having resonances (Sect. 8.10), which stops them from being C^∞-linearizable.

This leads us to work 'up to topological equivalence' and therefore to look for a possible 'topologically universal deformation' of X. Unfortunately, the existence of such an object is not clear, except in the few very simple cases that we shall see in this chapter.

10.2.4. The second obstacle is the following. It is natural to say that the family under consideration exhibits a bifurcation when the phase portrait changes, that is when the corresponding element of the family is not structurally stable. However, since the structurally stable fields do not form a dense subset, there exist generic families of vector fields in which *no* element is structurally stable; for such a family, all the values of the parameter would therefore be bifurcation values.

This time the objection is more serious, and it forces us to modify our ambitions considerably. One solution is to make a list of 'generic' properties of phase portraits, and to say that we have a bifurcation when one of the properties on the list is not satisfied. In this way we can at least be sure that for a generic family the set of bifurcation values will be 'thin'.

A reasonable list to choose is the one given by the Kupka-Smale Theorem 9.8.2. Hence we make this rather conservative definition:

Definition 10.2.5. The family of vector fields $X(\mu)$ *exhibits a bifurcation* for the value μ_0 of the parameter if $X(\mu_0)$ is not of KS type.

Thus the phase portrait of the vector field $X(\mu_0)$ exhibits one of: a non-hyperbolic singular point, a non-hyperbolic periodic orbit, or a stable and an unstable manifold that meet non-transversely. In the first two cases we often say that we are dealing with a *local bifurcation*.

10.2.6. It is natural to study these bifurcations in order of complexity given by their 'codimension', first of all keeping only those which are inevitable in a generic 1-parameter family, then passing to two parameters, and so on. It is reasonable to conjecture (and it can in fact be proved, with suitable regularity assumptions) that the codimension 1 bifurcations are given by the following list. We let X denote the vector field exhibiting the bifurcation, and we indicate the *exception* to the properties of the class KS that characterizes this bifurcation:

a) The linearization of X at a singular point has 0 as an eigenvalue with multiplicity 1, all the other eigenvalues having nonzero real part.

b) The linearization of X at a singular point has a pair of conjugate purely imaginary and nonzero eigenvalues of multiplicity 1, all the other eigenvalues having nonzero real part.

c) One of the periodic orbits of X has 1 as a characteristic multiplier of multiplicity 1, with none of the other multipliers lying on the unit circle.

d) One of the periodic orbits of X has -1 as a characteristic multiplier of multiplicity 1, with none of the other multipliers lying on the unit circle.

e) One of the periodic orbits of X has a pair of complex conjugate multipliers of multiplicity 1 belonging to the unit circle, with none of the other multipliers lying on the unit circle.

f) One of the orbits of X is a nontransverse intersection of a stable manifold and an unstable manifold.

We already observe from this list the extent to which the situation is more complicated than it is for functions, where there is only one codimension 1 bifurcation, namely the fold.

10.3 The Centre Manifold Theorem

Consider a vector field X on V of class C^r, $r \in [2, +\infty]$, and a singular point a on V. Let $L \in \mathrm{End}(T_a V)$ denote the linearization of X at a. By 7.3.5 we can decompose $T_a V$ into the direct sum of three subspaces T_a^s, T_a^u and T_a^0 which are invariant under L and on which the restrictions of L have all their eigenvalues with real part negative, positive or zero, respectively.

Theorem 10.3.1. (Centre Manifold Theorem.) *There exist connected C^1 submanifolds $V^s(a)$, $V^u(a)$ and $V^0(a)$ of V that are tangent at a to the corresponding subspaces of $T_a V$ and are invariant under the flow of X (so they are everywhere tangent to X). The submanifolds $V^s(a)$ and $V^u(a)$ are uniquely determined and are of class C^r. For every $s < r$ (hence for $s = r - 1$ if r is finite, and for all s if $r = \infty$) we may require that the submanifold $V^0(a)$ be of class C^s on some neighbourhood of a.* $\square$

Here we have precisely the same difficulties as in Theorem 8.8.5: either we talk about immersed submanifolds, or we replace V by a sufficiently small open set containing a so that we can then deal only with genuine submanifolds.

10.3.2. As in Definition 8.8.1 and 9.6.4 we say that $V^s(a)$ and $V^u(a)$ are the *stable and unstable manifolds* of the singular point a (with respect to X). Any connected submanifold tangent to T_a^0 at a and invariant under the flow of X is called a *centre manifold* of a. In general such a manifold is not uniquely determined; moreover, it can be chosen to be of class C^r when $r < \infty$, but cannot necessarily be chosen to be C^∞ when X is C^∞. See [RO] or [MV] for further technical details.

Retaining the notation of the previous theorem we have:

Theorem 10.3.3. *In a neighbourhood of the point a the field X is orbitally equivalent to a vector field on $V^0(a) \times V^s(a) \times V^u(a)$ that is the product of its restrictions to the three subspaces.* □

First note that the vector fields obtained by restricting X to the stable and unstable manifolds have as their respective linearizations at a the restrictions of L to T_a^s and T_a^u, so they have a as a hyperbolic (attracting or repelling) singular point. Therefore the vector fields are orbitally equivalent to $y \mapsto -y$ and to $z \mapsto z$ respectively (Theorem 8.5.3 and Corollary 8.5.4). Thus Theorem 3.3 is equivalent to saying that X is orbitally equivalent to the vector field $(x, y, z) \mapsto (X(x), -y, z)$ on $V^0(a) \times \mathbf{R}^q \times \mathbf{R}^r$, with $q = \dim(T_a^s)$, $r = \dim(T_a^u)$. If a is hyperbolic we have $T_a^0 = \{0\}$ and so $V^0(a) = \{a\}$ and the preceding theorems recover the Grobman-Hartman Linearization Theorem 8.7.2 and the Stable Manifold Theorem 8.8.5.

10.3.4. The above theorems act as a substitute for a non-existent theorem of the type of Corollary 4.5.5 for eliminating superfluous variables en route to constructing (possibly) a 'topologically universal deformation' for phase portraits in the neighbourhood of a singularity. Consider a family $\mu \mapsto X(\mu)$ of vector fields on V parametrized by a manifold M, and a point $\mu_0 \in M$ such that the vector field $X(\mu_0)$ has a singular point a_0. Let p, q and r denote the dimensions of the subspaces $T_{a_0}^s$, $T_{a_0}^u$ and $T_{a_0}^0$. Apply the usual technique of eliminating the parameters (see 6.2.5) and taking as base manifold the product $V \times M$ and as vector field the field $\overline{X}(x, \mu) = (X(\mu)(x), 0)$ which has a singular point $a = (a_0, \mu_0)$. In $T_a(V \times M) = T_{a_0}V \times T_{\mu_0}M$, the linearization of $\overline{X}$ is the endomorphism taking each pair (ξ, η) to the pair $((\Lambda_{a_0}X_0) \cdot \xi), 0)$; its associated subspaces are therefore $T_{a_0}^s \times \{0\}$, $T_{a_0}^u \times \{0\}$ and $T_{a_0}^0 \times T_{\mu_0}M$. It follows that the stable and unstable manifolds of a in $V \times M$ are $V^s(a_0) \times \{0\}$ and $V^u(a_0) \times \{0\}$.

Let W denote a centre manifold of a in $V \times M$; then $T_a W$ is equal to $T_{a_0}^0 \times T_{\mu_0}M$, and the projection π from W to M is a submersion in the neighbourhood of a. The fibre $W_0 = W \cap \pi^{-1}(\mu_0)$ is a submanifold of V, tangent to T_a^0 at a_0 and invariant under the flow of $X(\mu_0)$; it is therefore a centre manifold of a_0 with respect to $X(\mu_0)$. Moreover, in a neighbourhood of a we can identify W and $W_0 \times M$ by a diffeomorphism; transporting the restriction of $\overline{X}$ via this diffeomorphism we obtain a family $(Z(\mu))$ of vector fields on W_0, parametrized by (a neighbourhood of m_0 in) M. Then applying Theorem 3.3 (or rather a slightly improved version guaranteeing an orbital equivalence compatible with the projections to M) we conclude that the given family $(X(\mu))$ is orbitally equivalent in a neighbourhood of (a_0, μ_0) to the family $(x, y, z) \mapsto (Z(\mu)(x), -y, z)$ on the product $W_0 \times \mathbf{R}^q \times \mathbf{R}^r$. In particular, if a_0 is hyperbolic we recover Proposition 9.7.7: the phase portrait (local in a neighbourhood of a_0) remains orbitally constant as μ varies in a neighbourhood of μ_0. In the general case we see that *the study of the bifurcation of the local phase portrait of the field X on V reduces to the analogous study for the field Z on the centre manifold W_0.*

Naturally, this does not all go through without some 'concrete' difficulties (such as finding W, finding Z, ...) as well as theoretical ones (determining the differentiability class of W, ...).

10.4 The Saddle-Node Bifurcation

10.4.1. The simplest local bifurcation occurs when the linearized vector field at the singular point has 0 as an eigenvalue of multiplicity 1, all the others having nonzero real part. The centre manifold technique seen above reduces the discussion to the case when V is of dimension 1. We assume moreover that the family of vector fields is a generic 1-dimensional family. Thus V and M are 1-dimensional, and we consider a point a_0 of V and a point m_0 of M such that

$$X(m_0)(a_0) = 0, \quad \varLambda_{a_0} X(m_0) = 0. \tag{10.4.1}$$

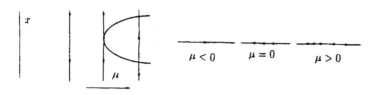

Fig. 10.1. Saddle-node bifurcation in one dimension

The equilibrium manifold is the subset C of the surface $V \times M$ defined by the equation $X(m)(a) = 0$. The genericity assumption and the above condition imply that C is a curve and that the projection of C on M has a fold at the point (a, m). We can therefore find local coordinates x on $V \times M$ and μ on M, valid in neighbourhoods of (a_0, m_0) and m_0 respectively, centred at these points and such that C is defined by the local equation $x^2 = \mu$. By changing the sign of X if necessary we can assume that for $\mu < 0$ the flow of $X(\mu)$ is in the direction of decreasing x, so in a suitable neighbourhood of (a_0, m_0) we have the phase portrait described in Fig. 10.1.

10.4.2. Consequently, the family we are studying is orbitally equivalent (on the centre manifold) to the model family $X(\mu)(x) = \mu - x^2$. Taking account of the 'transverse' variables we then obtain the general topological model for the *saddle-node bifurcation*

$$\dot{x} = \mu - x^2, \ \dot{y} = -y, \ \dot{z} = z, \quad (x, y, z) \in \mathbf{R} \times \mathbf{R}^q \times \mathbf{R}^r \qquad (10.4.2)$$

(to which we should add $\dot{\mu} = 0$), in a neighbourhood of the point $(x, y, z) = (0, 0, 0)$ and of the value 0 of the parameter μ. The surface $W : \{y = 0, z = 0\}$ is a centre manifold. The singular points are given by $x^2 = \mu$, $y = 0$, $z = 0$. They form a curve in W whose projection onto the parameter axis has a fold at the initial point $(0, 0, 0)$. In W the phase portrait is given by Fig. 10.1. For $\mu > 0$ there are two hyperbolic singular points; one of them $(x = -\sqrt{\mu})$ is repelling and the other $(x = \sqrt{\mu})$ is attracting. The name *saddle-node* comes from the fact that when $q + r = 1$ the bifurcation consists of the coincidence of a saddle and a node (Figs. 10.2, 10.3).

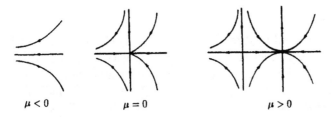

$\mu < 0$ $\mu = 0$ $\mu > 0$

Fig. 10.2. Saddle-node bifurcation in two dimensions (attracting)

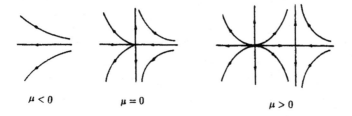

$\mu < 0$ $\mu = 0$ $\mu > 0$

Fig. 10.3. Saddle-node bifurcation in two dimensions (repelling)

Note that there exists a unique orbit connecting the two singular points. It lies on one of the separatrices of the saddle, which we see is the stable separatrix if the transverse direction is unstable, and conversely.

10.4.3. One of the features of this bifurcation is that for $\mu < 0$ there is no singular point for the field $X(\mu)$. In mechanics we often meet the closely related problem in which the given family is assumed to have a fixed singular point for all values of μ. We therefore replace the initial conditions (4.1) by

$$X(m)(a_0) = 0, \ m \in M; \quad \Lambda_{a_0} X(m_0) = 0. \qquad (10.4.3)$$

Arguing as before, we suppose that V and M are 1-dimensional and let x denote a local coordinate on V centred at a_0. By the Hadamard Lemma (Proposition 4.2.3) we have $X(m)(x) = xY(m, x)$, with $Y(m_0, a_0) = 0$. The genericity assumption this time implies that $\{Y(m, x) = 0\}$ is a curve that cuts both $\{m = m_0\}$ and $\{a = a_0\}$ transversely. This amounts to saying that this curve is the graph of a local diffeomorphism from M to V. Transporting the coordinate x via this diffeomorphism we obtain a local coordinate μ on M, centred at m_0, such that the equilibrium manifold $C = \{X(m)(a) = 0\}$ is defined by the equation $x(x - \mu) = 0$. Changing the sign of X if necessary we obtain the phase portraits in Fig. 10.4.

As before, when taking account of the 'transverse' variables we obtain the general topological model for the *transcritical bifurcation*

$$\dot{x} = \mu x - x^2, \; \dot{y} = -y, \; \dot{z} = z, \quad (x, y, z) \in \mathbf{R} \times \mathbf{R}^q \times \mathbf{R}^r, \qquad (10.4.4)$$

in a neighbourhood of $(x, y, z, \mu) = (0, 0, 0, 0)$. The surface $W : \{y = 0, \; z = 0\}$ is again a centre manifold. The singular points are now given by

$$x(x - \mu) = 0, \; y = 0, \; z = 0.$$

They form the union of two lines in W, where the phase portrait is given by Fig. 10.4. The adjective 'transcritical' refers to the fact that there is 'exchange of stability' between the singular points as they cross.

10.4.4. If a symmetry condition is imposed we find another variant of the saddle-node bifurcation called the *pitchfork bifurcation*, which we have already met (see in particular Figs. 5.7 and 5.8). The topological model (in the centre manifold) is

$$\dot{x} = \mu x - \varepsilon x^3, \quad \varepsilon = \pm 1; \qquad (10.4.5)$$

through symmetry-breaking it gives rise to the two-parameter family

$$\dot{x} = \nu + \mu x - \varepsilon x^3, \quad \varepsilon = \pm 1. \qquad (10.4.6)$$

10.5 The Hopf Bifurcation

10.5.1. We now move to the other case of local codimension 1 bifurcation of a singularity, in which the linearized vector field at the singular point has a conjugate pair of purely imaginary eigenvalues (say $i\omega$ and $-i\omega$ with $\omega > 0$) of multiplicity 1, all the others having nonzero real part. This time the centre manifold technique reduces the problem to the case where $V = \mathbf{R}^2$ and where $X(\mu_0)(0) = 0$ and $\Lambda_0 X(\mu_0) = \left(\begin{smallmatrix} 0 & -\omega \\ \omega & 0 \end{smallmatrix}\right)$. It is more convenient to take $V = \mathbf{C}$, so that $\Lambda_0 X(\mu_0)$ is the map $z \mapsto i\omega z$. The singular point 0 of $X(\mu_0)$ is nondegenerate and we can apply Proposition 9.7.7; therefore there exists (for μ sufficiently close to μ_0) a singular point $a(\mu)$ close to 0 for $X(\mu)$. By translation $z \mapsto z - a(\mu)$ we reduce to the case when $X(\mu)(0) = 0$ for all μ. The eigenvalues of $\Lambda_0 X(\mu)$ are of the form $\lambda(\mu), \bar{\lambda}(\mu)$ where λ is a C^r function such that $\lambda(\mu_0) = i\omega$. In a neighbourhood of μ_0 let us write $\lambda(\mu) = \alpha(\mu) + i\beta(\mu) = \beta(\mu)(\gamma(\mu) + i)$, with $\beta(\mu_0) = \omega$ and $\gamma(\mu_0) = 0$. For

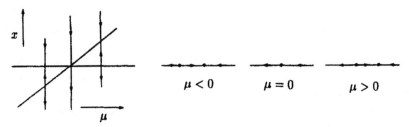

Fig. 10.4. Transcritical bifurcation

a generic family the derivative of γ does not vanish at μ_0 and we can take γ to be the first coordinate on the parameter manifold M. As elsewhere, we argue up to orbital equivalence and so we can replace $X(\mu)$ by $X(\mu)/\beta(\mu)$. Up to change of coordinates we have $V = \mathbf{C}$, $M = \mathbf{R}^k$ and

$$X(\mu)(0) = 0, \quad \Lambda_0 X(\mu) \cdot z = (i + \mu_1)z.$$

In particular, for a one-parameter family we obtain the expansion

$$X(\mu)(z) = (\mu + i)z + o(|z|). \tag{10.5.1}$$

10.5.2. We now try to find a family of local coordinate changes of the form $z \mapsto z'_\mu(z)$ which makes the higher order terms in (5.1) vanish. However, we immediately come up against the phenomenon of resonance (Sect. 8.10). The eigenvalues of $\Lambda_0 X(\mu)$ are $(\mu + i)$ and $-(\mu + i)$; they exhibit in particular the resonances $\lambda_1 = (n+1)\lambda_1 + n\lambda_2$, which implies difficulties with the 'resonant' monomials $z^{n+1}\bar{z}^n = z(z\bar{z})^n$, the first of which is $z(z\bar{z})$. It can be proved (exercise) that after a suitable coordinate-change of the type indicated, the Taylor expansion of $X(\mu)$ up to order 3 is of the form

$$X(\mu)(z) = (\mu + i)z + d(\mu)z(z\bar{z}) + o(|z|^3).$$

For a generic family the real part of $d(0)$ is nonzero. After carrying out a scalar multiplication $z \mapsto \alpha(\mu)z$ we replace d by $d\alpha\bar{\alpha}$ and thus we can reduce to the case where

$$X(\mu)(z) = (\mu + i)z - (\varepsilon + i\beta(\mu))z(z\bar{z}) + o(|z|^3), \quad \varepsilon = \pm 1.$$

Assume first that the remainder term of the Taylor series is zero. Letting $r^2 = z\bar{z}$ we calculate immediately

$$L_{X(\mu)}r = \mu r - \varepsilon r^3,$$

which proves that the circle of radius $\sqrt{\varepsilon\mu}$ (for $\varepsilon\mu > 0$) is invariant under $X(\mu)$ and is therefore one of its closed orbits. It can be proved that the presence of the remainder term does not alter this result qualitatively: for small μ and $\varepsilon\mu > 0$ there is a closed orbit close to the circle of radius $\sqrt{\varepsilon\mu}$.

More precisely, it can be proved that the given family is orbitally equivalent to the *reduced Poincaré-Andronov form*

$$\dot{z} = (\mu + i)z - \varepsilon z(z\bar{z}), \quad z \in \mathbf{C}, \quad \varepsilon = \pm 1, \tag{10.5.1}$$

or, if the real expression is preferred,

$$\begin{aligned}
\dot{x} &= \mu x - y - \varepsilon x(x^2 + y^2), \\
\dot{y} &= \mu y + x - \varepsilon y(x^2 + y^2), \quad (x, y) \in \mathbf{R}^2, \quad \varepsilon = \pm 1.
\end{aligned} \tag{10.5.2}$$

In polar coordinates (r, θ) we have

$$\dot{\theta} = 1, \quad \dot{r} = \mu r - \varepsilon r^3, \quad \varepsilon = \pm 1. \tag{10.5.3}$$

The orbits are therefore the solutions of the differential equation

$$\frac{dr}{d\theta} = \mu r - \varepsilon r^3.$$

Hence we recover the pitchfork bifurcation. This is the situation on the centre manifold; as in the previous section we have to 'multiply by a fixed hyperbolic singularity' in order to obtain the complete phase portrait.

10.5.3. In the polar form the description of the bifurcation is clear. According to the sign of ε we obtain one or other of Figs. 10.5a,b for the r variable, and one or other of Figs. 10.6a,b for $z = (x, y)$.

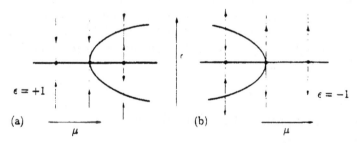

$\varepsilon = +1$ $\varepsilon = -1$

(a) μ (b) μ

Fig. 10.5. Hopf bifurcation: radius r of periodic orbit plotted against parameter μ. (a) $\varepsilon = +1$; (b) $\varepsilon = -1$

These figures are in place of a long explanation. Note the following salient features:

a) The point 0 is always a focus, hyperbolic for $\mu \neq 0$, attracting for $\mu < 0$ and repelling for $\mu > 0$. It therefore changes stability as μ passes through zero. At the precise moment of this bifurcation it is attracting for $\varepsilon = 1$ and repelling for $\varepsilon = -1$, but *not exponentially*.

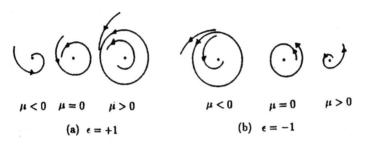

$\mu < 0$ $\mu = 0$ $\mu > 0$ $\mu < 0$ $\mu = 0$ $\mu > 0$

(a) $\epsilon = +1$ (b) $\epsilon = -1$

Fig. 10.6. Hopf bifurcation: (a) birth or (b) death of the periodic orbit as μ increases through 0

b) Far enough away from zero the phase portrait does not change, the orbits flowing towards the origin when $\epsilon = 1$ and away from it when $\epsilon = -1$. The reversal of stability of the focus (for $\epsilon\mu > 0$) has effect only in the disc of radius $R(\mu) = \sqrt{\epsilon\mu}$. The boundary of this disc is a periodic orbit which therefore appears (or disappears) at the moment of bifurcation. This appearance (or disappearance) is 'rapid' since $R(\mu)$ has infinite derivative for $\mu = 0$.

c) The periodic orbit (for $\epsilon\mu > 0$) is hyperbolic (exercise), attracting if $\epsilon = 1$, repelling if $\epsilon = -1$.

d) If we place ourselves at a point near the focus for a value of μ for which the focus is attracting, and let μ pass through the bifurcation value 0, then the focus becomes repelling and 'drives away' our point. If $\epsilon = 1$ the point will be captured by the closed orbit which is attracting and not very far away if μ is small; this a called a *weak* loss of stability. On the other hand, if $\epsilon = -1$ there if nothing to stop the escape and as soon as μ is nonzero our point flows away indefinitely (meaning that it leaves the domain of validity of this local model and will possibly be captured by another attractor); therefore in this case there is *strong* loss of stability.

10.6 Local Bifurcations of a Closed Orbit

10.6.1. Now we turn to local bifurcations of a periodic orbit Ω. Thus suppose that we are given a family $(X(\mu))$ of vector fields on V, parametrized by a manifold M, and a value μ_0 of the parameter such that $X(\mu_0)$ has a periodic orbit Ω. Fix a point $a \in \Omega$ and a transversal W to Ω at a, which enables us to construct the corresponding Poincaré map $p(\mu_0)$. The latter is a local diffeomorphism of W having a fixed point at a. For values μ of the parameter close to μ_0, the manifold W remains transverse to $X(\mu)$ in a neighbourhood of a and we can construct the Poincaré map $p(\mu)$ associated to $X(\mu)$ and to the transversal W at a. Naturally the map $\mu \mapsto p(\mu)$ is as regular as the family X, so that p is a family of diffeomorphisms defined in the neighbourhood of a

in W. Conversely, being given this family of diffeomorphisms we are able to reconstruct orbitally the family of phase portraits of $X(\mu)$ in a neighbourhood of Ω. Thus *the study of bifurcations of phase portraits of a vector field in the neighbourhood of a periodic orbit reduces to the study of the bifurcations of the corresponding Poincaré map.*

10.6.2. This study is facilitated by a Centre Manifold Theorem and an Orbital Equivalence Theorem for periodic orbits quite analogous to those in Sect. 3. Consider the Poincaré map p for a vector field with respect to a periodic orbit Ω and a transversal at a point $a \in \Omega$. Decompose the tangent space $T_a V$ into three subspaces (each containing $T_a \Omega = \mathbf{R} X(a)$) according to the positions of the eigenvalues of $T_a p$ (the characteristic multipliers of Ω) relative to the unit circle. It can then be proved that there exist three submanifolds of V remaining invariant under the flow of X and tangent to these three subspaces respectively. On the first two the Poincaré map (obtained by restricting p) has a hyperbolic fixed point at a, attracting or repelling, so that all the difficulty is concentrated on the third, the *centre manifold.*

10.6.3. Let $V^0(\Omega)$ denote the centre manifold. Let q and r be the numbers (counted with their multiplicities) of the characteristic multipliers of p that lie inside or outside the unit circle, respectively. Let $q' = q$ and $q'' = 0$, or $q' = q - 1$ and $q'' = 1$, according to whether $T_a p$ preserves or reverses the orientation of the stable tangent subspace. Define r' and r'' likewise. Then, just as Theorem 3.3 generalizes Theorem 8.7.1, we have the following generalization of Theorem 8.7.2: in a neighbourhood of the fixed point a the diffeomorphism p is topologically conjugate to the diffeomorphism

$$(x, y', y'', z', z'') \mapsto (p(x), \frac{y'}{2}, -\frac{y''}{2}, 2z', -2z'')$$

of $V^0(\Omega) \times \mathbf{R}^{q'} \times \mathbf{R}^{q''} \times \mathbf{R}^{r'} \times \mathbf{R}^{r''}$ in a neighbourhood of $(a, 0, 0, 0, 0)$.

We then deduce from 9.6.3 that *the orbital equivalence class of the flow in a neighbourhood of Ω depends only on its restriction to $V^0(\Omega)$, the integers q and r and the 'symbols' q'' and r''.*

In the same way, arguing as in 3.4, we can reduce the study of bifurcations of Ω to the case when $V = V^0(\Omega)$. This means that we have reduced the problem to studying the deformations of local diffeomorphisms in a neighbourhood of a fixed point when the spectrum of the tangent map is a subset of the unit circle.

10.7 Saddle-node Bifurcation for a Closed Orbit

10.7.1. The first case of bifurcation associated to a closed orbit is where at the bifurcation value μ_0 precisely one of the characteristic multipliers of the orbit Ω is equal to 1. As we have seen, the study reduces to the case when there is just one transverse direction, that is when V has dimension 2. Therefore we have to study deformations in a neighbourhood of 0 of a local diffeomorphism p from $\mathbf{R}$ to $\mathbf{R}$ such that $p(0) = 0$ and $p'(0) = 1$. An argument somewhat analogous to that in 4.1 shows that the one-parameter deformation

$$p(\mu) : x \mapsto x - x^2 + \mu \qquad (10.7.1)$$

is orbitally universal in a neighbourhood of $(\mu, x) = (0,0)$. We then have a discussion very similar to that of 4.2. The fixed points of $p(\mu)$ form the curve $x^2 = \mu$ whose projection to the parameter axis has a fold at the initial point $(0,0)$. For $\mu > 0$ the field $X(\mu)$ therefore has two hyperbolic closed orbits, one attracting and the other repelling, which bound a cylindrical region of 'width' $\sqrt{\mu}$. See Fig. 10.7.

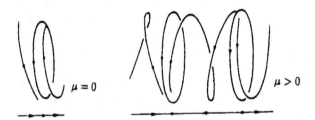

Fig. 10.7. Saddle-node bifurcation for a closed orbit

10.7.2. Similarly, as in 4.3 we can construct the *transcritical* version

$$p(\mu) : x \mapsto (1 + \mu)x - x^2. \qquad (10.7.2)$$

Here for all $\mu \neq 0$ we have two hyperbolic closed orbits corresponding to $x = 0$ and $x = \mu$ whose stabilities exchange for $\mu = 0$.

10.8 Period-doubling Bifurcation

10.8.1. The second case, which has no analogue for singular points, is where at the bifurcation value one of the multipliers becomes equal to -1 with the others not belonging to the unit circle. The centre manifold is then 1-dimensional and we can reduce to the case where V has dimension 2. Choosing a local coordinate x on the transversal at the relevant point we obtain as the initial condition for the Poincaré map

$$p(\mu_0)(0) = 0, \quad p(\mu_0)' = -1.$$

However, the fixed point 0 of the Poincaré map $p(\mu_0)$ is then nondegenerate and we can apply the Implicit Function Theorem to deduce that for μ close to μ_0 the map $p(\mu)$ has a unique fixed point close to 0. Making a suitable translation we can therefore assume that $p(\mu)(0) = 0$ for all μ and $p(\mu_0)'(0) = -1$. For a generic family the derivative of $p(\mu)'(0)$ with respect to μ does not vanish, and we can take $p(\mu)'(0)$ as the parameter for our 1-dimensional deformation, or in other words assume that $p(\mu)(x) = x(-1 + \mu) + o(x)$. If we now try to linearize p we see that we can eliminate the x^2 term but meet an obstacle with the x^3 term. For a generic family this term will not vanish; by a suitable rescaling $x \mapsto c(\mu)x$ we can reduce to the case where $p(\mu)(x) = x(-1+\mu) \pm x^3 + o(x^3)$. It can then be shown that the higher order terms do not affect the topology and we obtain the topological model that we are looking for:

$$p(\mu) : x \mapsto -x + \mu x - \varepsilon x^3, \quad \varepsilon = \pm 1, \tag{10.8.1}$$

where x and μ are close to 0.

The only fixed point close to 0 is 0; it corresponds to a closed orbit which is attracting for $\mu < 0$, repelling for $\mu > 0$. Such a change of stability in the neighbourhood of this orbit should, as in the Hopf bifurcation, correspond to the appearance of a frontier of the zone where it takes place. In fact the presence of the multiplier -1 for the original orbit implies a multiplier 1 for the orbit traversed twice and it is the latter that undergoes a bifurcation.

10.8.2. More precisely, the square of the map $p(\mu)$ has the form

$$p(\mu) \circ p(\mu) : x \mapsto x - 2\mu x + 2\varepsilon x^3 + \cdots,$$

where the unwritten terms are divisible by $\mu^2 x$ or by x^5. In a neighbourhood of $(0,0)$ the fixed points of $p(\mu) \circ p(\mu)$ form the union of the curve $x = 0$ and a curve $\varepsilon\mu - x^2 + \cdots = 0$ whose projection on the μ-axis has a fold at the point $(0,0)$. For $\varepsilon\mu > 0$ and μ small enough we thus obtain two fixed points for $p(\mu) \circ p(\mu)$ close to zero, say $a(\mu) > 0$ and $b(\mu) < 0$ with clearly $b(\mu) = p(\mu)(a(\mu))$. Therefore the $X(\mu)$-orbit common to $a(\mu)$ and $b(\mu)$ is closed, with period nearly twice that of Ω. It is hyperbolic, attracting if $\varepsilon > 0$, repelling if $\varepsilon < 0$. It bounds a Möbius band centred on Ω whose width is of order $\sqrt{\varepsilon\mu}$. According to the sign of ε we have a weak or strong loss of stability, just as in 5.3. See Fig. 10.8.

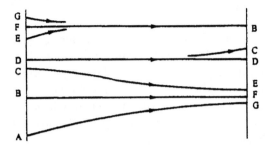

Fig. 10.8. Period-doubling takes place on a Möbius band

10.9 Hopf Bifurcation for a Closed Orbit

10.9.1. The last case of local bifurcation that we shall look at is the analogue for closed orbits of the Hopf bifurcation. Thus we assume that for the value μ_0 of the parameter the field X has a closed orbit for which a pair of complex conjugate multipliers belong to the unit circle, with the others being inside or outside. The centre manifold technique reduces the study to the following case: the ambient manifold V has dimension 3, the closed orbit Ω of $X(\mu_0)$ is of codimension 2 and the characteristic multipliers of Ω are λ and $\bar{\lambda}$ with $|\lambda| = 1$, $\lambda \neq \pm 1$. We can then identify the transverse tangent subspace with $\mathbf{C}$ so that for the bifurcation value the automorphism tangent to the Poincaré map is complex scalar multiplication by λ.

10.9.2. We therefore have to study deformations (in a neighbourhood of 0) of a local diffeomorphism p_0 of $\mathbf{C}$ such that

$$p_0(z) = \lambda z + o(z),$$

with $\lambda = e^{2\pi i\theta} \in \mathbf{S}^1$, $\theta \neq 0$, $\theta \neq \frac{1}{2}$; note that these two forbidden values correspond to a saddle-node and to period-doubling, respectively. As in the case of the saddle-node bifurcation, the natural method is to mimic the analysis of Sect. 5.

To start with, everything works nicely. As the fixed point 0 of $p(\mu_0)$ is nondegenerate the diffeomorphism $p(\mu)$ for μ close to μ_0 has a fixed point $a(\mu)$ close to 0 that depends regularly on μ. This is translated geometrically into the *persistence in the phase portrait* of $X(\mu)$ *of a periodic orbit* $\Omega(\mu)$ that depends regularly on μ. By the translation $z \mapsto z - a(\mu)$ we reduce to the case where $p(\mu)(0) = 0$ for all μ; thus $\Omega(\mu)$ is the orbit of 0 for $X(\mu)$. Next, there exists a regular function $\lambda(\mu)$, with $\lambda(\mu_0) = \lambda$, such that the multipliers of $\Omega(\mu)$, that is to say the eigenvalues of $T_0p(\mu)$, are $\lambda(\mu)$ and $\bar{\lambda}(\mu)$. Hence we obtain a first order Taylor expansion of the form

$$p_\mu(z) = \lambda(\mu)z + o(|z|).$$

Now assume that the family being studied is *generic*. If $\dim(M) = 1$ we may take $|\lambda(\mu)| - 1$ as a local coordinate on M and obtain

$$p_\mu(z) = (1 + \mu)e^{2\pi i\theta(\mu)}z + o(|z|), \quad \mu \in \mathbf{R}, \ \theta(0) = \theta, \ \mu_0 = 0. \qquad (10.9.1)$$

Note in passing that if we had $\dim(M) = 2$ then λ would be a local diffeomorphism and we would arrive at

$$p_\mu(z) = (1 + \mu)e^{2\pi i\theta(\mu)}z + o(|z|), \quad \mu \in \mathbf{C}, \ \theta(0) = \theta, \mu_0 = 0. \qquad (10.9.2)$$

10.9.3. Up to now there have been no snags, but we are about to meet some. The problem is that the structure of local diffeomorphisms in dimension 2 is much more complicated than that of vector fields in dimension 2 (the case of Sect. 5) or of local diffeomorphisms in dimension 1 (the cases of Sects. 7 and 8) and we even have difficulty in describing p_0 itself. When θ is *irrational* (that is the 'generic' assumption – in the measure-theoretical sense) only the monomials $z(z\bar{z})^k$ and $\bar{z}(z\bar{z})^k$ are resonant; in degree < 4 this gives only the monomials $z^2\bar{z}$ and $z\bar{z}^2$. After a suitable change of coordinates (exercise) we obtain a third order Taylor expansion of the form

$$p_\mu(z) = z\left((1 + \mu)e^{2\pi i\theta(\mu)} + \alpha(\mu)z\bar{z}\right) + o(|z|^3). \qquad (10.9.3)$$

On the other hand, when θ is *rational* there are other resonances. They appear in degree ≤ 3 when θ is a *third or fourth root of unity*. Thus we obtain two additional exceptional possibilities:

$$p_\mu(z) = (1+\mu)e^{2\pi i\theta(\mu)}z + a(\mu)z^2\bar{z} + b(\mu)\bar{z}^2 + o(|z|^3), \quad \theta(0) = \pm\frac{1}{3}; \quad (10.9.4)$$

$$p_\mu(z) = (1+\mu)e^{2\pi i\theta(\mu)}z + a(\mu)z^2\bar{z} + b(\mu)\bar{z}^3 + o(|z|^3), \quad \theta(0) = \pm\frac{1}{4}. \quad (10.9.5)$$

10.9.4. Let us disregard these last two cases and assume that the family has the expansion (9.3). We continue to mimic Sect. 5. Since the family is generic the real part of $a(0)$ is nonzero and as in 5.2 we can reduce to the case where $a(\mu)$ can be written $-(\varepsilon + i\beta(\mu))$ with $\varepsilon = \pm 1$. Suppose first of all that the Taylor remainder is identically zero. Then $p(\mu)$ transforms the circle $|z| = r$ into the circle $|z| = \rho(\mu, r)$ with

$$\rho(\mu, r) = r|(1 + \mu) + a(\mu)r^2| = r(1 + \mu - \varepsilon r^2 + o(r^2)).$$

Thus for small μ we see the appearance of a circle invariant under $p(\mu)$ and with radius equivalent to $\sqrt{\varepsilon\mu}$. As in 5.2, the next step is to prove that the presence of the Taylor remainder does not alter the qualitative conclusion: when μ is small, $p(\mu)$ has an invariant closed orbit $C(\mu)$ close to the circle of radius $\sqrt{\varepsilon\mu}$. In this way we obtain a direct analogue of Fig. 10.5.

10.9.5. However, it is here that the real difficulties begin. While in the case of Sect. 5 the invariant closed curve $C(\mu)$ has to be an orbit of the vector field $X(\mu)$ and that is the end of the story, the situation is not the same here. Each orbit of $p(\mu)$ is countable and so cannot fill $C(\mu)$.

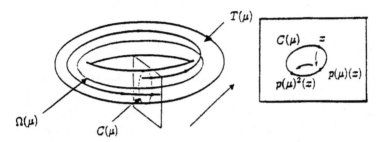

Fig. 10.9. Hopf bifurcation for a closed orbit: creation of an invariant torus

The description now is that the union of the orbits with respect to $X(\mu)$ of the different points of the curve $C(\mu)$ forms an *invariant torus* for the flow, with $C(\mu)$ being its section by the chosen transversal. For μ small and $\varepsilon\mu > 0$ the closed orbit $\Omega(\mu)$ is thus the core of an invariant torus $T(\mu)$ with 'small' radius approximately $\sqrt{\varepsilon\mu}$. As in Sect. 5, this torus is attracting or repelling according to the sign of ε. See Fig. 10.9.

10.9.6. As regards the dynamics of $X(\mu)$ on the torus or, what amounts to the same thing, the structure of the diffeomorphism induced by $p(\mu)$ on $C(\mu)$, the situation is much more complicated (even ignoring the fact that the precise differentiability class of $C(\mu)$ is already unclear). At this point we need to have recourse to the theory of *diffeomorphisms of the circle*. The first invariant of a diffeomorphism f of the circle is an angle (or equivalently a real number $n(f)$ defined modulo 1) called the *rotation number* of f that in a certain sense tells us to which rotation f is closest. The topological properties of f depend on *arithmetical* properties of $n(f)$: is it rational? irrational? Can it be approximated well or badly by rational numbers? ... In particular, it can be proved that structurally stable diffeomorphisms of the circle have rational rotation number and they form an open dense set on which the rotation number is locally constant.

10.9.7. Clearly the rotation number $n(\mu)$ of the diffeomorphism of $C(\mu)$ induced by $p(\mu)$ depends on μ. It is here that the most striking phenomenon occurs. Suppose the family is generic. On the one hand the $n(\mu)$ are almost all rational in the sense that the corresponding μ form an open dense set and on this set the phase portraits are locally constant. On the other hand, according to a theorem of Michel Herman, *the complement of this open set does not have measure zero!*

10.9.8. As for the two cases (9.4) and (9.5) that we have abandonned along the way, their analysis is even more difficult. We simply refer the reader to contemplation of Figs. 143,144,145,149, 150 and 152 in [A4]. See also Chapt. 5 of [AP].

10.10 An Example of a Codimension 2 Bifurcation

10.10.1. We consider the local codimension 2 bifurcation characterized by the fact that for the particular parameter value the vector field has a singularity whose linearized field has a double zero eigenvalue (but is not diagonalizable – since that would be a codimension 3 situation), while the other eigenvalues have nonzero real part as they must. The centre manifold technique reduces the study to the case where $V = \mathbf{R}^2$ and where, for the bifurcation value of the parameter, the field is of the form

$$X(x,y) = (y,0) + o(\|(x,y)\|).$$

10.10.2. Suppose as usual that the family is generic. Then we can show that an appropriate curvilinear coordinate change allows us to express the second order Taylor expansion in the form

$$X(x,y) = (y, ax^2 + bxy) + o(\|(x,y)\|^2).$$

Since we are in the generic case, a and b are nonzero. By suitable scalar multiplications of x, y and the time we reduce to one of the two cases $(a,b) = (-1,1)$ and $(a,b) = (-1,-1)$. Let us take the second case, so that

$$X(x,y) = (y, -x^2 - xy) + o(\|(x,y)\|^2).$$

We then show (by an *ad hoc* analysis) that it suffices to study the following 2-parameter family, which is thus a 'topologically versal deformation' of X:

$$\dot{x} = y, \quad \dot{y} = \mu + \nu y - x^2 - xy,$$

in a neighbourhood of the point $(x,y) = (0,0)$ and parameter $(\mu, \nu) = (0,0)$. In fact this is the first order system deduced from the second order equation

$$\ddot{x} = \mu + \nu \dot{x} - x^2 - x\dot{x}.$$

This system is often called the *Bogdanov-Takens singularity*: for details and references see Sect. 7.3 in [GH].

10.10.3. The singular points are given by $y = 0$ and $x^2 = \mu$. For $\mu < 0$ there are no singular points. For $\mu = 0$ the origin is the only singular point, the linearization being $(x, y) \mapsto (y, \nu y)$ with eigenvalues ν and 0. For $\mu = 0$ and $\nu \neq 0$ we therefore have a saddle-node bifurcation with 'transverse part' stable for $\nu < 0$ and unstable for $\nu > 0$, the two singular points describing the fold $y = 0$, $x^2 = \mu$. In particular, as we saw in studying that bifurcation, the orbit joining the two singular points is a stable (unstable) separatrix of the saddle when $\nu < 0$ ($\nu > 0$).

For $\mu = \nu = 0$ the system is

$$\dot{x} = y, \quad \dot{y} = -x^2 - x\dot{x}$$

which comes from the second order differential equation $\ddot{x} + x^2 + x\dot{x} = 0$. This can also be written $y\,dy + x^2\,dx + xy\,dx = 0$. It is not difficult to deduce that the orbits that have the origin as limit-point approach the branches of the curve with equation $y^2/2 + x^3/3 = 0$. From this, and from studying the signs of $\dot{x}$ and $\dot{y}$, we arrive at (exercise) the phase portrait given in Fig. 10.10.

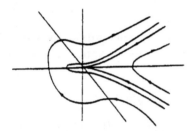

Fig. 10.10. Bogdanov-Takens singularity ($\mu = \nu = 0$)

10.10.4. Suppose $\mu > 0$. To simplify matters let $\mu = \lambda^2$, with $\lambda > 0$. The system becomes

$$\dot{x} = y, \quad \dot{y} = \lambda^2 + \nu y - x^2 - xy. \tag{10.10.2}$$

The singular points are $a = (-\lambda, 0)$ and $b = (\lambda, 0)$. At the point a the linearized system is $\dot{x} = y$, $\dot{y} = 2\lambda x + (\nu + \lambda)y$, with determinant -2λ; hence a is a saddle point. At the point b the linearized system is $\dot{x} = y$, $\dot{y} = -2\lambda x + (\nu - \lambda)y$, with determinant 2λ and trace $\nu - \lambda$. Hence:

a) if $\nu - \lambda$ is < 0, that is if $\nu < \sqrt{\mu}$, then b is hyperbolic and attracting;
b) if $\nu - \lambda$ is > 0, that is if $\nu > \sqrt{\mu}$, then b is hyperbolic and repelling;
c) if $\nu - \lambda$ is zero, that is if $\nu = \sqrt{\mu}$, the eigenvalues are $\pm i\sqrt{2\lambda} = \pm\sqrt{2}\mu^{1/4}$ and at b we have a Hopf bifurcation.

In particular for $\nu < 0$ and $\nu > \lambda$ we obtain the local phase portraits in Figs. 10.11a,b, justified on the one hand by the remark above concerning the orbit joining the two singular points, and on the other hand by the fact that along the orbits x increases when $y > 0$ and decreases when $y < 0$.

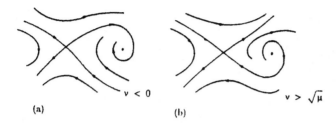

Fig. 10.11. Bifurcation from a Bogdanov-Takens singularity

10.10.5. To study the Hopf bifurcation when $\nu = \sqrt{\mu}$ we use the method that was followed in Sect. 5. To simplify later calculations put $\nu = \frac{1}{2}\lambda^2$ and $\mu = \frac{1}{4}\lambda^4$, and move b to the origin by putting $x = \nu - w/\lambda$. We obtain

$$\dot{w} = -\lambda y, \quad \dot{y} = \lambda w - \frac{w^2}{\lambda^2} + \frac{wy}{\lambda}.$$

A change of time-scale then gives the equivalent system

$$\dot{w} = -y, \quad \dot{y} = w - \frac{w^2}{\lambda^3} + \frac{wy}{\lambda^2}. \tag{10.10.3}$$

Next we have to eliminate the terms of degree 2, which we do by a coordinate transformation of the form $w \mapsto w + Q(w, y)$, $y \mapsto y + R(w, y)$ where Q and R are suitable quadratic forms (exercise). We arrive finally at a form of the type given in 5.2 and we see that the closed orbit appears for $\nu < \sqrt{\mu}$ and is therefore repelling. This enables us to sketch the phase portrait as in Fig. 10.12.

10.10.6. If we now fix μ (small and > 0) and vary ν we note that for ν slightly less than $\sqrt{\mu}$ the point b is surrounded by a repelling closed orbit which keeps the saddle separatrices apart, while for $\nu < 0$ one of these separatrices arrives at b. Therefore somewhere in between the two there is a bifurcation of a global nature (Fig. 10.13). As we shall see, this bifurcation entails the existence of a homoclinic orbit.

10.11 An Example of Non-local Bifurcation

10.11.1. To attack this problem we first of all use a technique from algebraic geometry called *blowing up*: we put

$$\mu = \varepsilon^4, \quad \nu = \varepsilon^2 \alpha, \quad x = \varepsilon^2 u, \quad y = \varepsilon^3 v, \quad \varepsilon \geq 0 \tag{10.11.1}$$

and replace the time t by εt. The new coordinates for phase space are u and v, and the new parameters are ε and α. The new system is

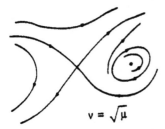

$$v = \sqrt{\mu}$$

Fig. 10.12. Creation of a periodic orbit

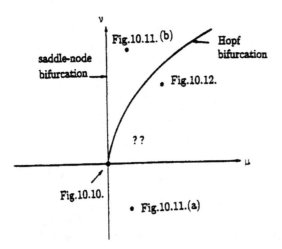

Fig. 10.13. Partial bifurcation diagram

$$\dot{u} = v, \quad \dot{v} = 1 - u^2 + \varepsilon(\alpha v - uv). \tag{10.11.2}$$

Its singular points are fixed; they are $a = (-1, 0)$ and $b = (1, 0)$.

The parameter change induces a bijection of the half-space $\{\mu > 0\}$ onto the half-space $\{\varepsilon > 0\}$, but it transforms the special point $(\mu, \nu) = (0, 0)$ into the whole straight line $\varepsilon = 0$ (hence the term 'blowing up').

10.11.2. The main advantage of this transformation is the following: for $\varepsilon = 0$ we obtain the vector field

$$X_0(u, v) = (v, 1 - u^2)$$

which is associated to the second order equation $\ddot{u} = 1 - u^2$ and is therefore *conservative*. Explicitly, it has the first integral

$$E(u, v) = \frac{v^2}{2} - u + \frac{u^3}{3}.$$

The phase portrait for X_0 can be deduced immediately (Fig. 10.14); all the orbits in the region $E(u, v) < 2/3$ are closed; this region is bounded by

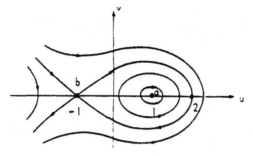

Fig. 10.14. Phase portrait of the conservative system X_0

the homoclinic separatrix S which lies on the closed loop of the curve with equation $E(u, v) = 2/3$ and which in particular passes through the point $m_0 = (2, 0)$.

10.11.3. After the transformation the system turns out to be a perturbation of a conservative system, which was not the case for the original system. More precisely, letting $X(\varepsilon, \alpha)$ denote the vector field (11.2) we have

$$X(\varepsilon, \alpha)(u, v) = X_0(u, v) + \varepsilon Z(u, v),$$

with $Z(u, v) = (0, \alpha v - uv)$. The function E is a first integral of X_0 and hence we have:

$$L_{X(\varepsilon, \alpha)} E = \varepsilon L_Z E = \varepsilon v^2 (\alpha - u). \tag{10.11.3}$$

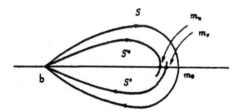

Fig. 10.15. The homoclinic separatrix S for X_0 may not persist for X

Let $\gamma_0 : t \mapsto (u_0(t), v_0(t))$ be the integral curve of X_0 such that $\gamma_0(0)$ is the point $m_0 = (2, 0)$. The map $\gamma_0 : \mathbf{R} \to \mathbf{R}^2$ is an immersion whose image is S and $\gamma_0(t)$ tends to b as t tends to $+\infty$ or to $-\infty$. For ε small enough the two separatrices S^s and S^u of the saddle b of $X(\varepsilon, \alpha)$ are close to S and hence cut $v = 0$ (which is transverse to S at m_0) at two points m^s and m^u close to m_0. To say that the saddle b of X has a homoclinic saddle-connection (close to S) is to say that $m^s = m^u$ or indeed that $E(m^s) = E(m^u)$. See Fig. 10.15. We shall solve this equation by a perturbation method.

10.11.4. Let $\gamma^s = (u^s, v^s)$ be the integral curve of $X(\varepsilon, \alpha)$ such that $\gamma^s(0) = m^s$ and whose image is S^s. By (11.3) we have

$$E(b) - E(m^s) = \varepsilon \int_0^\infty v^s(t)^2(\alpha - u^s(t))dt,$$

and an analogous formula for the unstable separatrix. From this and the theorem on the regularity of integral curves (suitably formulated, since the interval is infinite) we deduce

$$E(m^s) - E(m^u) = \varepsilon M(\alpha) + O(\varepsilon^2), \tag{10.11.4}$$

where $M(\alpha)$ is the *Melnikov integral*

$$M(\alpha) = \int_{-\infty}^\infty v_0(t)^2(\alpha - u_0(t))dt. \tag{10.11.5}$$

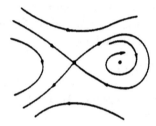

Fig. 10.16. Homoclinic orbit for X when $\nu = 5\sqrt{\mu}/7$ approx.

The function M is of the form $a\alpha - b$. Explicit calculation of the solution γ_0 and then of the integral gives $b/a = 5/7$. Hence $\alpha = 5/7$ is a simple root of M. From this and the Implicit Function Theorem we obtain the following result: there exists a regular function $\alpha(\varepsilon)$, with $\alpha(0) = 5/7$, such that for small enough ε the saddle b of the field $X(\alpha, \varepsilon)$ has a homoclinic saddle-connection for $\alpha = \alpha(\varepsilon)$ and only in this case.

10.11.5. Returning to the original coordinates, we have therefore shown the appearance of a homoclinic orbit (Fig. 10.16) along a curve in parameter space that has third order contact with the semi-parabola $\nu = 5\sqrt{\mu}/7$ at the point $(\mu, \nu) = (0, 0)$. This completes the list of bifurcations of the two-parameter field (10.1). However, a great deal of detailed verification is still required in order to prove that this is indeed the case.

References

[A1]   Arnol'd, V.I.: Lectures on bifurcations in versal families, Russian Math. Surveys 27 (1972), 54-123.

[A2]   Arnol'd, V.I.: Critical points of smooth functions, Proc. Int. Congr. Math. Vancouver 1974, pp. 19-39.

[A3]   Arnol'd, V.I.: Ordinary Differential Equations, MIT Press, Cambridge Mass. 1973.

[A4]   Arnol'd, V.I.: Geometrical Methods in the Theory of Ordinary Differential Equations, Springer-Verlag, New York 1983.

[A5]   Arnol'd, V.I.: Mathematical Methods of Classical Mechanics, Springer-Verlag, New York 1978.

[A6]   Arnol'd, V.I.: Catastrophe Theory, Springer-Verlag, Berlin-Heidelberg-New York 1984 (3rd edition 1992).

[A7]   Arnol'd, V.I.: Catastrophe Theory, Section II in Dynamical Systems V, Encyclopaedia of Mathematical Sciences Vol.5, Springer-Verlag, Berlin-Heidelberg-New York 1994.

[AA]   Anosov, D.V., Aranson, S.Kh., Arnold, V.I., Bronshtein, I.U., Grines, V.Z., Il'yashenko, Yu.S.: Ordinary Differential Equations and Smooth Dynamical Systems, Springer-Verlag, Berlin Heidelberg 1997.

[AG]   Arnol'd, V.I., Gussein-Zade, S.M., Varchenko, A.N.: Singularities of Differentiable Maps, Vol.1, Birkhäuser, Basel 1985.

[AK]   Arnol'd, V.I., Kozlov, V.V., Neishtadt, A.I.: Dynamical Systems III: Mathematical Aspects of Classical and Celestial Mechanics, Encyclopaedia of Math. Sciences Vol 3, Springer-Verlag, Berlin Heidelberg 1988.

[AM]   Abraham, R., Marsden, J.E.: Foundations of Mechanics (2nd Ed.), Benjamin/Cummings, Reading, Mass. 1978.

[AP]   Arrowsmith, D.K., Place, C.M.: An Introduction to Dynamical Systems, Cambridge University Press, 1990.

[AR]   Abraham, R., Robbin, J.: Transversal Mappings and Flows, Benjamin-Cummings, Reading, Mass. 1967.

[AS]   Abraham, R.S., Shaw, C.D.: Dynamics: The Geometry of Behavior, Ariel Press, Santa Cruz 1988.

[BE]   Berlinski, D.: On Systems Analysis, An Essay Concerning the Limitations of Some Mathematical Methods in the Social, Political and Biological Sciences, MIT Press, Cambridge, Mass. 1976.

[BG]   Bruce, J.W., Giblin, P.J.: Curves and Singularities, Cambridge University Press 1984 (2nd ed. 1992).

[BN]   Bennequin, D.: Caustique mystique (d'après Arnold et al.), Séminaire Bourbaki 1984/85, Exposé 634, pp. 19-56, Astérisque 133-134, Soc. Mat. de France 1986.

[BO]   Bost, J.-B.: Tores invariants des systèmes dynamiques hamiltoniens (d'après Kolmogorov, Arnold, Moser, . . .), Séminaire Bourbaki 1984/85, Exposé 639, pp. 113-157, Astérisque 133-134, Soc. Mat. de France 1986.

[BL]   Bröcker, Th., Lander, L.: Differentiable Germs and Catastrophes, London Math. Soc. Lecture Notes 17, Cambridge University Press 1975.

[CH]   Castrigiano, D.P.L., Hayes, S.A.: Catastrophe Theory, Addison Wesley, Reading, Mass. 1993.

[CO]   Cohen, R.L.: The immersion conjecture for differentiable manifolds, Annals of Math. 122 (1985), 237-328.

[DO]   Dombrowski, P.: 150 years after Gauss' *Disquisitiones generales circa superficies curvas*, Astérisque 62, Soc. Mat. de France 1979.

[GE]   Gilmore, R.: Catastrophe Theory for Scientists and Engineers, Wiley, New York 1981.

[GG]   Golubitsky, M., Guillemin, V.: Stable Mappings and Their Singularities, Springer-Verlag, New York 1973.

[GH]   Guckenheimer, J., Holmes, P.: Nonlinear Oscillations, Dynamical Systems, and Bifurcations of Vector Fields, Springer-Verlag, New York 1983.

[GI]   Gitler, S.: Immersion and embedding of manifolds, Proc. Symp. in Pure Math., Amer. Math. Soc., 22 (1971), 87-96.

[GS]   Golubitsky, M., Schaeffer, D.G., Stewart, I.: Singularities and Groups in Bifurcation Theory, Vols.I,II, Springer-Verlag, New York 1985, 1988.

[GW]   Gibson, C.G., Wirthmüller, K., du Plessis, A.A., Looijenga, E.J.N.: Topological Stability of Smooth Mappings, Lecture Notes in Math. 552, Springer-Verlag, Berlin-Heidelberg-New York 1976.

[HA]   Hartman, P.: Ordinary Differential Equations, Wiley, New York 1967.

[HM]   Dieudonné, J. (ed.): Abrégé d'histoire des mathématiques 1700-1900, Hermann, Paris 1978. Hermann, Paris 1978.

[HS]   Hirsch, M., Smale, S.: Differential Equations, Dynamical Systems and Linear Algebra, Academic Press, New York 1974.

[IR]   Irwin, M.: Smooth Dynamical Systems, Academic Press, New York 1980.

[KH]   Katok, A., Hasselblatt, B.: Introduction to the Modern Theory of Dynamical Systems, Cambridge University Press 1995.

[MA]   Manheim, J.H.: The Genesis of Point Set Topology, Pergamon Press, Oxford 1964.

[MR]   Mather, J.: How to stratify mappings and jet spaces, pp.128-176 in Singularités d'Applications Différentiables, Plans-sur-Bex 1975, Lecture Notes in Math. 535, Springer-Verlag, Berlin-Heidelberg-New York 1976.

[MV]   Medved', M.: Fundamentals of Dynamical Systems and Bifurcation Theory, Adam Hilger, Bristol 1992.

[NE]   Nelson, E.: Topics in Dynamics – I: Flows, Math. Notes 9, Princeton Univ. Press, NJ 1969.

[PM]   Palis, J., de Melo, W.: Geometric Theory of Dynamical Systems, an Introduction, Springer-Verlag, New York 1982.

[PS]   Poston,T., Stewart, I.: Catastrophe Theory and its Applications, Pitman, London 1978.

[RO]   Robinson, C.: Dynamical Systems, CRC Press Inc., Boca Raton 1995.

[T2]   Thom, R.: Structral Stability and Morphogenesis, Benjamin, Reading, Mass. 1975 (orig. French edition 1972).

[ZE]   Zeeman, E.C.: Catastrophe Theory, Selected Papers 1972-77, Addison-Wesley, London 1977.

Subject Index

Notation

In linear algebra we shall use the following conventions and notation:

a) Unless indicated otherwise, all vector spaces are over the field of real numbers. They will often be finite-dimensional, but this will always be explicitly mentioned.

b) A vector space E is the *direct sum* of subspaces $E_1, \ldots, E_p$ if every element of E can be written in a unique way as a sum $x_1 + \cdots + x_p$, where x_i belongs to E_i for each i. When $p = 2$ we also say that the subspaces E_1 and E_2 are *complementary*.

c) If E and F are two vector spaces, $L(E; F)$ denotes the vector space of all linear maps from E to F. The linear maps from E to E are called *endomorphisms* of E and we also write $\operatorname{End}(E)$ instead of $L(E; E)$.

d) For all $u \in L(E; F)$ and $x \in E$ we denote the image $u(x) \in F$ of x under u also by $u \cdot x$. Thus we have $v \cdot (u \cdot x) = (v \circ u) \cdot x$. Likewise we let $u \cdot E'$ denote the *image* $\operatorname{Im}(E') = u(E')$ of a vector subspace E' of E. Recall in passing that the *kernel* $\operatorname{Ker}(u)$ of u is the vector subspace of E consisting of those x for which $u \cdot x = 0$. For u and v in $\operatorname{End}(E)$ we often write uv instead of $u \circ v$ and u^2 instead of $u \circ u$, etc.

e) Assume from now on that E is finite-dimensional. Let E^* denote the *dual* $L(E; \mathbf{R})$ of E. For $x \in E$ and $\xi \in E^*$, the real number $\xi(x) = \xi \cdot x$ will also be denoted $\langle \xi, x \rangle$ or $\langle x, \xi \rangle$.

f) A (euclidean) *scalar product* on E is a symmetric bilinear form $(x|y)$ on E such that $(x|x) > 0$ for all nonzero x. A finite-dimensional vector space equipped with a scalar product is called a *euclidean space*. In a euclidean space E the elements x of the space and ξ of its dual are in bijective correspondence: associated to $x \in E$ is the linear form $\xi \in E^*$ such that

$$(x|y) = \langle \xi, y \rangle, \quad y \in E.$$

g) A norm on E is said to be *euclidean* if it is of the form $\|x\| = \sqrt{(x|x)}$, where $(x|y)$ is a scalar product on E. We then have the *polarisation formula*

$$(x|y) = \frac{1}{2}(\|x + y\|^2 - \|x\|^2 - \|y\|^2),$$

which allows the scalar product to be reconstructed from the associated norm.

Printing: Druckhaus Beltz, Hemsbach
Binding: Buchbinderei Schäffer, Grünstadt